Geometric Theory of Dynamical Systems
An Introduction

Jacob Palis, Jr.
Welington de Melo

Geometric Theory of Dynamical Systems
An Introduction

Translated by A. K. Manning

With 114 Illustrations

Springer-Verlag
New York   Heidelberg   Berlin

Jacob Palis, Jr.
Welington de Melo
Instituto de Matemática Pura e Aplicada
Estrada Dona Castorina 110
Jardim Botânico 22460
Rio de Janeiro-RJ
Brazil

A. K. Manning (*Translator*)
Mathematics Institute
University of Warwick
Coventry CV4 7AL
England

AMS Subject Classifications (1980): 58-01, 58F09, 58F10, 34C35, 34C40

Library of Congress Cataloging in Publication Data
Palis Junior, Jacob.
  Geometric theory of dynamical systems.

  Bibliography: p.
  Includes index.
  1. Global analysis (Mathematics) 2. Differentiable
dynamical systems. I. Melo, Welington de. II. Title.
QA614.P2813   514'.74   81-23332
                   AACR2

© 1982 by Springer-Verlag New York Inc.
All rights reserved. No part of this book may be translated or reproduced in any form without written permission from Springer-Verlag, 175 Fifth Avenue, New York, New York 10010, U.S.A.

Printed in the United States of America.

9 8 7 6 5 4 3 2 1

ISBN 0-387-90668-1 Springer-Verlag New York Heidelberg Berlin
ISBN 3-540-90668-1 Springer-Verlag Berlin Heidelberg New York

Acknowledgments

This book grew from courses and seminars taught at IMPA and several other institutions both in Brazil and abroad, a first text being prepared for the Xth Brazilian Mathematical Colloquium. With several additions, it later became a book in the Brazilian mathematical collection *Projeto Euclides*, published in Portuguese. A number of improvements were again made for the present translation.

We are most grateful to many colleagues and students who provided us with useful suggestions and, above all, encouragement for us to present these introductory ideas on Geometric Dynamics. We are particularly thankful to Paulo Sad and, especially to Alcides Lins Neto, for writing part of a first set of notes, and to Anthony Manning for the translation into English.

Introduction

... cette étude qualitative (des équations différentielles) aura par elle-même un intérêt du premier ordre ...

HENRI POINCARÉ, 1881.

We present in this book a view of the Geometric Theory of Dynamical Systems, which is introductory and yet gives the reader an understanding of some of the basic ideas involved in two important topics: structural stability and genericity.

This theory has been considered by many mathematicians starting with Poincaré, Liapunov and Birkhoff. In recent years some of its general aims were established and it experienced considerable development.

More than two decades passed between two important events: the work of Andronov and Pontryagin (1937) introducing the basic concept of structural stability and the articles of Peixoto (1958–1962) proving the density of stable vector fields on surfaces. It was then that Smale enriched the theory substantially by defining as a main objective the search for generic and stable properties and by obtaining results and proposing problems of great relevance in this context. In this same period Hartman and Grobman showed that local stability is a generic property. Soon after this Kupka and Smale successfully attacked the problem for periodic orbits.

We intend to give the reader the flavour of this theory by means of many examples and by the systematic proof of the Hartman–Grobman and the Stable Manifold Theorems (Chapter 2), the Kupka–Smale Theorem (Chapter 3) and Peixoto's Theorem (Chapter 4). Several of the proofs we give are simpler than the original ones and are open to important generalizations. In Chapter 4, we also discuss basic examples of stable diffeomorphisms with infinitely many periodic orbits. We state general results on the structural stability of dynamical systems and make some brief comments on other topics, like bifurcation theory. In the Appendix to Chapter 4, we present the important concept of rotation number and apply it to describe a beautiful example of a flow due to Cherry.

Prerequisites for reading this book are only a basic course on Differential Equations and another on Differentiable Manifolds the most relevant results of which are summarized in Chapter 1. In Chapter 2 little more is required than topics in Linear Algebra and the Implicit Function Theorem and Contraction Mapping Theorem in Banach Spaces. Chapter 3 is the least elementary but certainly not the most difficult. There we make systematic use of the Transversality Theorem. Formally Chapter 4 depends on Chapter 3 since we make use of the Kupka–Smale Theorem in the more elementary special case of two-dimensional surfaces.

Many relevant results and varied lines of research arise from the theorems proved here. A brief (and incomplete) account of these results is presented in the last part of the text. We hope that this book will give the reader an initial perspective on the theory and make it easier for him to approach the literature.

Rio de Janeiro, September 1981.   JACOB PALIS, JR.
                                        WELINGTON DE MELO

Contents

List of Symbols   xi

Chapter 1
Differentiable Manifolds and Vector Fields   1

§0 Calculus in $\mathbb{R}^n$ and Differentiable Manifolds   1
§1 Vector Fields on Manifolds   10
§2 The Topology of the Space of C^r Maps   19
§3 Transversality   23
§4 Structural Stability   26

Chapter 2
Local Stability   39

§1 The Tubular Flow Theorem   39
§2 Linear Vector Fields   41
§3 Singularities and Hyperbolic Fixed Points   54
§4 Local Stability   59
§5 Local Classification   68
§6 Invariant Manifolds   73
§7 The λ-lemma (Inclination Lemma). Geometrical Proof of Local Stability   80

Chapter 3
The Kupka–Smale Theorem   91

§1 The Poincaré Map   92
§2 Genericity of Vector Fields Whose Closed Orbits Are Hyperbolic   99
§3 Transversality of the Invariant Manifolds   106

Chapter 4
Genericity and Stability of Morse–Smale Vector Fields   115

§1 Morse–Smale Vector Fields; Structural Stability   116
§2 Density of Morse–Smale Vector Fields on Orientable Surfaces   132
§3 Generalizations   150
§4 General Comments on Structural Stability. Other Topics   153
    Appendix: Rotation Number and Cherry Flows   181

References   189

Index   195

List of Symbols

$\mathbb{R}$	real line
$\mathbb{R}^n$	Euclidean n-space
$\mathbb{C}^n$	complex n-space
C^n	differentiability class of mappings having n continuous derivatives
C^∞	infinitely differentiable
C^ω	real analytic
$df(p), df_p$ or $Df(p)$	derivative of f at p
$(\partial/\partial t)f, \partial f/\partial t$	partial derivative
$D_2 f(x, y)$	partial derivative with respect to the second variable
$d^n f(p)$	nth derivative of f at p
$L(\mathbb{R}^n, \mathbb{R}^m)$	space of linear mappings
$L^r(\mathbb{R}^m; \mathbb{R}^k)$	space of r-linear mappings
$\|\ \|$	norm
$g \circ f$	composition of the mappings g and f
$\emptyset$	empty set
$f\mid M$	restriction of map f to subset M
$\overline{U}$	closure of set U
TM_p	tangent space of M at p
TM	tangent bundle of M
$\mathfrak{X}^r(M)$	space of C^r vector fields on M
$f_* X$	vector field induced on the range of f by X
X_t	diffeomorphism induced by flow of X at time t
$\mathcal{O}(p)$	orbit of p
$\omega(p)$	ω-limit set of p
$\alpha(p)$	α-limit set of p
S^n	unit n-sphere

T^2	two-dimensional torus
grad f	gradient field of f
$\int f$	integral of f
id_M	identity map of M
$\langle \, , \, \rangle$	Riemannian metric
$\langle \, , \, \rangle_p$	inner product in the tangent space of p defined by Riemannian metric
$C^r(M, N)$	space of C^r mappings
$\| \ \|_r$	C^r-norm
$\text{Diff}^r(M)$	space of C^r diffeomorphisms
$f \pitchfork S$	f is transversal to S
$\mathcal{O}_X(p)$	orbit of X through p
$\mathcal{O}_+(p)$	positive orbit of p
$\alpha'(t)$	derivative at t of map of interval
T^n	n-dimensional torus
$\mathscr{L}(\mathbb{R}^n)$	space of linear operators on $\mathbb{R}^n$
$\mathscr{L}(\mathbb{C}^n)$	complex vector space of linear operators on $\mathbb{C}^n$
L^k	$L \circ L \circ \cdots \circ L$
$\text{Exp}(L), e^L$	exponential of L
$GL(\mathbb{R}^n)$	group of invertible linear operators of $\mathbb{R}^n$
$H(\mathbb{R}^n)$	space of hyperbolic linear isomorphisms of $\mathbb{R}^n$
$\mathcal{H}(\mathbb{R}^n)$	space of hyperbolic linear vector fields of $\mathbb{R}^n$
$\text{Sp}(L)$	spectrum of L
$\mathscr{G}_0$	space of vector fields having all singularities simple
$\det(A)$	determinant of A
$\mathscr{G}_1$	space of vector fields having all singularities hyperbolic
G_0	space of diffeomorphisms having all fixed points elementary
G_1	space of diffeomorphisms whose fixed points are all hyperbolic
$C_b^0(\mathbb{R}^m)$	space of continuous bounded maps from $\mathbb{R}^m$ to $\mathbb{R}^m$
$\dim M$	dimension of M
$W^s(p)$	stable manifold of p
$W^u(p)$	unstable manifold of p
$W^s_\beta(p)$	stable manifold of size β
$W^u_\beta(p)$	unstable manifold of size β
$W^s_{\text{loc}}(0)$	local stable manifold
$W^u_{\text{loc}}(0)$	local unstable manifold
$\mathscr{G}_{12}$	space of vector fields in $\mathscr{G}_1$ whose closed orbits are all hyperbolic
$\mathfrak{X}(T)$	space of vector fields in $\mathscr{G}_1$ whose closed orbits of period $\leq T$ are all hyperbolic
$L_\alpha(X)$	union of the α-limit sets of orbits of X
$L_\omega(X)$	union of the ω-limit sets of orbits of X
$\Omega(X)$	set of nonwandering points of X
M–S	set of Morse–Smale vector fields
∂M	boundary of M
$\text{int } A$	interior of set A

Chapter 1
Differentiable Manifolds and Vector Fields

This chapter establishes the concepts and basic facts needed for understanding later chapters.

First we set out some classical results from Calculus in $\mathbb{R}^n$, Ordinary Differential Equations and Submanifolds of $\mathbb{R}^n$. Next we define vector fields on manifolds and we apply the local results of the Theory of Differential Equations in $\mathbb{R}^n$ to this case. We introduce the qualitative study of vector fields, with the concepts of α- and ω-limit sets, and prove the important Poincaré–Bendixson Theorem.

In Section 2 we define the C^r topology on the set of differentiable maps between manifolds. We show that the set of C^r maps with the C^r topology is a separable Baire space and that the C^∞ maps are dense in it. From this we obtain topologies with the same properties for the spaces of vector fields and diffeomorphisms.

Section 3 is devoted to the Transversality Theorem, which we shall use frequently.

We conclude the chapter by establishing the general aims of the Geometric or Qualitative Theory of Dynamical Systems. In particular we discuss the concepts of topological equivalence and structural stability for differential equations defined on submanifolds of $\mathbb{R}^n$.

§0 Calculus in $\mathbb{R}^n$ and Differentiable Manifolds

In this section we shall state some concepts and basic results from Calculus in $\mathbb{R}^n$, Differential Equations and Differentiable Manifolds. The proofs of the facts set out here on Calculus in $\mathbb{R}^n$ can be found in [46], [48]; on Differential

Equations in the much recommended introductory texts [4], [41], [116] or the more advanced ones [33], [35] and also [47]; on Differentiable Manifolds in [29], [38], [49].

Let $f: U \subset \mathbb{R}^m \to \mathbb{R}^k$ be a map defined on the open subset U of $\mathbb{R}^m$. We say that f is *differentiable at a point p of U* if there exists a linear transformation $T: \mathbb{R}^m \to \mathbb{R}^k$ such that, for small v, $f(p + v) = f(p) + T(v) + R(v)$ with $\lim_{v \to 0} R(v)/\|v\| = 0$. We say that the linear map T is the *derivative of f at p* and write it as $df(p)$ or sometimes df_p or $Df(p)$. The existence of the derivative of f at p implies, in particular, that f is continuous at p. If f is differentiable at each point of U we have a map $df: L(\mathbb{R}^m, \mathbb{R}^k)$ which to each p in U associates the derivative of f at p. Here $L(\mathbb{R}^m, \mathbb{R}^k)$ denotes the vector space of linear maps from $\mathbb{R}^m$ to $\mathbb{R}^k$ with the norm $\|T\| = \sup\{\|Tv\|; \|v\| = 1\}$. If df is continuous we say that f is *of class C^1* in U. It is well known that f is C^1 if and only if the partial derivatives of the coordinate functions of f, $\partial f^i/\partial x_j: U \to \mathbb{R}$, exist and are continuous. The matrix of $df(p)$ with respect to the canonical bases of $\mathbb{R}^m$ and $\mathbb{R}^k$ is $[(\partial f^i/\partial x_j)(p)]$. Analogously we define $d^2f(p)$ as the derivative of df at p. Thus $d^2f(p)$ belongs to the space $L(\mathbb{R}^m, L(\mathbb{R}^m, \mathbb{R}^k))$, which is isomorphic to the space $L^2(\mathbb{R}^m; \mathbb{R}^k)$ of bilinear maps from $\mathbb{R}^m \times \mathbb{R}^m$ to $\mathbb{R}^k$. The norm induced on $L^2(\mathbb{R}^m; \mathbb{R}^k)$ by this isomorphism is $\|B\| = \sup\{\|B(u, v)\|; \|u\| = \|v\| = 1\}$. We say that f is *of class C^2* in U if $d^2f: U \to L^2(\mathbb{R}^m; \mathbb{R}^k)$ is continuous. By induction we define $d^rf(p)$ as the derivative at p of $d^{r-1}f$. We have $d^rf(p) \in L^r(\mathbb{R}^m; \mathbb{R}^k)$, where $L^r(\mathbb{R}^m; \mathbb{R}^k)$ is the space of r-linear maps with the norm $\|C\| = \sup\{\|C(v_1, \ldots, v_r)\|; \|v_1\| = \cdots = \|v_r\| = 1\}$. Then we say that f is *of class C^r* in U if $d^rf: U \to L^r(\mathbb{R}^m; \mathbb{R}^k)$ is continuous. Finally, f is *of class C^∞* in U if it is of class C^r for all $r \geq 0$. We remark that f is of class C^r if and only if all the partial derivatives up to order r of the coordinate functions of f exist and are continuous. Let U, V be open sets in $\mathbb{R}^m$ and $f: U \to V$ a surjective map of class C^r. We say that f is a *diffeomorphism of class C^r* if there exists a map $g: V \to U$ of class C^r such that $g \circ f$ is the identity on U.

0.0 Proposition. *Let $U \subset \mathbb{R}^m$ be an open set and $f_n: U \to \mathbb{R}^k$ be a sequence of maps of class C^1. Suppose that f_n converges pointwise to $f: U \to \mathbb{R}^k$ and that the sequence df_n converges uniformly to $g: U \to L(\mathbb{R}^m, \mathbb{R}^k)$. Then f is of class C^1 and $df = g$.* □

0.1 Proposition (Chain Rule). *Let $U \subset \mathbb{R}^m$ and $V \subset \mathbb{R}^n$ be open sets. If $f: U \to \mathbb{R}^n$ is differentiable at $p \in U, f(U) \subset V$ and $g: V \to \mathbb{R}^k$ is differentiable at $q = f(p)$, then $g \circ f: U \to \mathbb{R}^k$ is differentiable at p and $d(g \circ f)(p) = dg(f(p)) \circ df(p)$.* □

Corollary 1. *If f and g are both of class C^r, then $g \circ f$ is of class C^r.* □

Corollary 2. *If $f: U \to \mathbb{R}^k$ is differentiable at $p \in U$ and $\alpha: (-1, 1) \to U$ is a curve such that $\alpha(0) = p$ and $(d/dt)\alpha(0) = v$, then $f \circ \alpha$ is a curve which is differentiable at 0 and $(d/dt)(f \circ \alpha)(0) = df(p)v$.* □

§0 Calculus in $\mathbb{R}^n$ and Differentiable Manifolds

0.2 Theorem (Inverse Function). *Let $f: U \subset \mathbb{R}^m \to \mathbb{R}^m$ be of class C^r, $r \geq 1$. If $df(p): \mathbb{R}^m \to \mathbb{R}^m$ is an isomorphism, then f is a local diffeomorphism at $p \in U$ of class C^r; that is, there exist neighbourhoods $V \subset U$ of p and $W \subset \mathbb{R}^m$ of $f(p)$ and a C^r map $g: W \to V$ such that $g \circ f = I_V$ and $f \circ g = I_W$, where I_V denotes the identity map of V and I_W the identity of W.* □

0.3 Theorem (Implicit Function). *Let $U \subset \mathbb{R}^m \times \mathbb{R}^n$ be an open set and $f: U \to \mathbb{R}^n$ a C^r map, $r \geq 1$. Let $z_0 = (x_0, y_0) \in U$ and $c = f(z_0)$. Suppose that the partial derivative with respect to the second variable, $D_2 f(z_0): \mathbb{R}^n \to \mathbb{R}^n$, is an isomorphism. Then there exist open sets $V \subset \mathbb{R}^m$ containing x_0 and $W \subset U$ containing z_0 such that, for each $x \in V$, there exists a unique $\xi(x) \in \mathbb{R}^n$ with $(x, \xi(x)) \in W$ and $f(x, \xi(x)) = c$. The map $\xi: V \to \mathbb{R}^n$, defined in this way, is of class C^r and its derivative is given by $d\xi(x) = [D_2 f(x, \xi(x))]^{-1} \circ D_1 f(x, \xi(x))$.* □

Remark. These theorems are also valid in Banach spaces.

0.4 Theorem (Local Form for Immersions). *Let $U \subset \mathbb{R}^m$ be open and $f: U \to \mathbb{R}^{m+n}$ a C^r map, $r \geq 1$. Suppose that, for some $x_0 \in U$, the derivative $df(x_0): \mathbb{R}^m \to \mathbb{R}^{m+n}$ is injective. Then there exist neighbourhoods $V \subset U$ of x_0, $W \subset \mathbb{R}^n$ of the origin and $Z \subset \mathbb{R}^{m+n}$ of $f(x_0)$ and a C^r diffeomorphism $h: Z \to V \times W$ such that $h \circ f(x) = (x, 0)$ for all $x \in V$.* □

0.5 Theorem (Local Form for Submersions). *Let $U \subset \mathbb{R}^{m+n}$ be open and $f: U \to \mathbb{R}^n$ a C^r map, $r \geq 1$. Suppose that, for some $z_0 \in U$, the derivative $df(z_0)$ is surjective. Then there exist neighbourhoods $Z \subset U$ of z_0, $W \subset \mathbb{R}^n$ of $c = f(z_0)$ and $V \subset \mathbb{R}^m$ of the origin and a C^r diffeomorphism $h: V \times W \to Z$ such that $f \circ h(x, w) = w$ for all $x \in V$ and $w \in W$.* □

Let $f: U \subset \mathbb{R}^m \to \mathbb{R}^n$ be a C^r map, $r \geq 1$. A point $x \in U$ is *regular* if $df(x)$ is surjective; otherwise x is a *critical point*. A point $c \in \mathbb{R}^n$ is a *regular value* if every $x \in f^{-1}(c)$ is a regular point; otherwise c is a *critical value*. A subset of $\mathbb{R}^n$ is *residual* if it contains a countable intersection of open dense subsets. By Baire's Theorem every residual subset of $\mathbb{R}^n$ is dense.

0.6 Theorem (Sard [64]). *If $f: U \subset \mathbb{R}^m \to \mathbb{R}^n$ is of class C^∞ then the set of regular values of f is residual in $\mathbb{R}^n$.* □

We should remark here that if $f^{-1}(c) = \emptyset$ then c is a regular value. For the existence of a regular point $x \in U$ we need $m \geq n$. If $m < n$ all the points of U are critical and, therefore, $f(U)$ is "meagre" in $\mathbb{R}^n$, that is $\mathbb{R}^n - f(U)$ is residual.

We are going now to state some basic results on differential equations. A *vector field* on an open set $U \subset \mathbb{R}^m$ is a map $X: U \to \mathbb{R}^m$. We shall only consider fields of class C^r, $r \geq 1$. An *integral curve* of X, through a point

$p \in U$, is a differentiable map $\alpha \colon I \to U$, where I is an open interval containing 0, such that $\alpha(0) = p$ and $\alpha'(t) = X(\alpha(t))$ for all $t \in I$. We say that α is a *solution* of the differential equation $dx/dt = X(x)$ with initial condition $x(0) = p$.

0.7 Theorem (Existence and Uniqueness of Solutions). *Let X be a vector field of class C^r, $r \geq 1$, on an open set $U \subset \mathbb{R}^m$ and let $p \in U$. Then there exists an integral curve of X, $\alpha \colon I \to U$, with $\alpha(0) = p$. If $\beta \colon J \to U$ is another integral curve of X with $\beta(0) = p$ then $\alpha(t) = \beta(t)$ for all $t \in I \cap J$.* □

A *local flow* of X at $p \in U$ is a map $\varphi \colon (-\varepsilon, \varepsilon) \times V_p \to U$, where V_p is a neighbourhood of p in U, such that, for each $q \in V_p$, the map $\varphi_q \colon (-\varepsilon, \varepsilon) \to U$, defined by $\varphi_q(t) = \varphi(t, q)$, is an integral curve through q; that is, $\varphi(0, q) = q$ and $(\partial/\partial t)\varphi(t, q) = X(\varphi(t, q))$ for all $(t, q) \in (-\varepsilon, \varepsilon) \times V_p$.

0.8 Theorem. *Let X be a vector field of class C^r in U, $r \geq 1$. For all $p \in U$ there exists a local flow, $\varphi \colon (-\varepsilon, \varepsilon) \times V_p \to U$, which is of class C^r. We also have*

$$D_1 D_2 \varphi(t, q) = DX(\varphi(t, q)) \cdot D_2 \varphi(t, q)$$

and $D_2 \varphi(0, q)$ is the identity map of $\mathbb{R}^m$, where D_1 and D_2 denote the partial derivatives with respect to the first and second variables. □

We can also consider vector fields that depend on a parameter and the dependence of their solutions on the parameter. Let E be a Banach space and $F \colon E \times U \to \mathbb{R}^m$ a C^r map, $r \geq 1$. For each $e \in E$ the map $F_e \colon U \to \mathbb{R}^m$, defined by $F_e(p) = F(e, p)$, is a vector field on U of class C^r. The following theorem shows that the solutions of this field F_e depend continuously on the parameter $e \in E$.

0.9 Theorem. *For every $e \in E$ and $p \in U$ there exist neighbourhoods W of e in E and V of p in U and a C^r map $\varphi \colon (-\varepsilon, \varepsilon) \times V \times W \to U$ such that*

$$\varphi(0, q, \lambda) = q,$$
$$D_1 \varphi(t, q, \lambda) = F(\lambda, \varphi(t, q, \lambda))$$

for every $(t, q, \lambda) \in (-\varepsilon, \varepsilon) \times V \times W$. □

Next we introduce the concept of differentiable manifold. To simplify the exposition we define manifolds as subsets of $\mathbb{R}^k$. At the end of this section we discuss the abstract definition.

Let M be a subset of Euclidean space $\mathbb{R}^k$. We shall use the induced topology on M; that is, $A \subset M$ is open if there exists an open set $A' \subset \mathbb{R}^k$ such that $A = A' \cap M$. We say that $M \subset \mathbb{R}^k$ is a *differentiable manifold* of dimension m if, for each point $p \in M$, there exists a neighbourhood $U \subset M$ of p and a homeomorphism $x \colon U \to U_0$, where U_0 is an open subset of $\mathbb{R}^m$, such that

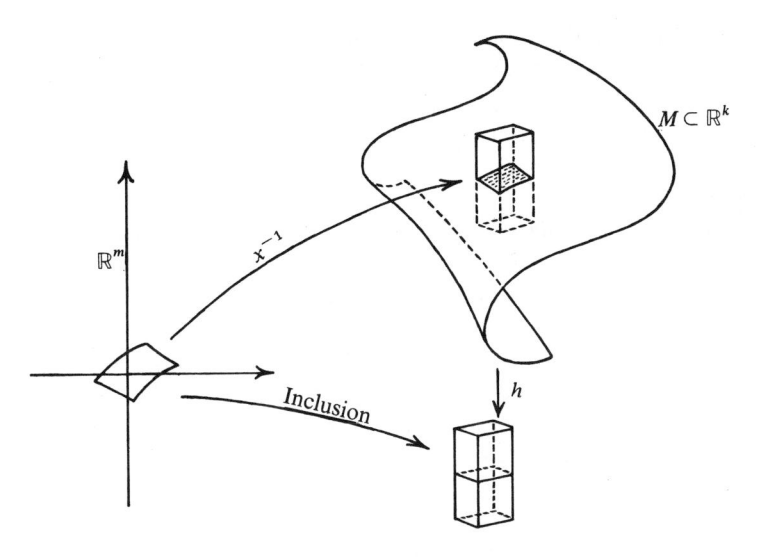

Figure 1

the inverse homeomorphism $x^{-1}: U_0 \to U \subset \mathbb{R}^k$ is an immersion of class C^∞. That is, for each $u \in U_0$, the derivative $dx^{-1}(u): \mathbb{R}^m \to \mathbb{R}^k$ is injective. In this case we say that (x, U) is a *local chart around* p and U is a *coordinate neighbourhood* of p. If the homeomorphisms x^{-1} above are of class C^r we say that M is a manifold of class C^r. What we have called a differentiable manifold corresponds to one of class C^∞. It follows from the Local Form for Immersions 0.4 that, if (x, U) is a local chart around $p \in M$, then there exist neighbourhoods A of p in $\mathbb{R}^k$, V of $x(p)$ and W of the origin in $\mathbb{R}^{k-m}$ and a C^∞ diffeomorphism $h: A \to V \times W$ such that, for all $q \in A \cap M$, we have $h(q) = (x(q), 0)$. In particular, a local chart is the restriction of a C^∞ map of an open subset of $\mathbb{R}^k$ into $\mathbb{R}^m$ (Figure 1). From this remark we obtain the following proposition.

0.10 Proposition. *Let* $x: U \to \mathbb{R}^m$ *and* $y: V \to \mathbb{R}^m$ *be local charts in* M. *If* $U \cap V \neq \emptyset$ *then the change of coordinates* $y \circ x^{-1}: x(U \cap V) \to y(U \cap V)$ *is a* C^∞ *diffeomorphism* (Figure 2). □

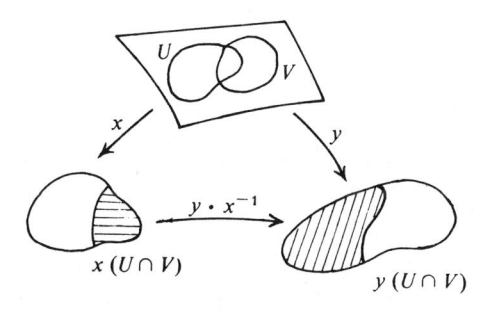

Figure 2

We shall now define differentiable maps between manifolds. Let M^m and N^n be manifolds and $f: M^m \to N^n$ a map. We say that f is *of class C^r* if, for each point $p \in M$, there exist local charts $x: U \to \mathbb{R}^m$ around p and $y: V \to \mathbb{R}^n$ with $f(U) \subset V$ such that $y \circ f \circ x^{-1}: x(U) \to y(V)$ is of class C^r. As the changes of coordinates are of class C^∞ this definition is independent of the choice of charts.

Let us consider a differentiable curve $\alpha: (-\varepsilon, \varepsilon) \to M \subset \mathbb{R}^k$ with $\alpha(0) = p$. It is easy to see that α is differentiable according to the above definition if and only if α is differentiable as a curve in $\mathbb{R}^k$. Hence there exists a tangent vector $(d\alpha/dt)(0) = \alpha'(0)$. The set of vectors tangent to all such curves α is called the tangent space to M at p and denoted by TM_p. Let us consider a local chart $x: U \to \mathbb{R}^m$, $x(p) = 0$. It is easy to see that the image of the derivative $dx^{-1}(0)$ coincides with TM_p. Thus TM_p is a vector space of dimension m.

Let $f: M \to N$ be a differentiable map and $v \in TM_p$, $p \in M$. Consider a differentiable curve $\alpha: (-\varepsilon, \varepsilon) \to M$ with $\alpha(0) = p$ and $\alpha'(0) = v$. Then $f \circ \alpha: (-\varepsilon, \varepsilon) \to N$ is a differentiable curve, so we can define $df(p)v = (d/dt)(f \circ \alpha)(0)$. This definition is independent of the curve α.

The map $df(p): TM_p \to TN_{f(p)}$ is linear and is called the *derivative of f at p*.

As a differentiable manifold is locally an open subset of a Euclidean space all the theorems from Calculus that we listed earlier extend to manifolds.

0.11 Proposition (Chain Rule). *Let $f: M \to N$ and $g: N \to P$ be maps of class C^r between differentiable manifolds. Then $g \circ f: M \to P$ is of class C^r and $d(g \circ f)(p) = dg(f(p)) \circ df(p)$.* □

A map $f: M \to N$ is a *C^r diffeomorphism* if it is of class C^r and has an inverse f^{-1} of the same class. In this case, for each $p \in M$, $df(p): TM_p \to TN_{f(p)}$ is an isomorphism whose inverse is $df^{-1}(f(p))$. In particular, M and N have the same dimension. We say that $f: M \to N$ is a *local diffeomorphism* at $p \in M$ if there exist neighbourhoods $U(p) \subset M$ and $V(f(p)) \subset N$ such that the restriction of f to U is a diffeomorphism onto V.

0.12 Proposition (Inverse Function). *If $f: M \to N$ is of class C^r, $r \geq 1$, and $df(p)$ is an isomorphism for some $p \in M$ then f is a local diffeomorphism of class C^r at p.* □

Now consider a subset S of a manifold M. S is a *submanifold of class C^r* of M of dimension s if, for each $p \in S$, there exist open sets $U \subset M$ containing p, $V \subset \mathbb{R}^s$ containing 0 and $W \subset \mathbb{R}^{m-s}$ containing 0 and a diffeomorphism of class C^r $\varphi: U \to V \times W$ such that $\varphi(S \cap U) = V \times \{0\}$ (Figure 3).

We remark that $\mathbb{R}^k$ is a differentiable manifold and that, if $M \subset \mathbb{R}^k$ is a manifold as defined above, then M is a submanifold of $\mathbb{R}^k$. The submanifolds of $M \subset \mathbb{R}^k$ are those submanifolds of $\mathbb{R}^k$ that are contained in M.

§0 Calculus in $\mathbb{R}^n$ and Differentiable Manifolds

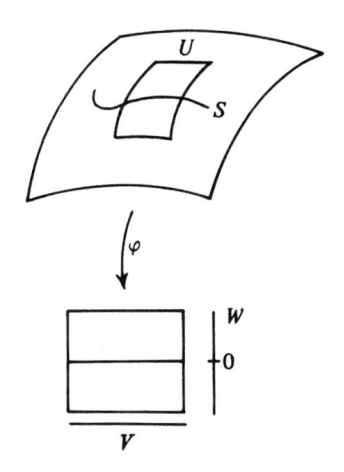

Figure 3

0.13 Proposition (Local Form for Immersions). *Let $f: M^m \to N^{m+n}$ be a map of class C^r, $r \geq 1$, and $p \in M$ a point for which $df(p)$ is injective. Then there exist neighbourhoods $U(p)$, $V(f(p))$, $U_0(0)$ in $\mathbb{R}^m$ and $V_0(0)$ in $\mathbb{R}^n$ and C^r diffeomorphisms $\varphi: U \to U_0$ and $\psi: V \to U_0 \times V_0$ such that $\psi \circ f \circ \varphi^{-1}(x) = (x, 0)$.* □

A C^r map $f: M \to N$ is an *immersion* if $df(p)$ is injective for all $p \in M$. An injective immersion $f: M \to N$ is an *embedding* if $f: M \to f(M) \subset N$ is a homeomorphism, where $f(M)$ has the induced topology. In this case $f(M)$ is a submanifold of N. If $f: M \to N$ is only an injective immersion we say that $f(M)$ is an *immersed submanifold*. The examples of Figure 4 show submanifolds that are immersed but not embedded.

0.14 Proposition (Local Form for Submersions). *Let $f: M^{m+n} \to N^n$ be a map of class C^r, $r \geq 1$, and $p \in M$ a point for which $df(p)$ is surjective. There exist neighbourhoods $U(p)$, $V(f(p))$, $U_0(0)$ in $\mathbb{R}^m$ and $V_0(0)$ in $\mathbb{R}^n$ and diffeomorphisms $\varphi: U \to U_0 \times V_0$ and $\psi: V \to V_0$ such that $\psi \circ f \circ \varphi^{-1}(x, y) = y$.* □

A point $q \in N$ is a *regular value* of $f: M^m \to N^n$, where f is a C^r map with $r \geq 1$, if, for all $p \in M$ with $f(p) = q$, $df(p)$ is surjective. It follows from the last proposition that $f^{-1}(q)$ is a C^r submanifold of M of dimension $m - n$.

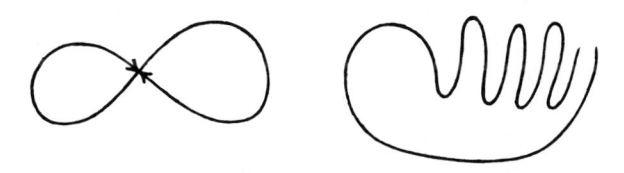

Figure 4

0.15 Proposition (Sard). *Let $f: M \to N$ be a C^∞ map. The set of regular values of f is residual; in particular, it is dense in N.*

The proof of Proposition 0.15 follows from Sard's Theorem by taking local charts and using the fact that every open cover of a manifold admits a countable subcover. □

We remark that, if M is compact, the set of regular values of $f: M \to N$ is open and dense in N.

Let us consider a countable cover $\{U_n\}$ of a manifold M by open sets. We say that this cover is *locally finite* if, for every $p \in M$, there exists a neighbourhood V of p which intersects only a finite number of elements of the cover. A *partition of unity* subordinate to the cover $\{U_n\}$ is a countable collection $\{\varphi_n\}$ of nonnegative real functions of class C^∞ such that:

(a) for each index n the support of φ_n is contained in U_n; recall that the support of φ_n is the closure of the set of points where φ_n is positive;
(b) $\sum_n \varphi_n(p) = 1$ for all $p \in M$.

0.16 Proposition. *Given a locally finite countable cover of M there exists a partition of unity subordinate to this cover.* □

Corollary 1. *Let $K \subset M$ be a closed subset. There exists a map $f: M \to \mathbb{R}$ of class C^∞ such that $f^{-1}(0) = K$.* □

Corollary 2. *Let $f: M \to \mathbb{R}^s$ be a C^r map where $M \subset \mathbb{R}^k$ is a closed manifold. Then there exists a C^r map $\tilde{f}: \mathbb{R}^k \to \mathbb{R}^s$ such that $\tilde{f}|M = f$.* □

It follows from this proposition that, given open sets U and V in M with $\overline{U} \subset V$, there exists a C^∞ real-valued function $\varphi \geq 0$ with $\varphi = 1$ on U and $\varphi = 0$ on $M - V$. Such a function is called a bump function.

We shall now define the *tangent bundle* TM of a manifold $M^m \subset \mathbb{R}^k$. Put $TM = \{(p, v) \in \mathbb{R}^k \times \mathbb{R}^k; p \in M, v \in TM_p\}$. Give TM the induced topology as a subset of $\mathbb{R}^k \times \mathbb{R}^k$; then the natural projection $\pi: TM \to M$, $\pi(p, v) = p$, is continuous. Let us show that TM is a differentiable manifold and that π is actually of class C^∞. Let $x: U \to \mathbb{R}^m$ be a local chart for M. We define the map $Tx: \pi^{-1}(U) \to \mathbb{R}^m \times \mathbb{R}^m$ by $Tx(p, v) = (x(p), dx(p)v)$. It is easy to see that $(Tx, \pi^{-1}(U))$ is a local chart for TM and, therefore, $TM \subset \mathbb{R}^k \times \mathbb{R}^k$ is a manifold. Note that the expression for π using the local charts $(Tx, \pi^{-1}(U))$ is simply the natural projection of $\mathbb{R}^m \times \mathbb{R}^m$ on the first factor; thus π is C^∞. It is also easy to see that, if $f: M \to N$ is of class C^{r+1}, then $df: TM \to TN$, $df(p, v) = (f(p), df(p)v)$, is C^r.

As we remarked earlier there is an abstract definition of manifolds which, *a priori*, is more general than the one we have just presented. For this let M be a Hausdorff topological space with a countable basis. A *local chart* in M is a pair (x, U), where $U \subset M$ is open and $x: U \to U_0 \subset \mathbb{R}^m$ is a homeomorphism onto an open subset U_0 of $\mathbb{R}^m$. We say that U is a *parametrized* neighbourhood in M. If (x, U) and (y, V) are local charts in M, with $U \cap V \neq \emptyset$, the change of coordinates $y \circ x^{-1} \colon x(U \cap V) \to y(U \cap V)$ is a homeomorphism. A *differentiable manifold of class C^r, $r \geq 1$*, is a topological space together with a family of local charts such that (a) the parametrized neighbourhoods cover M and (b) the changes of coordinates are C^r diffeomorphisms. Such a family of local charts is called a C^r *atlas for M*.

Using the local charts we can define differentiability of maps between these manifolds just as we did before. In particular, a curve $\alpha \colon (-\varepsilon, \varepsilon) \to M$ is differentiable if $x \circ \alpha \colon (-\varepsilon, \varepsilon) \to \mathbb{R}^m$ is differentiable, where (x, U) is a local chart with $\alpha(-\varepsilon, \varepsilon) \subset U$. The *tangent vector to α at $p = \alpha(0)$* is defined as the set of differentiable curves $\beta \colon (-\varepsilon, \varepsilon) \to M$ such that $\beta(0) = p$ and $d(x \circ \beta)(0) = d(x \circ \alpha)(0)$. This definition does not depend on the choice of the local chart (x, U). The *tangent space to M at p*, TM_p, is the set of tangent vectors to differentiable curves passing through p. It follows that TM_p has a natural m-dimensional vector space structure. If $f \colon M \to N$ is a differentiable map between manifolds with $f(p) = q$, we define $df(p) \colon TM_p \to TN_q$ as the map which takes the tangent vector at p to the curve $\alpha \colon (-\varepsilon, \varepsilon) \to M$ to the tangent vector at q to the curve $f \circ \alpha \colon (-\varepsilon, \varepsilon) \to N$. It is easy to see that this definition does not depend on the choice of the curve α and that $df(p)$ is a linear map. We say that $f \colon M \to N$ is an *immersion* if $df(p)$ is injective for all $p \in M$. An *embedding* is an injective immersion $f \colon M \to N$ which has a continuous inverse $f^{-1} \colon f(M) \subset N \to M$. If $f \colon M \to \mathbb{R}^k$ is an embedding of class C^∞ then $f(M) \subset \mathbb{R}^k$ is a submanifold of $\mathbb{R}^k$ according to the definition given earlier.

The next theorem relates the abstract definition of manifold to that of submanifold of Euclidean space.

0.17 Theorem (Whitney). *If M is a differentiable manifold of dimension m there exists a proper embedding $f \colon M \to \mathbb{R}^{2m+1}$.* □

Let M be a differentiable manifold and $S \subset M$ a submanifold. A *tubular neighbourhood* of S is a pair (V, π) where V is a neighbourhood of S in M and $\pi \colon V \to S$ is a submersion of class C^∞ such that $\pi(p) = p$ for $p \in S$.

0.18 Theorem. *Every submanifold $S \subset M$ has a tubular neighbourhood.* □

Lastly, every manifold of class C^r, $r \geq 1$, can be considered in a natural way as a manifold of class C^∞.

0.19 Theorem (Whitney). *Let M be a manifold of class C^r, $r \geq 1$. Then there exists a C^r embedding $f \colon M \to \mathbb{R}^{2m+1}$ such that $f(M)$ is a closed C^∞ submanifold of $\mathbb{R}^{2m+1}$.* □

By Theorem 0.17 this result is equivalent to the following: if $\mathscr{A}$ is a C^r atlas on M there exists a C^∞ atlas $\tilde{\mathscr{A}}$ on M such that, if $(x, U) \in \mathscr{A}$ and $(\tilde{x}, \tilde{U}) \in \tilde{\mathscr{A}}$ with $U \cap \tilde{U} \neq \emptyset$, then $\tilde{x} \circ x^{-1}$ and $x \circ \tilde{x}^{-1}$ are of class C^r.

§1 Vector Fields on Manifolds

We begin here the qualitative study of differential equations. As this study has both local and global aspects the natural setting for it is a differentiable manifold. One of the first basic results which is global in character is the Poincaré–Bendixson Theorem with which we shall close this section.

Let $M^m \subset \mathbb{R}^k$ be a differentiable manifold. A *vector field* of class C^r on M is a C^r map $X: M \to \mathbb{R}^k$ which associates a vector $X(p) \in TM_p$ to each point $p \in M$. This corresponds to a C^r map $X: M \to TM$ such that πX is the identity on M where π is the natural projection from TM to M. We denote by $\mathfrak{X}^r(M)$ the set of C^r vector fields on M.

An *integral curve* of $X \in \mathfrak{X}^r(M)$ through a point $p \in M$ is a C^{r+1} map $\alpha: I \to M$, where I is an interval containing 0, such that $\alpha(0) = p$ and $\alpha'(t) = X(\alpha(t))$ for all $t \in I$. The image of an integral curve is called an *orbit* or *trajectory*.

If $f: M \to N$ is a C^{r+1} diffeomorphism and $X \in \mathfrak{X}^r(M)$ then $Y = f_*X$, defined by $Y(q) = df(p)(X(p))$ with $q = f(p)$, is a C^r vector field on N, since $f_*X = df \circ X \circ f^{-1}$. If $\alpha: I \to M$ is an integral curve of X then $f \circ \alpha: I \to N$ is an integral curve of Y. In particular, f takes trajectories of X onto trajectories of Y. Thus, if $x: U \to U_0 \subset \mathbb{R}^m$ is a local chart, then $Y = x_*X$ is a C^r vector field in U_0; we say that Y is the expression of X in the local chart (x, U). By these remarks the local theorems on existence, uniqueness and differentiability of solutions extend to vector fields on manifolds as in the following proposition.

1.1 Proposition. *Let E be a Banach space and $F: E \times M \to TM$ a C^r map, $r \geq 1$, such that $\pi F(\lambda, p) = p$, where $\pi: TM \to M$ is the natural projection. For every $\lambda_0 \in E$ and $p_0 \in M$ there exist neighbourhoods W of λ_0 in E and V of p_0 in M, a real number $\varepsilon > 0$ and a C^r map $\varphi: (-\varepsilon, \varepsilon) \times V \times W \to M$ such that $\varphi(0, p, \lambda) = p$ and $(\partial/\partial t)\varphi(t, p, \lambda) = F(\lambda, \varphi(t, p, \lambda))$ for all $t \in (-\varepsilon, \varepsilon)$, $p \in V$ and $\lambda \in W$. Moreover, if $\alpha: (-\varepsilon, \varepsilon) \to M$ is an integral curve of the vector field $F_\lambda = F(\lambda, \cdot)$ with $\alpha(0) = p$ then $\alpha = \varphi(\cdot, p, \lambda)$.* □

1.2 Proposition. *Let I, J be open intervals and let $\alpha: I \to M$, $\beta: J \to M$ be integral curves of $X \in \mathfrak{X}^r(M)$, $r \geq 1$. If $\alpha(t_0) = \beta(t_0)$, for some $t_0 \in I \cap J$, then $\alpha(t) = \beta(t)$ for all $t \in I \cap J$. Hence there exists an integral curve $\gamma: I \cup J \to M$ which coincides with α on I and with β on J.*

PROOF. By the local uniqueness, if $\alpha(t_1) = \beta(t_1)$ there exists $\varepsilon > 0$ such that $\alpha(t) = \beta(t)$ for $|t - t_1| < \varepsilon$. Therefore the set $\tilde{I} \subset I \cap J$ where α coincides with β is open. As the complement of $\tilde{I}$ is also open and $I \cap J$ is connected we have $\tilde{I} = I \cap J$. □

§1 Vector Fields on Manifolds

1.3 Proposition. *Let M be a compact manifold and $X \in \mathfrak{X}^r(M)$. There exists on M a global C^r flow for X. That is, there exists a C^r map $\varphi \colon \mathbb{R} \times M \to M$ such that $\varphi(0, p) = p$ and $(\partial/\partial t)\varphi(t, p) = X(\varphi(t, p))$.*

PROOF. Consider an arbitrary point p in M. We shall show that there exists an integral curve through p defined on the whole of $\mathbb{R}$. Let $(a, b) \subset \mathbb{R}$ be the domain of an integral curve $\alpha \colon (a, b) \to M$ with $0 \in (a, b)$ and $\alpha(0) = p$. We say that (a, b) is maximal if for every interval J with the same property we have $J \subset (a, b)$. We claim that, if (a, b) is maximal, then $b = +\infty$. If this is not the case consider a sequence $t_n \to b$, $t_n \in (a, b)$. As M is compact, we may suppose (by passing to a subsequence) that $\alpha(t_n)$ converges to some $q \in M$. Let $\varphi \colon (-\varepsilon, \varepsilon) \times V_q \to M$ be a local flow for X at q. Take n_0 such that $b - t_{n_0} < \varepsilon/2$ and $\alpha(t_{n_0}) \in V_q$. Define $\gamma \colon (a, t_{n_0} + \varepsilon) \to M$ by $\gamma(t) = \alpha(t)$ if $t \le t_{n_0}$ and $\gamma(t) = \varphi(t - t_{n_0}, \alpha(t_{n_0}))$ if $t \ge t_{n_0}$. It follows that γ is an integral curve of X, which is a contradiction because $(a, t_{n_0} + \varepsilon) \supset (a, b]$. In the same way we may show that $a = -\infty$ and, therefore, there exists an integral curve $\alpha \colon \mathbb{R} \to M$ with $\alpha(0) = p$. By Proposition 1.2 this integral curve is unique. We define $\varphi(t, p) = \alpha(t)$. It is clear that $\varphi(0, p) = p$ and $(\partial/\partial t)\varphi(t, p) = X(\varphi(t, p))$. We claim that $\varphi(t + s, p) = \varphi(t, \varphi(s, p))$ for $t, s \in \mathbb{R}$ and $p \in M$. Indeed, let $\beta(t) = \varphi(t + s, p)$ and $\gamma(t) = \varphi(t, \varphi(s, p))$. We see that β and γ are integral curves of X and $\beta(0) = \gamma(0) = \varphi(s, p)$, which proves the claim. Lastly we show that φ is of class C^r. Let $p \in M$ and $\psi \colon (-\varepsilon_p, \varepsilon_p) \times V_p \to M$ be a local flow for X, which is of class C^r by Proposition 1.1. Also, by the uniqueness of solutions ψ is the restriction of φ to $(-\varepsilon_p, \varepsilon_p) \times V_p$. In particular, $\varphi_t = \varphi(t, \cdot)$ is of class C^r on V_p for $|t| < \varepsilon_p$. By the compactness of M there exists $\varepsilon > 0$ such that φ_t is of class C^r on M for $|t| < \varepsilon$. Moreover, for any $t \in \mathbb{R}$, we can choose an integer n so that $|t/n| < \varepsilon$ and deduce that $\varphi_t = \varphi_{t/n} \circ \cdots \circ \varphi_{t/n}$ is of class C^r. For any $t_0 \in \mathbb{R}$ and $p_0 \in M$, φ is C^r on a neighbourhood of (t_0, p_0). For if $|t - t_0| < \varepsilon_{p_0}$ and $p \in V_{p_0}$, then $\varphi(t, p) = \varphi_{t_0} \circ \varphi(t - t_0, p)$ is C^r since φ_{t_0} and $\varphi|(-\varepsilon_{p_0}, \varepsilon_{p_0}) \times V_{p_0}$ are C^r. That completes the proof. $\square$

Corollary. *Let $X \in \mathfrak{X}^r(M)$ and let $\varphi \colon \mathbb{R} \times M \to M$ be the flow determined by X. For each $t \in \mathbb{R}$ the map $X_t \colon M \to M$, $X_t(p) = \varphi(t, p)$, is a C^r diffeomorphism. Moreover, $X_0 = $ identity and $X_{t+s} = X_t \circ X_s$ for all $t, s \in \mathbb{R}$.* $\square$

Let $X \in \mathfrak{X}^r(M)$ and let X_t, $t \in \mathbb{R}$, be the flow of X. The *orbit* of X through $p \in M$ is the set $\mathcal{O}(p) = \{X_t(p); t \in \mathbb{R}\}$. If $X(p) = 0$ the orbit of p reduces to p. In this case we say that p is a *singularity* of X. Otherwise, the map $\alpha \colon \mathbb{R} \to M$, $\alpha(t) = X_t(p)$, is an immersion. If α is not injective there exists $\omega > 0$ such that $\alpha(\omega) = \alpha(0) = p$ and $\alpha(t) \ne p$ for $0 < t < \omega$. In this case the orbit of p is diffeomorphic to the circle S^1 and we say that it is a *closed orbit* with *period* ω. If the orbit is not singular or closed it is called *regular*. Thus a regular orbit is the image of an injective immersion of the line.

The *ω-limit set* of a point $p \in M$, $\omega(p)$, is the set of those points $q \in M$ for which there exists a sequence $t_n \to \infty$ with $X_{t_n}(p) \to q$. Similarly, we define

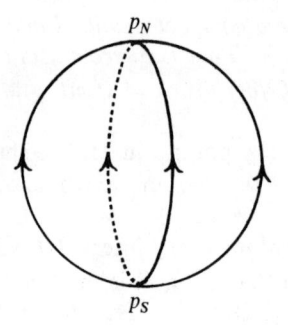

        Figure 5

the α-*limit set* of p, $\alpha(p) = \{q \in M; X_{t_n}(p) \to q$ for some sequence $t_n \to -\infty\}$. We note that the α-limit of p is the ω-limit of p for the vector field $-X$. Also, $\omega(p) = \omega(\tilde{p})$ if $\tilde{p}$ belongs to the orbit of p. Indeed, $\tilde{p} = X_{t_0}(p)$ and so, if $X_{t_n}(p) \to q$ where $t_n \to \infty$, then $X_{t_n - t_0}(\tilde{p}) \to q$ and $t_n - t_0 \to \infty$. Thus we can define the ω-limit of the orbit of p as $\omega(p)$. Intuitively $\alpha(p)$ is where the orbit of p "is born" and $\omega(p)$ is where it "dies".

EXAMPLE 1. We shall consider the unit sphere $S^2 \subset \mathbb{R}^3$ with centre at the origin and use the standard coordinates (x, y, z) in $\mathbb{R}^3$. We call $p_N = (0, 0, 1)$ the north pole and $p_S = (0, 0, -1)$ the south pole of S^2. We define the vector field X on S^2 by $X(x, y, z) = (-xz, -yz, x^2 + y^2)$. It is clear that X is of class C^∞ and that the singularities of X are p_N, p_S. As X is tangent to the meridians of S^2 and points upwards, $\omega(p) = p_N$ and $\alpha(p) = p_S$ if $p \in S^2 - \{p_N, p_S\}$ (Figure 5).

EXAMPLE 2. Rational and irrational flows on the torus. Let $\varphi: \mathbb{R}^2 \to T^2 \subset \mathbb{R}^3$ be given by

$$\varphi(u, v) = ((2 + \cos 2\pi v) \cos 2\pi u, (2 + \cos 2\pi v) \sin 2\pi u, \sin 2\pi v).$$

We see that φ is a local diffeomorphism and takes horizontal lines in $\mathbb{R}^2$ to parallels of latitude in T^2, vertical lines to meridians and the square $[0, 1] \times [0, 1]$ onto T^2. Moreover, $\varphi(u, v) = \varphi(\tilde{u}, \tilde{v})$ if and only if $u - \tilde{u} = m$ and $v - \tilde{v} = n$ for some integers m and n. For each $\alpha \in \mathbb{R}$ we consider the vector field in $\mathbb{R}^2$ given by $X^\alpha(u, v) = (1, \alpha)$. It is easy to see that $Y^\alpha = \varphi_* X^\alpha$ is well defined and is a C^∞ vector field on T^2. The orbits of Y^α are the images by φ of the orbits of X^α and these are the lines of slope α in $\mathbb{R}^2$. We shall show that, for α rational, every orbit of Y^α is closed and, for α irrational, every orbit of Y^α is dense in T^2. For each $c \in \mathbb{R}$ let Δ_c denote the straight line in $\mathbb{R}^2$ through $(0, c)$ with slope α; $\Delta_c = \{(u, c + \alpha u); u \in \mathbb{R}\}$. As we have already observed, $\varphi(\Delta_c)$ is an orbit of Y^α. If α is rational this orbit is closed for each $c \in \mathbb{R}$. For if $\alpha = n/m$ then $(m, c + (n/m)m) \in \Delta_c$ and $\varphi(m, c + n) = \varphi(0, c)$. Suppose now that α is irrational and $c \in \mathbb{R}$. We claim that $C = \{\bar{c} \in \mathbb{R}; \varphi(\Delta_c) = \varphi(\Delta_{\bar{c}})\}$ is dense in $\mathbb{R}$. It follows that $\bigcup_{\bar{c} \in C} \Delta_{\bar{c}}$ is dense in $\mathbb{R}^2$ and, therefore,

§1 Vector Fields on Manifolds

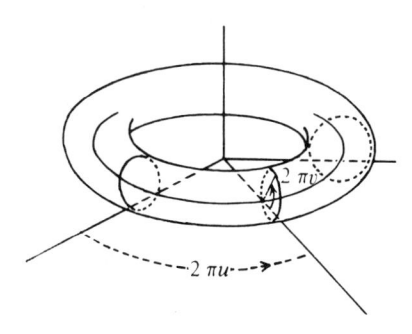

Figure 6

$\varphi(\Delta_{\bar{c}}) = \varphi(\bigcup_{c \in C} \Delta_c)$ is dense in T^2. To show that C is dense in $\mathbb{R}$ it is enough to prove that $G = \{m\alpha + n; m, n \in \mathbb{Z}\}$ is dense in $\mathbb{R}$, because $c \in C$ if and only if $c - \bar{c} \in G$. As G is a subgroup of the additive group $\mathbb{R}$ we know that G is either dense or discrete. It remains, therefore, to show that G is not discrete. But for each $m \in \mathbb{Z}$, there exists $n \in \mathbb{Z}$ such that $u_m = m\alpha + n$ belongs to the interval $[0, 1]$. The sequence u_m has a cluster point and, as α is irrational, its terms are distinct. Thus G is dense.

The vector field Y^α above is called the *rational* or *irrational field* on T^2 according as to whether α is rational or not. If α is rational the ω-limit of any orbit is itself. If α is irrational, the ω-limit of any orbit is the whole torus T^2.

EXAMPLE 3 (Gradient Vector Fields). Consider a manifold $M^m \subset \mathbb{R}^k$. At each point $p \in M$ we take in TM_p the inner product $\langle\ ,\ \rangle_p$ induced by $\mathbb{R}^k$. We denote the norm induced by this inner product by $\|\ \|_p$ or, simply, by $\|\ \|$. If X and Y are C^∞ vector fields on M then the function $g: M \to \mathbb{R}$, $g(p) = \langle X(p), Y(p)\rangle_p$ is of class C^∞. Let $f: M \to \mathbb{R}$ be a C^{r+1} map. For each $p \in M$ there exists a unique vector $X(p) \in TM_p$ such that $df_p v = \langle X(p), v\rangle_p$ for all $v \in TM_p$. This defines a vector field X which is of class C^r. It is called the *gradient* of f and written as $X = \text{grad } f$. We shall now indicate some basic properties of gradient fields. Firstly, grad $f(p) = 0$ if and only if $df_p = 0$. Along nonsingular orbits of $X = \text{grad } f$ we have f strictly increasing because $df_p X(p) = \|X(p)\|^2$. In particular grad f does not have closed orbits. Moreover, the ω-limit of any orbit consists of singularities. For let us suppose that $X(q) \neq 0$ and $q \in \omega(p)$ for some $p \in M$. Let S be the intersection of $f^{-1}(f(q))$ with a small neighbourhood of q. We see that S is a submanifold of dimension $m - 1$ orthogonal to $X = \text{grad } f$ and, by the continuity of the flow, the orbit through any point near q intersects S. As $q \in \omega(p)$ there exists a sequence p_n in the orbit of p converging to q. Thus the orbit of p intersects S in more than one point (in fact, in infinitely many points) which is absurd since f is increasing along orbits. On the other hand, it is clear that, if the ω-limit of an orbit of a gradient vector field contains more than one singularity, it must contain infinitely many. We are going to show that this can in fact occur.

Figure 7   Figure 8

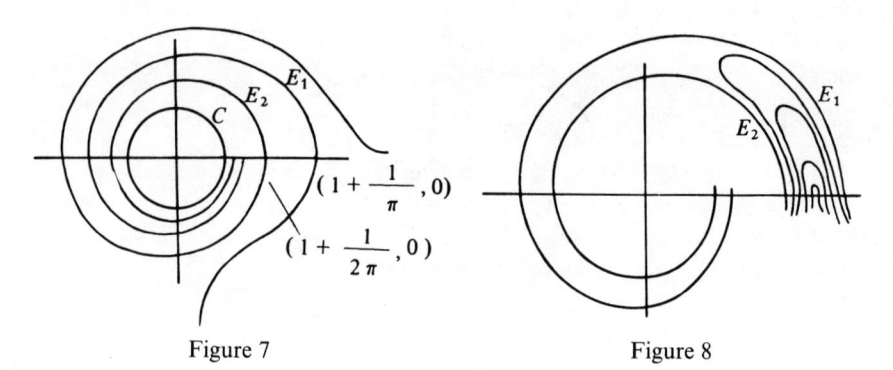

Let $f: \mathbb{R}^2 \to \mathbb{R}$ be defined by

$$f(r\cos\theta, r\sin\theta) = \begin{cases} e^{1/(r^2-1)}, & \text{if } r < 1; \\ 0, & \text{if } r = 1; \\ e^{-1/(r^2-1)}\sin(1/(r-1) - \theta), & \text{if } r > 1. \end{cases}$$

Let $X = \operatorname{grad} f$. We have $X(r\cos\theta, r\sin\theta) = 0$ if and only if $r = 0$ or $r = 1$. We are going to show that there exists an orbit of X whose ω-limit is the circle C with centre at the origin and radius 1. Note that $f^{-1}(0) = C \cup E_1 \cup E_2$, where E_1 and E_2 are the spirals (Figure 7) defined by

$$E_1 = \{(r\cos\theta, r\sin\theta); r = 1 + 1/(\pi + \theta), -\pi < \theta < \infty\},$$

$$E_2 = \{(r\cos\theta, r\sin\theta); r = 1 + 1/(2\pi + \theta), -2\pi < \theta < \infty\}.$$

Let us consider the region $U = \{(r\cos\theta, r\sin\theta); 1 + 1/(2\pi + \theta) < r < 1 + 1/(\pi + \theta), \theta \geq 0\}$ and let I be the interval $\{(x, 0); 1 + 1/2\pi \leq x \leq 1 + 1/\pi\}$. We shall show the existence of a point $p_0 \in I$ whose positive orbit remains in the region U. Hence the ω-limit of p_0 will be the circle C. In Figure 8 we draw some level curves of the function f on U.

The intersection of the level curve through a point $p \in I$ with U is a compact segment whose ends are in I. The length of this segment tends to infinity as p approaches the ends of I.

Let $q \in E_1$. As $X(q)$ is orthogonal to E_1 and points out of U (because f is negative in U), we see that the negative orbit of q intersects one of the level curves through a point in the interior of I. So the negative orbit of q intersects I. Therefore, the set $J = \{p \in I; X_t(p) \in U$ for $0 \leq t < s$ and $X_s(p) \in E_1\}$ is nonempty. Moreover, given $q \in E_1$, there exists $p \in J$ such that the positive orbit of p contains q and the segment of the orbit between p and q is in U. On the other hand, given $q \in E_2$, the negative orbit of q also intersects I so that $J \neq I$.

Let p_0 be the infimum of J. We claim that the positive orbit of p_0 remains in U. For if this is not the case there exists a point q in the positive orbit of p_0 such that the segment of the orbit between p_0 and q is contained in U and $X_t(q) \notin U$ for sufficiently small $t > 0$. Thus $q \in E_1$ or $q \in E_2$ or $q \in I$. If $q \in E_1$ then each positive orbit through a point of J intersects E_1 in a point of the segment between $(1 + 1/\pi, 0)$ and q. This is absurd because the negative

§1 Vector Fields on Manifolds

orbit through any point of E_1 intersects I and therefore J. If $q \in E_2$ or $q \in I$ then the positive orbit of each point near p_0 leaves U without meeting E_1 which is absurd since p_0 is the infimum of J. Thus the positive orbit of p_0 is contained in U, which proves our claim.

We remark that the vector field on S^2 in Example 1 is the gradient of the height function that measures height above the plane tangent to the sphere S^2 at p_S. Other simple examples can be obtained by considering the function on a surface in $\mathbb{R}^3$ which measures the distance from its points to a plane. Some of these examples will be considered later.

Next we shall discuss some general properties of ω-limit sets.

1.4 Proposition. *Let $X \in \mathfrak{X}^r(M)$ where M is a compact manifold and let $p \in M$. Then*

(a) $\omega(p) \neq \emptyset$,
(b) $\omega(p)$ *is closed*,
(c) $\omega(p)$ *is invariant by the flow of X, that is $\omega(p)$ is a union of orbits of X, and*
(d) $\omega(p)$ *is connected.*

PROOF. Let $t_n \to \infty$ and $p_n = X_{t_n}(p)$. As M is compact p_n has a convergent subsequence whose limit belongs to $\omega(p)$. Thus $\omega(p) \neq \emptyset$. Suppose now that $q \notin \omega(p)$. Then it has a neighbourhood $V(q)$ disjoint from $\{X_t(p); t \geq T\}$ for some $T > 0$. This implies that the points of $V(q)$ do not belong to $\omega(p)$ and so $\omega(p)$ is closed. Next suppose that $q \in \omega(p)$ and $\tilde{q} = X_s(q)$. Take $t_n \to \infty$ with $X_{t_n}(p) \to q$. Then $X_{t_n + s}(p) = X_s X_{t_n}(p)$ converges to $X_s(q) = \tilde{q}$ and so $\tilde{q} \in \omega(p)$. This shows that $\omega(p)$ is invariant by the flow. Suppose that $\omega(p)$ is not connected. Then we can choose open sets V_1 and V_2 such that $\omega(p) \subset V_1 \cup V_2$, $\omega(p) \cap V_1 \neq \emptyset$, $\omega(p) \cap V_2 \neq \emptyset$ and $\overline{V}_1 \cap \overline{V}_2 = \emptyset$. The orbit of p accumulates on points of both V_1 and V_2 so, given $T > 0$, there exists $t > T$ such that $X_t(p) \in M - (V_1 \cup V_2) = K$, say. Thus there exists a sequence $t_n \to \infty$ with $X_{t_n}(p) \in K$. Passing to a subsequence, if necessary, we have $X_{t_n}(p) \to q$ for some $q \in K$. But this implies that $q \in \omega(p) \subset V_1 \cup V_2$ which is absurd. □

Remark. Clearly the properties above are also true for the α-limit set. On the other hand, if the manifold were not compact we should have to restrict attention to an orbit contained in a compact set for positive time (or for negative time). Figure 9 shows an orbit of a vector field on $\mathbb{R}^2$ whose ω-limit is not connected.

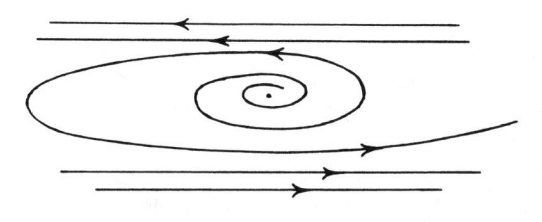

Figure 9

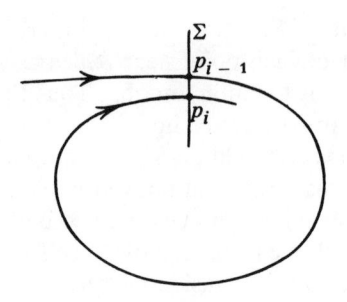

Figure 10

As we have already seen, the ω-limit of an orbit of an irrational flow on the torus T^2 is the whole torus. There are more complex examples of vector fields on T^2 with rather complicated ω-limit sets as in Example 13 of Chapter 4. Meanwhile for the sphere S^2 the situation is much simpler because of the following topological fact: every continuous closed curve without self-intersections separates S^2 into two regions that are homeomorphic to discs (the Jordan Curve Theorem). The structure of an ω-limit set of a vector field on S^2 is described by the Poincaré–Bendixson Theorem whose proof we develop now through a sequence of lemmas.

Let $X \in \mathfrak{X}^r(S^2)$, $r \geq 1$.

1.5 Lemma. *Let $\Sigma \subset S^2$ be an arc transversal to X. The positive orbit through a point $p \in S^2$, $\mathcal{O}_+(p)$, intersects Σ in a monotonic sequence; that is, if p_i is the ith intersection of $\mathcal{O}_+(p)$ with Σ, then $p_i \in [p_{i-1}, p_{i+1}] \subset \Sigma$.*

PROOF. Consider the piece of orbit from p_{i-1} to p_i together with the segment $[p_{i-1}, p_i] \subset \Sigma$. These make a closed curve which bounds a disc D and, as Σ is transversal to X which points inwards into D, the positive orbit of p_i is contained in D. Thus $p_i \in [p_{i-1}, p_{i+1}]$. □

Corollary. *The ω-limit of a trajectory γ intersects Σ in at most one point.*

PROOF. Suppose that $\omega(\gamma)$ contains two points q_1 and q_2 of Σ. Let p_n be the sequence of intersections of γ with Σ. Then there exist subsequences of p_n converging to q_1 and q_2, which leads to a contradiction because of the monotonic property of p_n. □

1.6 Lemma. *If the ω-limit set of a trajectory γ does not contain singularities then $\omega(\gamma)$ is a closed orbit and the orbits through points close enough to a point p of γ have the same closed orbit as their ω-limits.*

PROOF. Let $q \in \omega(\gamma)$. We show that the orbit of q is closed. Take $x \in \omega(q)$, which cannot, therefore, be a singularity. Consider a segment Σ transversal

§1 Vector Fields on Manifolds   17

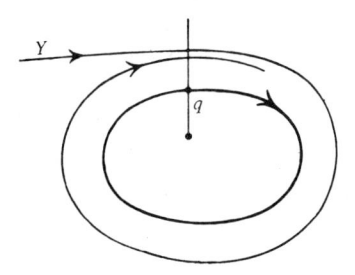

Figure 11

to X containing x. By the previous lemma the positive orbit of q intersects Σ in a monotonic sequence $q_n \to x$. As $q_n \in \omega(\gamma)$ the above corollary shows that $q_n = x$ for all n. Thus the orbit of q is closed. By taking a transversal segment containing q we conclude as in Lemma 1.5 that $\omega(\gamma)$ reduces to the orbit of q. The proof of the second statement is immediate. □

1.7 Lemma. *Let p_1 and p_2 be distinct singularities of the vector field contained in the ω-limit of a point $p \in S^2$. Then there exists at most one orbit $\gamma \subset \omega(p)$ such that $\alpha(\gamma) = p_1$ and $\omega(\gamma) = p_2$.*

PROOF. To get a contradiction suppose there exist two orbits $\gamma_1, \gamma_2 \subset \omega(p)$ such that $\alpha(\gamma_i) = p_1$ and $\omega(\gamma_i) = p_2$ for $i = 1, 2$. The curve C_1, consisting of the orbits γ_1, γ_2 and the points p_1, p_2, separates S^2 into two discs, one of which contains p as shown in Figure 12. Let Σ_1 and Σ_2 be segments transversal to X through the points $q_1 \in \gamma_1$ and $q_2 \in \gamma_2$ respectively. As $\gamma_1, \gamma_2 \subset \omega(p)$ the positive orbit of p intersects Σ_1 in a point a and later intersects Σ_2 in a point b. Consider the curve C_2 consisting of the arcs $ab \subset \mathcal{O}(p)$, $bq_2 \subset \Sigma_2$, $q_2 p_2 \subset \gamma_2$, $p_2 q_1 \subset \gamma_1$, $q_1 a \subset \Sigma_1$ and the point p_2. We see that C_2 separates S^2 into two discs A and B. The positive orbit of the point b remains entirely in A which gives a contradiction since $\gamma_1, \gamma_2 \subset \omega(p)$. □

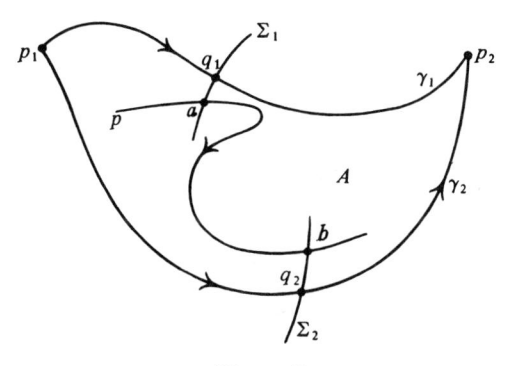

Figure 12

1.8 Theorem (Poincaré–Bendixson). *Let $X \in \mathfrak{X}^r(S^2)$ be a vector field with a finite number of singularities. Take $p \in S^2$ and let $\omega(p)$ be the ω-limit set of p. Then one of the following possibilities holds:*

(1) *$\omega(p)$ is a singularity;*
(2) *$\omega(p)$ is a closed orbit;*
(3) *$\omega(p)$ consists of singularities $p_1, \ldots, p_n$ and regular orbits such that if $\gamma \subset \omega(p)$ then $\alpha(\gamma) = p_i$ and $\omega(\gamma) = p_j$.*

PROOF. If $\omega(p)$ does not contain a singularity then, by Lemma 1.6, $\omega(p)$ is a closed orbit. If $\omega(p)$ does not contain regular points then $\omega(p)$ is a unique singularity since X only has a finite number of singularities and $\omega(p)$ is connected.

Suppose, then, that $\omega(p)$ contains regular points and singularities. Let γ be a regular trajectory contained in $\omega(p)$. We claim that $\omega(\gamma)$ is a singularity. If $\omega(\gamma)$ contains some regular point q take a segment Σ through q transversal to X. As $\gamma \subset \omega(p)$ the Corollary to Lemma 1.5 says that γ intersects Σ in only one point. By Lemma 1.6 γ is a closed trajectory and $\omega(p) = \gamma$. This is absurd because $\omega(p)$ contains singularities. Thus $\omega(\gamma)$ is a singularity. Similarly $\alpha(\gamma)$ is a singularity. □

In the following examples we illustrate some facts about this theorem.

EXAMPLE. Let X be a vector field on S^2 as in Figure 13. The north and south poles are singularities and the equator is a closed orbit. The other orbits are born at a pole and die at the equator.

Let $\varphi \colon S^2 \to \mathbb{R}$ be a nonnegative C^∞ function which vanishes precisely on the equator of the sphere. Consider the vector field $Y = \varphi \cdot X$. Each point of the equator is a singularity of Y and the ω-limit set of a point p which is neither a pole nor on the equator is the whole of the equator. This example shows that the Poincaré–Bendixson Theorem is not valid without the hypothesis of a finite number of singularities.

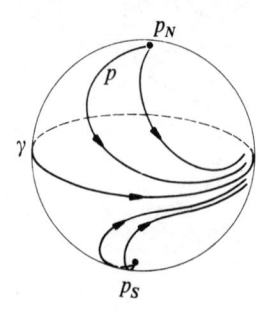

Figure 13

§2 The Topology of the Space of C^r Maps

Southern hemisphere   Northern hemisphere

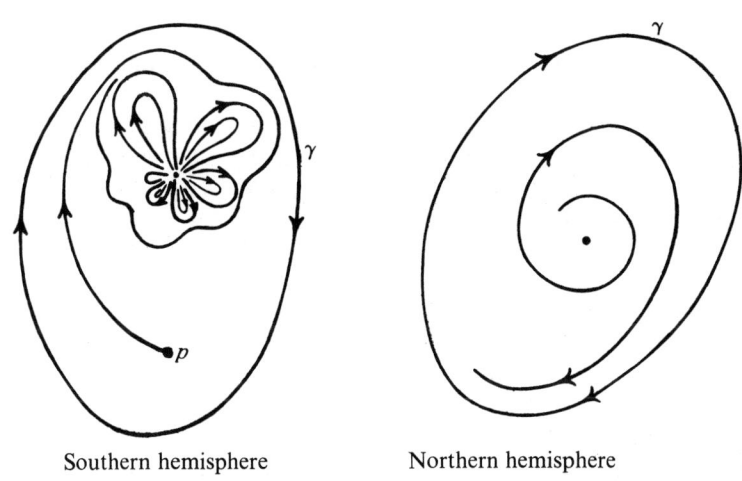

Figure 14

EXAMPLE. Let X be a vector field on S^2 as in Figure 14. The vector field X has two singularities p_S and p_N and one closed orbit γ. The orbits in the northern hemisphere have p_N as α-limit and γ as ω-limit. In the southern hemisphere we have the singularity p_S which is the centre of a rose with infinitely many petals each bounded by an orbit which is born in p_S and dies in p_S. In the interior of each petal the situation is as in Figure 15.

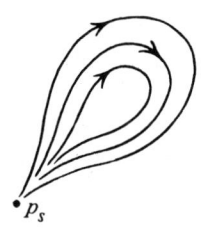

Figure 15

The other orbits in the southern hemisphere have γ as α-limit and the edge of the rose as ω-limit. Therefore the ω-limit of an orbit can contain infinitely many regular orbits, which shows that Lemma 1.7 is not valid if $p_1 = p_2$.

§2 The Topology of the Space of C^r Maps

We introduce here a natural topology on the space $\mathfrak{X}^r(M)$ of C^r vector fields on a compact manifold M. In this topology, two vector fields $X, Y \in \mathfrak{X}^r(M)$ will be close if the vector fields and their derivatives up to order r are close at all points of M.

Let us consider first the space $C^r(M, \mathbb{R}^s)$ of C^r maps, $0 \leq r < \infty$, defined on a compact manifold M. We have a natural vector space structure on $C^r(M, \mathbb{R}^s)$: $(f + g)(p) = f(p) + g(p)$, $(\lambda f)(p) = \lambda f(p)$ for $f, g \in C^r(M, \mathbb{R}^s)$ and $\lambda \in \mathbb{R}$. Let us take a finite cover of M by open sets $V_1, \ldots, V_k$ such that each V_i is contained in the domain of a local chart (x_i, U_i) with $x_i(U_i) = B(2)$ and $x_i(V_i) = B(1)$, where $B(1)$ and $B(2)$ are the balls of radii 1 and 2 with centre at the origin in $\mathbb{R}^m$. For $f \in C^r(M, \mathbb{R}^s)$ we write $f^i = f \circ x_i^{-1} : B(2) \to \mathbb{R}^s$. We define

$$\|f\|_r = \max_i \sup\{\|f^i(u)\|, \|df^i(u)\|, \ldots, \|d^r f^i(u)\| ; u \in B(1)\}.$$

2.1 Proposition. $\| \ \|_r$ *is a complete norm on* $C^r(M, \mathbb{R}^s)$.

PROOF. It is immediate that $\| \ \|_r$ is a norm on $C^r(M, \mathbb{R}^s)$. It remains to prove that every Cauchy sequence converges. Let $f_n : M \to \mathbb{R}^s$ be a Cauchy sequence in the norm $\| \ \|_r$. If $p \in M$ then $f_n(p)$ is a Cauchy sequence in $\mathbb{R}^s$ and so converges. We put $f(p) = \lim f_n(p)$. In particular, $f_n^i(u) \to f^i(u)$ for $u \in B(1)$ and $i = 1, \ldots, k$. On the other hand, for each $u \in B(1)$, $df_n^i(u)$ is a Cauchy sequence in $L(\mathbb{R}^m, \mathbb{R}^s)$ and so converges to a linear transformation $T^i(u)$. We claim that the convergence $df_n^i \to T^i$ is uniform. For notice that

$$\|df_n^i(u) - T^i(u)\| \leq \|df_n^i(u) - df_{n'}^i(u)\| + \|df_{n'}^i(u) - T^i(u)\|.$$

Given $\varepsilon > 0$, there exists n_0 such that, if $n, n' \geq n_0$, then $\|df_n^i(u) - df_{n'}^i(u)\| < \varepsilon/2$ for all $u \in B(1)$. On the other hand, for each $u \in B(1)$ there exists $n' \geq n_0$, which depends on u, such that $\|df_{n'}^i(u) - T^i(u)\| < \varepsilon/2$. Thus, for $n \geq n_0$, we have $\|df_n^i(u) - T^i(u)\| < \varepsilon$ for all $u \in B(1)$. By Proposition 0.0, f^i is of class C^1 and $df^i = T^i$. It follows that $f_n \to f$ in the norm $\| \ \|_1$. With the same argument we can show by induction that f is of class C^r and $f_n \to f$ in the norm $\| \ \|_r$. □

It is easy to see that the topology defined on $C^r(M, \mathbb{R}^s)$ by the norm $\| \ \|_r$ does not depend on the cover $V_1, \ldots, V_k$ of M.

Next we describe some important properties of the space $C^r(M, \mathbb{R}^s)$ with the C^r topology.

A subset of a topological space is *residual* if it contains a countable intersection of open dense sets. A topological space is a *Baire space* if every residual subset is dense. As $C^r(M, \mathbb{R}^s)$ is a complete metric space we immediately obtain the following proposition.

2.2 Proposition. $C^r(M, \mathbb{R}^s)$ *is a Baire space*. □

Let us show that $C^r(M, \mathbb{R}^s)$ contains a countable dense subset. For $f \in C^r(M, \mathbb{R}^s)$ consider $f^i = f \circ x_i^{-1} : B(2) \subset \mathbb{R}^m \to \mathbb{R}^s$. Note that the map $j^r f^i : B(2) \to B(2) \times \mathbb{R}^s \times L(\mathbb{R}^m, \mathbb{R}^s) \times \cdots \times L^r(\mathbb{R}^m; \mathbb{R}^s) = E$ defined by

§2 The Topology of the Space of C^r Maps

$j^r f^i(u) = (u, f^i u, df^i(u), \ldots, d^r f^i(u))$ is continuous. Thus $J^r(f^i) = j^r f^i(\overline{B(1)})$ is a compact subset of E. It is easy to see that, if $\mathcal{W}$ is a neighbourhood of f in $C^r(M, \mathbb{R}^s)$, there exists a neighbourhood W of $J^r(f) = J^r(f^1) \times \cdots \times J^r(f^k)$ in $E \times E \times \cdots \times E$ such that, if $g \in C^r(M, \mathbb{R}^s)$ and $J^r g = J^r g^1 \times \cdots \times J^r g^k \subset W$, then $g \in \mathcal{W}$.

2.3 Proposition. $C^r(M, \mathbb{R}^s)$ *is separable; that is, it has a countable base of open sets.*

PROOF. As $E^k = E \times \cdots \times E$ is an open set in a Euclidean space, there exists a countable base of open sets $E_1, \ldots, E_i, \ldots$ for the topology of E^k. Let $\tilde{E}_1, \ldots, \tilde{E}_j, \ldots$ be the collection of those open subsets of E^k that are finite unions of the E_i. Let $\mathscr{E}_j = \{g \in C^r(M, \mathbb{R}^s); J^r(g) \subset \tilde{E}_j\}$ for each j. It is clear that $\mathscr{E}_j$ is open in $C^r(M, \mathbb{R}^s)$. Let $\mathcal{W}$ be a neighbourhood of f in $C^r(M, \mathbb{R}^s)$ and W a neighbourhood of $J^r(f)$ such that $g \in \mathcal{W}$ if $J^r(g) \subset W$. As $J^r(f)$ is compact there exists a finite cover of $J^r(f)$ by open sets E_i contained in W. Let $\tilde{E}_j$ be the union of these E_i; it is clear that $J^r(f) \subset \tilde{E}_j \subset W$. Therefore, $\mathscr{E}_j$ contains f and is contained in $\mathcal{W}$. This shows that $\{\mathscr{E}_1, \ldots, \mathscr{E}_j, \ldots\}$ is a countable base for the topology of $C^r(M, \mathbb{R}^s)$. □

Next we show that every C^r map can be approximated in the C^r topology by a C^∞ map.

2.4 Lemma. *Let $f: U \subset \mathbb{R}^m \to \mathbb{R}^s$ be a C^r map with U an open set. Let $K \subset U$ be compact. Given $\varepsilon > 0$ there exists a C^∞ map $g: \mathbb{R}^m \to \mathbb{R}^s$ such that $\|f - g\|_r < \varepsilon$ on K.*

PROOF. Let us consider a bump function $\varphi: \mathbb{R}^m \to \mathbb{R}$ which takes the value 1 on K and is 0 outside a neighbourhood of K contained in U. Taking $h = \varphi f$ we have $h = f$ in K and $h = 0$ outside U. As h is of class C^r and K is compact, there exists $\delta > 0$ such that

$$\sup\{\|d^j h(u + v) - d^j h(u)\|; u \in K, \|v\| < \delta\} < \varepsilon,$$

where d^j denotes the derivative of order j, for $j = 1, \ldots, r$. Let $\varphi_\delta: \mathbb{R}^m \to \mathbb{R}$ be a bump function such that $\varphi_\delta(v) = 0$ if $\|v\| > \delta$ and $\int \varphi_\delta(v) \, dv = 1$. We define $g: \mathbb{R}^m \to \mathbb{R}^s$ by $g(u) = \int \varphi_\delta(v) h(u + v) \, dv = \int \varphi_\delta(z - u) h(z) \, dz$. It follows that

$$d^j g(u) = \int \varphi_\delta(v) d^j h(u + v) \, dv$$

and

$$d^j g(u) = (-1)^j \int d^j \varphi_\delta(z - u) h(z) \, dz.$$

From the second expression it follows that g is C^∞. On the other hand, from the first expression we have

$$\|d^j g(u) - d^j h(u)\| = \left\|\int \varphi_\delta(v) d^j h(u+v)\, dv - \int \varphi_\delta(v)\, d^j h(u)\, dv\right\|$$

$$= \left\|\int \varphi_\delta(v)(d^j h(u+v) - d^j h(u))\, dv\right\| < \varepsilon \quad \text{for } u \in K.$$

As $h = f$ on K, g satisfies the required condition. $\square$

2.5 Proposition. *The subset of C^∞ maps is dense in $C^r(M, \mathbb{R}^s)$.*

PROOF. Let (x_i, U_i), $i = 1, \ldots, k$ be local charts with $x_i(U_i) = B(2)$ and $M = \bigcup V_i$ where $V_i = x_i^{-1}(B(1))$. Take a partition of unity $\{\varphi_i : M \to \mathbb{R}\}$ subordinate to the cover $\{V_i\}$. Let $f \in C^r(M, \mathbb{R}^s)$ and $\varepsilon > 0$. By the previous lemma, given $\delta > 0$, there exists $\tilde{g}^i : \mathbb{R}^m \to \mathbb{R}^s$ of class C^∞ such that $\|f^i - \tilde{g}^i\|_r < \delta$ on $\overline{B(1)}$, where $f^i = f \circ x_i^{-1}$. Taking δ sufficiently small we have $\|\varphi_i f - \varphi_i \tilde{g}^i \circ x_i\|_r < \varepsilon/k$. Thus, $g = \sum \varphi_i \tilde{g}^i \circ x_i$ is of class C^∞ and $\|f - g\|_r = \|\sum(\varphi_i f - \varphi_i \tilde{g}^i \circ x_i)\|_r < \varepsilon/k + \cdots + \varepsilon/k = \varepsilon$. $\square$

Let us now consider a closed manifold N. By Whitney's Theorem, we may assume that N is a closed submanifold of $\mathbb{R}^s$, for some $s > 0$. As N is a closed subset of $\mathbb{R}^s$, $C^r(M, N)$ is closed in $C^r(M, \mathbb{R}^s)$. Therefore, $C^r(M, N)$, with the topology induced from $C^r(M, \mathbb{R}^s)$, is a separable Baire space.

Let $N_1 \subset \mathbb{R}^{s_1}$, $N_2 \subset \mathbb{R}^{s_2}$ be closed manifolds and $\Phi : N_1 \to N_2$ a C^l map, $r \le l \le \infty$. Define $\Phi_* : C^r(M, N_1) \to C^r(M, N_2)$ by $\Phi_* f = \Phi \circ f$.

2.6 Proposition. *The map Φ_* is continuous.*

PROOF. By Corollary 2 of Proposition 0.16 there exists a C^l map, $\tilde{\Phi} : \mathbb{R}^{s_1} \to \mathbb{R}^{s_2}$, such that $\tilde{\Phi} = \Phi$ on N_1. Let (x_i, U_i), $i = 1, \ldots, k$, be local charts in M as before. It is easy to see that, given $\varepsilon > 0$, there exists $\delta > 0$ such that, if $\|f \circ x_i^{-1} - g \circ x_i^{-1}\|_r < \delta$ on $\overline{B(1)}$, then $\|(\tilde{\Phi} \circ f - \tilde{\Phi} \circ g) \circ x_i^{-1}\|_r < \varepsilon$ on $\overline{B(1)}$. Therefore, if $\|f - g\|_r < \delta$ then we have $\|\Phi \circ f - \Phi \circ g\|_r = \|\tilde{\Phi} \circ f - \tilde{\Phi} \circ g\|_r < \varepsilon$, which shows the continuity of Φ_*. $\square$

Now let M and N be abstract manifolds with M compact. We can define a C^r topology on $C^r(M, N)$. For this it is sufficient to embed N in a Euclidean space $\mathbb{R}^2$ (Theorem 0.17). The proposition above shows that this topology is independent of the embedding.

2.7 Proposition. *The subset of maps of class C^∞ is dense in $C^r(M, N)$.*

PROOF. Suppose $N \subset \mathbb{R}^s$. Let $V \subset \mathbb{R}^s$ be a tubular neighbourhood of N and $\pi : V \to N$ the associated projection. Let $f \in C^r(M, N)$; by Proposition 2.5 we can approximate f by a C^∞ map $g : M \to \mathbb{R}^s$. Thus, $\pi \circ g : M \to N$ is C^∞ and $\pi \circ g$ approximates f because π_* is continuous. $\square$

2.8 Proposition. *The set $\text{Diff}^r(M)$ consisting of the C^r diffeomorphisms of M is open in $C^r(M, M)$.*

PROOF. Let $f \in \text{Diff}^r(M)$. We can suppose that $M \subset \mathbb{R}^s$. If $p \in M$ there exist, by the Inverse Function Theorem, neighbourhoods V_p of p in M and $\mathscr{V}_p$ of f in $C^r(M, M)$ such that, if $g \in \mathscr{V}_p$ then $g|V_p$ is a diffeomorphism. Let $V_{p_1}, \ldots, V_{p_j}$ be a finite subcover of M and put $\mathscr{V} = \bigcap_{i=1}^{j} \mathscr{V}_{p_i}$. If δ is the Lebesgue number of this cover then $d(p, q) < \delta$ and $p \neq q$ imply $g(p) \neq g(q)$ for every $g \in \mathscr{V}$. On the other hand, $\rho = \inf\{d(f(p), f(q)); p, q \in M$ and $d(p, q) \geq \delta\}$ is positive. By reducing $\mathscr{V}$ we can therefore suppose that if $g \in \mathscr{V}$ then g is injective. Since g is a local diffeomorphism it is a diffeomorphism. □

It follows from the previous proposition that $\text{Diff}^r(M)$ is a separable Baire space and that the subset of C^∞ diffeomorphisms is dense in it.

Finally let us consider the space $\mathfrak{X}^r(M)$ of C^r vector fields on a compact manifold M. Supposing that $M \subset \mathbb{R}^s$ we can easily see that $\mathfrak{X}^r(M)$ is a closed subspace of $C^r(M, \mathbb{R}^s)$. Thus, $\mathfrak{X}^r(M)$ is a separable Baire space. Let us show that every vector field $X \in \mathfrak{X}^r(M)$ can be approximated by a C^∞ vector field. In fact, X can be approximated by a C^∞ map $Y: M \to TM$. Let $\pi: TM \to M$ be the natural projection. As π_* is continuous, $\pi \circ Y$ is C^r close to $\pi \circ X = id_M$. By Proposition 2.8, $\varphi = \pi \circ Y$ is a diffeomorphism. Let $Z = Y \circ \varphi^{-1}$; Z is a C^∞ vector field since φ is C^∞ and $\pi \circ Z = id_M$. Moreover Z approximates X. (See Exercise 15 at the end of the chapter.)

§3 Transversality

Let $S \subset N$ be a C^r submanifold and let $f: M \to N$ be a C^k map where k, $r \geq 1$. We say that f is *transversal to S at a point* $p \in M$ if either $f(p) \notin S$ or $df_p(TM_p) + TS_{f(p)} = TN_{f(p)}$; that is, if the image of TM_p by df_p contains a subspace of $TN_{f(p)}$ that is complementary to $TS_{f(p)}$. We say that f is *transversal to S*, $f \pitchfork S$, if it is transversal to S at each point $p \in M$. Note that if the dimension of M is less than the codimension of S then f is transversal to S if and only if $f(M) \cap S = \emptyset$.

An interesting special case happens when $f: M \to N$ is a submersion; here f is transversal to every submanifold $S \subset N$. We define transversality between two submanifolds $S_1, S_2 \subset N$ in the following way: S_1 *is transversal to* S_2 if the inclusion map $i: S_1 \to N$ is transversal to S_2.

We recall that every submanifold is locally the inverse image of a regular value. More precisely, if $q \in S \subset N^n$ then there exist a neighbourhood V_q of q in N and a diffeomorphism $\varphi: V_q \to \mathbb{R}^s \times \mathbb{R}^{n-s}$ of class C^r such that

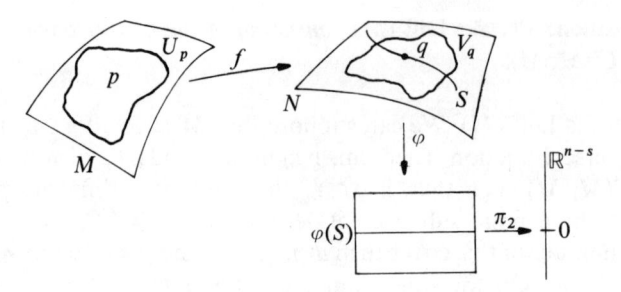

Figure 16

$\varphi(S \cap V_q) = \mathbb{R}^s \times \{0\}$; thus, $S \cap V_q = (\pi_2 \circ \varphi)^{-1}(0)$ where $\pi_2 \colon \mathbb{R}^s \times \mathbb{R}^{n-s} \to \mathbb{R}^{n-s}$ is the natural projection. Now let $f \colon M \to N$ and let $U_p \subset M$ be a neighbourhood of p with $f(U_p) \subset V_q$ where $q = f(p)$. Consider the map $\pi_2 \circ \varphi \circ f | U_p$ as in Figure 16.

The next proposition is immediate.

3.1 Proposition. *The map $f \colon M \to N$ is transversal to S at $p \in f^{-1}(S)$ if and only if 0 is a regular value of $\pi_2 \circ \varphi \circ f | U_p$ for some neighbourhood U_p as above.* □

Corollary. *Let $f \in C^k(M, N^n)$ and let S^s be a C^r submanifold of N with k, $r \geq 1$. If f is transversal to S then $f^{-1}(S)$ is either empty or a C^l submanifold of codimension $n - s$, where $l = \min(k, r)$.* □

3.2 Proposition. *If M is compact and $S \subset N$ is closed then those maps in $C^k(M, N)$ that are transversal to S form an open subset.*

PROOF. Let $f \in C^k(M, N)$ be transversal to S. For each $q \in S$ we take a neighbourhood V_q and a diffeomorphism $\varphi_q \colon V_q \to \mathbb{R}^s \times \mathbb{R}^{n-s}$ such that $\varphi_q(S \cap V_q) = \mathbb{R}^s \times \{0\}$. For each $p \in f^{-1}(q)$ let us consider a neighbourhood $\overline{U}_p$ such that $f(\overline{U}_p) \subset V_q$ and the derivative of $\pi_2 \circ \varphi_q \circ f$ is surjective at every point of $\overline{U}_p$. There exists a neighbourhood $\mathscr{V}_p(f) \subset C^k(M, N)$ such that the same happens for $\pi_2 \circ \varphi_q \circ g$ on $\overline{U}_p$ for every $g \in \mathscr{V}_p$. Let $U_{p_1}, \ldots, U_{p_m}$ be a finite subcover of the compact set $f^{-1}(S)$ and put $U = U_{p_1} \cup \cdots \cup U_{p_m}$ and $\mathscr{V}(f) = \mathscr{V}_{p_1} \cap \cdots \cap \mathscr{V}_{p_m}$. It is clear that if $g \in \mathscr{V}$ then g is transversal to S at all points of U. As $M - U$ is compact and $f(M - U) \cap S = \varnothing$ we can reduce $\mathscr{V}$ if necessary and obtain $g(M - U) \cap S = \varnothing$ for all $g \in \mathscr{V}$. Thus, every $g \in \mathscr{V}$ is transversal to S, which proves the proposition. □

We remark that, if $S \subset N$ is not closed, the proposition remains true for closed subsets of S. That is, if $\tilde{S} \subset S$ is closed in N, then the set of maps $f \colon M \to N$ that are transversal to S on $f^{-1}(\tilde{S})$ is open in $C^k(M, N)$.

Let Λ, M, N be manifolds and let $F \colon \Lambda \times M \to N$ be a C^∞ map. For $\lambda \in \Lambda$ we denote by $F_\lambda \colon M \to N$ the map defined by $F_\lambda(p) = F(\lambda, p)$. Let $S \subset N$ be a C^∞ submanifold and let $T_S \subset \Lambda$ be the set of points λ such that F_λ is transversal to S.

3.3 Proposition. *If $F: \Lambda \times M \to N$ is transversal to $S \subset N$ then T_S is residual in Λ.*

PROOF. Let $\pi: \Lambda \times M \to \Lambda$ be the natural projection. Since F is transversal to S, $\tilde{S} = F^{-1}(S)$ is a submanifold of $\Lambda \times M$ and $\pi_S = \pi | \tilde{S}: \tilde{S} \to \Lambda$ is a C^∞ map. It is easy to see that F_λ is transversal to S if and only if λ is a regular value of π_S. The proposition now follows from Sard's Theorem. □

Corollary 1. *Let $f: M \to \mathbb{R}^n$ be of class C^∞ and let $S \subset \mathbb{R}^n$ be a submanifold. The set of vectors $v \in \mathbb{R}^n$ such that $f + v$ is transversal to S is residual.*

PROOF. The map $F: \mathbb{R}^n \times M \to \mathbb{R}^n$ defined by $F(v, p) = f(p) + v$ is a submersion and, therefore, transversal to S. The corollary now follows immediately from the proposition. □

Corollary 2. *If M is a compact manifold then the set $T_S \subset C^k(M, \mathbb{R}^n)$ of maps that are transversal to a closed submanifold $S \subset \mathbb{R}^n$ is open and dense.*

PROOF. The openness of T_S follows from Proposition 3.2 while the density comes from Corollary 1 and the density of C^∞ maps in $C^k(M, R^n)$. □

We emphasize that the manifolds Λ, M and N in Proposition 3.3 are not necessarily compact. Note too that in Proposition 3.3 and Corollary 1 we require the maps and the submanifold S to be of class C^∞. This is because we use Sard's Theorem in the proof.

3.4 Theorem (Thom). *Suppose that M is compact and that $S \subset N$ is a closed submanifold. The set of maps $f \in C^k(M, N)$ that are transversal to S is open and dense.*

PROOF. As we have already shown the openness it remains to prove the density of maps transversal to S. Take $f \in C^k(M, N)$. We have to show that, for each $p \in M$, there exist neighbourhoods U_p of p in M and $\mathscr{V}_p$ of f in $C^k(M, N)$ such that the set of $g \in \mathscr{V}_p$ that are transversal to S on U_p is open and dense in $\mathscr{V}_p$. In fact, if $U_{p_1}, \ldots, U_{p_m}$ is a finite subcover of M and $\mathscr{V}(f) = \mathscr{V}_{p_1} \cap \cdots \cap \mathscr{V}_{p_m}$, then the set of $g \in \mathscr{V}$ that are transversal to S at every point of M is open and dense. In particular we can approximate f by a map transversal to S. Let us then construct these neighbourhoods U_p and $\mathscr{V}_p$. Let $y: V \to \mathbb{R}^n$ be a local chart around $f(p)$. We take a neighbourhood U_p of p such that $f(\bar{U}_p) \subset V$ and a neighbourhood $\mathscr{V}_p$ of f such that $g(\bar{U}_p) \subset V$ for all $g \in \mathscr{V}_p$. The argument in the proof of Proposition 3.2 also shows that the set of those $g \in \mathscr{V}_p$ that are transversal to S on $\bar{U}_p$ is open. In proving density we can start from a C^∞ map $g \in \mathscr{V}_p$ because C^∞ maps are dense in $C^k(M, N)$. It is easy to see that, if $v \in \mathbb{R}^n$ has small norm, there is $g_v \in \mathscr{V}_p$ near g with $y \circ g_v = y \circ g + v$ on $\bar{U}_p$ and $g_v = g$ outside a neighbourhood of $\bar{U}_p$. By Corollary 1 of Proposition 3.3 there exists $v \in \mathbb{R}^n$ of small norm with $y \circ g + v$ transversal to $y(S) \subset \mathbb{R}^n$. Thus g_v is transversal to S on $\bar{U}_p$, which completes the proof. □

Remark. In Theorem 3.4 we required S to be a C^∞ submanifold. However, the result remains true for S of class C^r, $r \geq 1$. In fact it suffices, in the proof above, to take a C^r local chart y such that $y(S)$ is a C^∞ submanifold of $\mathbb{R}^n$ and then to approximate g by a map $\tilde{g}$ that makes $y \circ \tilde{g}$ of class C^∞ on U_p.

§4 Structural Stability

The qualitative study of a differential equation consists of a geometric description of its orbit space. Thus it is natural to ask when do two orbit spaces have the same description, the same qualitative features; this means establishing an equivalence relation between differential equations. An equivalence relation that captures the geometric structure of the orbits is what we shall define below as topological equivalence.

Let $\mathfrak{X}^r(M)$ be the space of C^r vector fields on a compact manifold M with the C^r topology, $r \geq 1$. Two vector fields X, $Y \in \mathfrak{X}^r(M)$ are *topologically equivalent* if there exists a homeomorphism $h: M \to M$ which takes orbits of X to orbits of Y preserving their orientation; this last condition means that if $p \in M$ and $\delta > 0$ there exists $\varepsilon > 0$ such that, for $0 < t < \delta$, $hX_t(p) = Y_{t'}(h(p))$ for some $0 < t' < \varepsilon$. We say that h is a *topological equivalence* between X and Y. Here we have defined an equivalence relation on $\mathfrak{X}^r(M)$. Another stronger relation is conjugacy of the flows of the vector fields. Two vector fields X and Y are *conjugate* if there exists a topological equivalence h that preserves the parameter t; that is, $hX_t(p) = Y_t h(p)$ for all $p \in M$ and $t \in \mathbb{R}$.

The next proposition, whose proof is immediate, shows some of the qualitative features of the orbit space that must be the same for two equivalent vector fields.

4.1 Proposition. *Let h be a topological equivalence between X, $Y \in \mathfrak{X}^r(M)$. Then*

(a) *$p \in M$ is a singularity of X if and only if $h(p)$ is a singularity of Y,*
(b) *the orbit of p for the vector field X, $\mathcal{O}_X(p)$, is closed if and only if $\mathcal{O}_Y(h(p))$ is closed,*
(c) *the image of the ω-limit set of $\mathcal{O}_X(p)$ by h is the ω-limit set of $\mathcal{O}_Y(h(p))$ and similarly for the α-limit set.* □

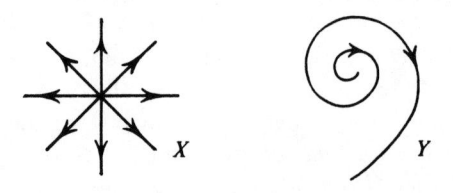

Figure 17

§4 Structural Stability

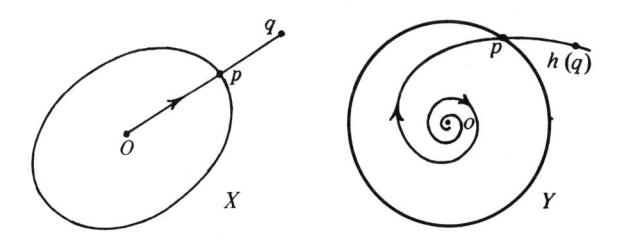

Figure 18

EXAMPLE 1. Let us consider the linear vector fields X and Y on $\mathbb{R}^2$ defined by $X(x, y) = (x, y)$ and $Y(x, y) = (x + y, -x + y)$. The corresponding flows are $X_t(x, y) = e^t(x, y)$ and $Y_t(x, y) = e^t(x \cos t + y \sin t, -x \sin t + y \cos t)$.

We shall construct a homeomorphism h of $\mathbb{R}^2$ conjugating X_t and Y_t. As 0 is the only singularity of X and Y we must have $h(0) = 0$. It is easy to see that the unit circle S^1 is transversal to X and Y. Moreover, all the trajectories of X and Y except 0 intersect S^1. We define $h(p) = p$ for $p \in S^1$. If $q \in \mathbb{R}^2 - \{0\}$ there exists a unique $t \in \mathbb{R}$ such that $X_t(q) = p \in S^1$. We put $h(q) = Y_{-t}(p) = Y_{-t} X_t(q)$. It is immediate that h is continuous and has continuous inverse on $\mathbb{R}^2 - \{0\}$. The continuity of h and its inverse at 0 can be checked using the flows of X and Y.

EXAMPLE 2. Let X and Y be linear vector fields on $\mathbb{R}^2$ whose matrices with respect to the standard basis are

$$X = \begin{pmatrix} 1 & 1 \\ -1 & 1 \end{pmatrix}, \quad Y = \begin{pmatrix} 0 & -1 \\ 1 & 0 \end{pmatrix}.$$

These vector fields are not equivalent since all the orbits of Y are closed and this is not true for X.

A vector field is structurally stable if the topological behaviour of its orbits does not change under small perturbations of the vector field. Formally we say that $X \in \mathfrak{X}^r(M)$ is *structurally stable* if there exists a neighbourhood $\mathscr{V}$ of X in $\mathfrak{X}^r(M)$ such that every $Y \in \mathscr{V}$ is topologically equivalent to X.

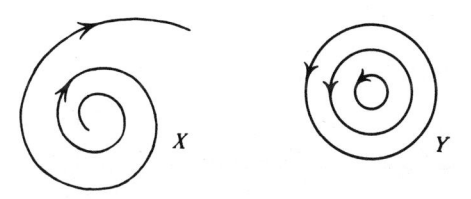

Figure 19

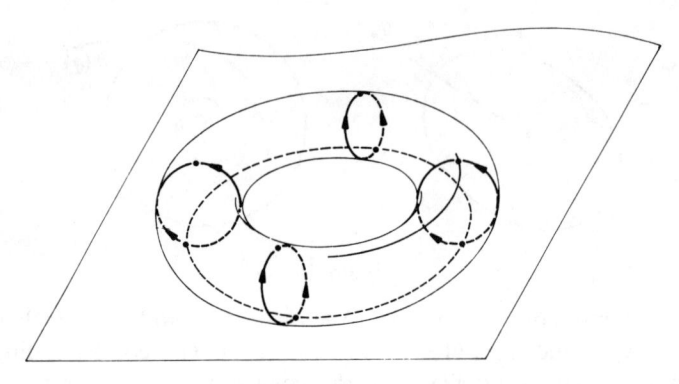

Figure 20

The zero vector field on any manifold is obviously unstable. On the other hand the linear field $X(p) = p$ considered in Example 1 is structurally stable in the space of linear vector fields on $\mathbb{R}^2$. In order to motivate the necessary conditions for structural stability that we shall introduce in later chapters we present next some examples of unstable vector fields.

EXAMPLE 3. Let us consider a rational vector field on the torus T^2, as in Example 2 of Section 1. This vector field is unstable in $\mathfrak{X}^r(T^2)$. In fact all its orbits are closed, whereas it can be approximated by an irrational vector field, which does not possess closed orbits. Actually, on (compact) manifolds of dimension two every vector field with infinitely many closed orbits is unstable. This is because we can approximate it by a vector field with only a finite number of closed orbits, as we shall see in Chapter 4.

EXAMPLE 4. Let π be a horizontal plane tangent to the torus T^2 which is embedded in the usual way in $\mathbb{R}^3$ so that π meets T^2 in a "parallel" or horizontal circle as in Figure 20. Let $f: T^2 \to \mathbb{R}$ be the function which to each point of T^2 associates its distance from π. We take $X = \operatorname{grad} f$. The parallel of T^2 contained in π is composed entirely of singularities of X. Now let $X' = \operatorname{grad} f'$, where f' is distance from a plane π' obtained from π by a small rotation. As only four of the planes parallel to π' are tangent to T^2 and each is only tangent at one point it follows that X' has only four singularities. Thus X is not equivalent to X' and so X is unstable. We shall show in Chapter 2 that every vector field with infinitely many singularities is unstable, since it can be approximated by another with a finite number of singularities.

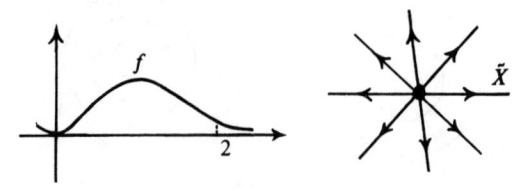

Figure 21

§4 Structural Stability

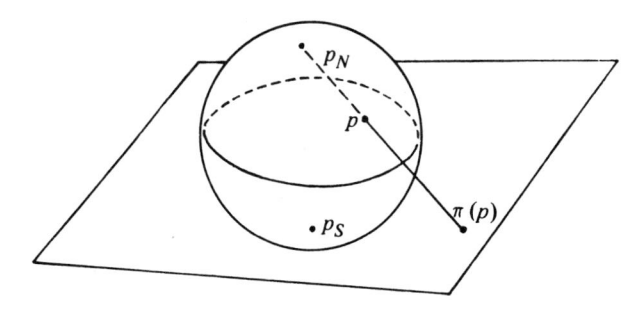

Figure 22

EXAMPLE 5. We now describe a vector field on S^2 which is unstable, even though it is topologically equivalent to the north pole–south pole vector field (Example 1 of Section 1), which is stable. Let $f : \mathbb{R} \to \mathbb{R}$ be a C^∞ map satisfying the following conditions: $f(t) > 0$ for $t \neq 0$; $f(t) = 1/t$ for $t > 1$; $f(0) = df/dt(0) = \cdots = d^r f/dt^r(0) = \cdots = 0$. Consider the vector field $\tilde{X}$ on $\mathbb{R}^2$ defined by $\tilde{X}(r \cos \theta, r \sin \theta) = (rf(r) \cos \theta, rf(r) \sin \theta)$. The vector field $\tilde{X}$ is radial and the origin is its only singularity, $d\tilde{X}(0) = 0$ and $\|\tilde{X}(p)\| = 1$ if $\|p\| \geq 1$.

Let $\pi: S^2 - \{p_N\} \to \mathbb{R}^2$ be the stereographic projection as shown in Figure 22. We define the vector field X on S^2 by $X(p) = d\pi^{-1}_{\pi(p)} \tilde{X}(\pi(p))$ if $p \neq p_N$ and $X(p_N) = 0$. We see that X is a C^∞ vector field and has two singularities p_N and p_S. Note that the identity map is a topological equivalence between X and the north pole–south pole vector field of Example 1, Section 1. Let us show that there exists a vector field C^r close to X having a closed orbit. Let $\tilde{Y}$ be the vector field on $\mathbb{R}^2$ defined by $\tilde{Y}(r \cos \theta, r \sin \theta) = (rl(r) \cos \theta + rg(r) \sin \theta, -rg(r) \cos \theta + rl(r) \sin \theta)$, where $l, g: \mathbb{R} \to \mathbb{R}$ are C^∞ maps with graphs as in Figure 23.

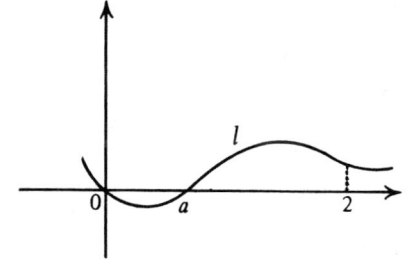

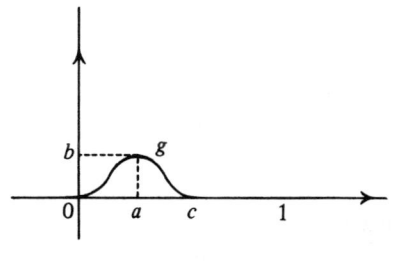

$l(0) = l(a) = 0,$ $\qquad\qquad$ $g(0) = g(t) = 0,\quad$ if $t \geq c$

$l(t) = 1/t,\quad$ if $t \geq 1$ $\qquad\qquad$ $g(t) > 0,\qquad\qquad$ if $0 < t < c$

$l(t) < 0,\quad$ if $0 < t < a$ $\qquad\quad$ $g'(a) = 0,$

$l(t) > 0,\quad$ if $t > a$ $\qquad\qquad\quad$ $g(a) = b.$

Figure 23

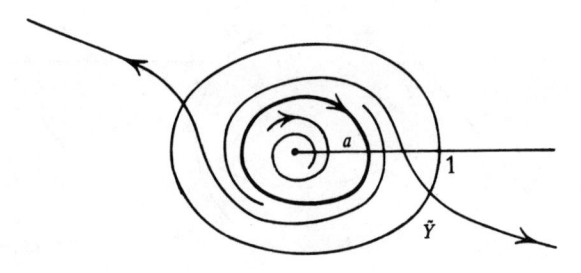

Figure 24

The circle S_a with centre at the origin and radius a is a closed orbit of $\tilde{Y}$ since $\tilde{Y}$ is tangent to S_a at each of its points. Outside the disc of radius 1, $\tilde{Y} = \tilde{X}$. Thus $\tilde{Y}$ defines a vector field Y on S^2 which has the north and south poles as "attracting" singularities, and a "repelling" closed orbit $\gamma = \pi^{-1}(S_a)$. If we choose l to be C^r close to f and g C^r close to the zero function then Y will be C^r close to X. As X is not topologically equivalent to Y, X is not structurally stable in $\mathfrak{X}^r(S^2)$.

We emphasize that, out of these examples, only the vector field of Example 5 is equivalent to a stable vector field. The basic reason for the instability in this case is that the derivative of the vector field at the singularity p_S is degenerate.

It is, in general, a delicate problem to prove the stability of a vector field. Many examples will be given in Chapter 4. We shall next analyse the stability of vector fields on S^1. This is a very simple case but it offers an insight into the general aims of the Theory of Dynamical Systems.

Let X^0 be one of the two unit vector fields on S^1. Any $X \in \mathfrak{X}^r(S^1)$ can be written in a unique way as $X(p) = f(p)X^0(p)$, $p \in S^1$, with $f \in C^r(S^1, \mathbb{R})$. It is clear that $X(p) = 0$ if and only if $f(p) = 0$. As we have already remarked, given any compact set $K \subset S^1$ there exists $f \in C^r(S^1, \mathbb{R})$ with $f^{-1}(0) = K$. Thus K is the set of singularities of $X = fX^0$. As topological equivalence preserves singularities we have at least as many equivalence classes of vector fields as homeomorphism classes of compact subsets of S^1. This shows that it is impossible to describe and classify the orbit structures of all the vector fields on S^1. It is, therefore, natural to restrict ourselves to a residual subset of $\mathfrak{X}^r(S^1)$, or preferably to an open dense one.

A singularity p of $X \in \mathfrak{X}^r(S^1)$ is *nondegenerate* (or *hyperbolic*) if $dX(p) \neq 0$, that is $df(p) \neq 0$ where $X = fX^0$. If $df(p) < 0$ then p is a *sink* (or *attracting singularity*) and if $df(p) > 0$ then p is a *source* (or *repelling singularity*). Let $G \subset \mathfrak{X}^r(S^1)$ be the subset consisting of those vector fields whose singularities are all hyperbolic; as these singularities are isolated it follows that the number of them is finite (possibly zero!). We claim that G is open and dense. In fact, let $X = fX^0$ and let $\tilde{f}: S^1 \to S^1 \times \mathbb{R}$ be defined by $\tilde{f}(p) = (p, f(p))$. It is clear that $X \in G$ if and only if $\tilde{f}$ is transversal to $S^1 \times \{0\}$. But the set of $f \in C^r(S^1, \mathbb{R})$ such that $\tilde{f}$ is transversal to $S^1 \times \{0\}$ is open. This set is also dense, since for any f the set of $v \in \mathbb{R}$ such that $\tilde{f} + v$ is transversal to $S^1 \times \{0\}$ is residual and so we can choose v small with $(f + v)X^0 \in G$.

§4 Structural Stability

If $X \in G$ and $X = fX^0$ we see from the graph of $f: S^1 \to \mathbb{R}$ that the sinks and sources of X must alternate round S^1. In particular, the number of singularities is even. From this it follows that if $X, Y \in G$ have the same number of singularities then X and Y are topologically conjugate. In fact, let $a_1, b_1, a_2, b_2, \ldots, a_s, b_s$ be the sinks and sources of X in order on S^1. Similarly, let $a'_1, b'_1, a'_2, b'_2, \ldots, a'_s, b'_s$ be the sinks and sources of Y in order on S^1. Define $h(a_i) = a'_i$ and $h(b_i) = b'_i$. Choose points $p_i \in (a_i, b_i)$, $q_i \in (b_i, a_{i+1})$ and $p'_i \in (a'_i, b'_i)$, $q'_i \in (b'_i, a'_{i+1})$. Define $h(p_i) = p'_i$ and $h(q_i) = q'_i$. If $p \in (a_i, b_i)$ there exists a unique $t \in \mathbb{R}$ such that $X_t(p) = p_i$; define $h(p) = Y_{-t}(p'_i) = Y_{-t}hX_t(p)$ and proceed similarly for points of (b_i, a_{i+1}). It is now clear that h is a homeomorphism that conjugates the flows of X and Y.

If X does not have singularities the only orbit of X is the whole of S^1. If $X = fX^0$ then either $f > 0$ or $f < 0$ on S^1. If $f > 0$ the identity is a topological equivalence between X and X^0. If $f < 0$ we take an orientation-reversing homeomorphism as our topological equivalence. Finally we claim that if $X \in \mathfrak{X}^r(S^1)$ is stable then $X \in G$. First we remark that the number of singularities of X is finite. This is because G is dense and so X must be equivalent to a vector field $Y \in G$ near X. We leave it to the reader to show that these singularities are hyperbolic; if not, we can perturb X in a way that increases the number of singularities contradicting the hypothesis that X is stable. Thus $X \in \mathfrak{X}^r(S^1)$ is stable if and only if $X \in G$. Therefore, the structurally stable vector fields in $\mathfrak{X}^r(S^1)$ form an open dense set and, as we have seen, it is possible to classify them.

The development of the geometric theory of differential equations led naturally to a parallel study of diffeomorphisms. We next introduce some basic concepts in the study of the orbit structure of a diffeomorphism.

Let $f \in \text{Diff}^r(M)$. The *orbit* of a point $p \in M$ is the set $\mathcal{O}(p) = \{f^n(p); n \in \mathbb{Z}\}$. When $\mathcal{O}(p)$ is finite we say that p is *periodic* and the least integer $n > 0$ such that $f^n(p) = p$ is called the *period* of p. If $f(p) = p$ we say that p is a *fixed point*. A point q belongs to the ω-*limit set* of p, $\omega(p)$, when there exists a sequence of integers $n_i \to \infty$ such that $f^{n_i}(p) \to q$. If $x \in \mathcal{O}(p)$ then $\omega(x) = \omega(p)$. Also $\omega(p)$ is nonempty, closed and invariant. *Invariant* means that $\omega(p)$ is a union of orbits of f. For p periodic $\omega(p) = \mathcal{O}(p)$; thus $\omega(p)$ is not connected if the period of p is greater than one. Analogously we define the α-*limit set* of p, $\alpha(p)$, as the ω-limit set of p for f^{-1}. The above properties of $\omega(p)$ hold for $\alpha(p)$ too.

Equivalence of the orbit structures of two diffeomorphisms is expressed by conjugacy. A *conjugacy* between $f, g \in \text{Diff}^r(M)$ is a homeomorphism $h: M \to M$ such that $h \circ f = g \circ h$. It follows that $h \circ f^n = g^n \circ h$ for any integer n and so $h(\mathcal{O}_f(p)) = \mathcal{O}_g(q)$ if $q = h(p)$. That is, h takes orbits of f onto orbits of g and, in particular, it takes periodic points onto periodic points of the same period. Also $h(\omega_f(p)) = \omega_g(q)$ and $h(\alpha_f(p)) = \alpha_g(q)$.

EXAMPLE 6. Let us consider two linear contractions in $\mathbb{R}$, $f(x) = \frac{1}{2}x$ and $g(x) = \frac{1}{3}x$. We shall show that f and g are conjugate. Take two points with coordinates $a > 0$, $b < 0$ and consider the intervals $[f(a), a]$, $[b, f(b)]$ and

$[g(a), a]$, $[b, g(b)]$. Define a homeomorphism $h: [f(a), a] \cup [b, f(b)] \to [g(a), a] \cup [b, g(b)]$ such that $h(a) = a$, $h(b) = b$, $h(f(a)) = g(a)$ and $h(f(b)) = g(b)$. For each $x \in \mathbb{R}$, $x \neq 0$, there exists an integer n such that $f^n(x) \in [f(a), a] \cup [b, f(b)]$. We put $h(x) = g^{-n}hf^n(x)$ and $h(0) = 0$. It is easy to see that h is well defined and is a conjugacy between f and g. On the other hand, the contractions $f(x) = \frac{1}{2}x$ and $g(x) = -\frac{1}{3}x$ are not conjugate.

By this argument we can also see that two contractions of $\mathbb{R}$ are conjugate if and only if they both preserve or both reverse the orientation of $\mathbb{R}$.

EXAMPLE 7. The linear transformations of $\mathbb{R}^2$, $f(x, y) = (\frac{1}{2}x, 2y)$ and $g(x, y) = (\frac{1}{3}x, 4y)$ are conjugate. We construct, as in Example 6, a conjugacy h_1 between $f|\mathbb{R} \times \{0\}$ and $g|\mathbb{R} \times \{0\}$ and a conjugacy h_2 between $f|\{0\} \times \mathbb{R}$ and $g|\{0\} \times \mathbb{R}$. The conjugacy between f and g is given by $h(x, y) = (h_1(x), h_2(y))$.

EXAMPLE 8. The linear transformations X_1, Y_1 induced at time 1 by the vector fields X, Y of Example 2 are not conjugate. It is sufficient to observe that Y_1 leaves invariant a family of concentric circles and X_1 does not.

Conjugacy gives rise naturally to the concept of structural stability for diffeomorphisms. Thus $f \in \text{Diff}^r(M)$ is *structurally stable* if there exists a neighbourhood $\mathscr{V}$ of f in $\text{Diff}^r(M)$ such that any $g \in \mathscr{V}$ is conjugate to f.

The identity map is obviously unstable. Also, the diffeomorphisms induced at time $t = 1$ by the vector fields X in Examples 3, 4 and 5 of this section are unstable.

EXAMPLE 9. Let us take a vector field $X \in \mathfrak{X}^r(S^1)$ which is stable and has singularities. As we have already seen, X has an even number of singularities alternately sinks and sources a_1, b_1, a_2, b_2, ..., a_s, b_s. Choose points $p_i \in (a_i, b_i)$ and $q_i \in (b_i, a_{i+1})$. Now consider the diffeomorphism $f = X_1$

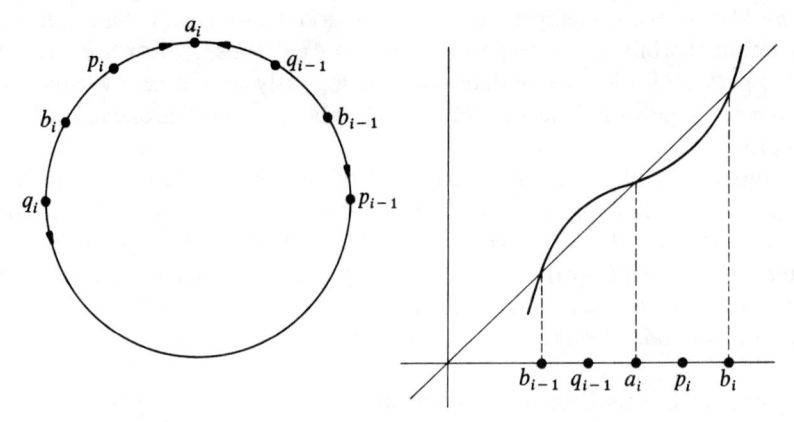

Figure 25

§4 Structural Stability

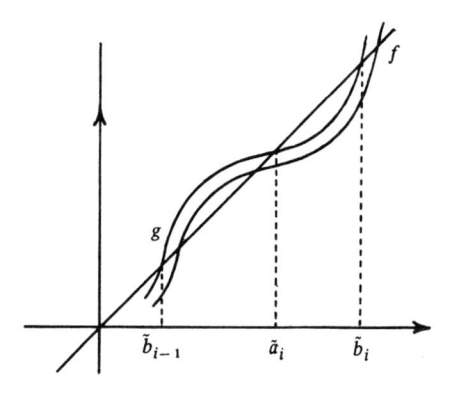

Figure 26

induced by X at time $t = 1$. We shall prove that $f \in \text{Diff}^r(S^1)$ is structurally stable. We know that f is a contraction on $[q_{i-1}, p_i]$ with fixed point a_i and f is an expansion on $[p_i, q_i]$ with fixed point b_i. If g is C^r close to f then g is a contraction in $[q_{i-1}, p_i]$ with a single fixed point $\tilde{a}_i$ close to a_i. In addition, g is an expansion on $[p_i, q_i]$ with a single fixed point $\tilde{b}_i$ near b_i. We put $h(a_i) = \tilde{a}_i$, $h(b_i) = \tilde{b}_i$, $h(p_i) = p_i$, $h(q_i) = q_i$, $h(f(p_i)) = g(p_i)$ and $h(f(q_i)) = g(q_i)$. We define h to be any homeomorphism from the interval $[p_i, f(p_i)]$ to $[p_i, g(p_i)]$ and from $[q_i, f(q_i)]$ to $[q_i, g(q_i)]$ and then we extend h to $[a_i, b_i]$ and $[b_{i-1}, a_i]$ as in Example 6. We obtain a conjugacy between f and g, which shows that f is structurally stable.

We emphasize that the stable diffeomorphisms in $\text{Diff}^r(S^1)$ form an open dense subset. The proof of this result is much more elaborate and will be done in Section 4 of Chapter 4. We also remark that, in the example above, we started from a stable vector field in $\mathfrak{X}^r(S^1)$ and showed that the diffeomorphism it induced at time $t = 1$ was stable in $\text{Diff}^r(S^1)$. The next example shows that this is not always the case.

EXAMPLE 10. Consider the unit vector field X^0 on S^1. Then S^1 is a closed orbit of X^0 of period 2π. The diffeomorphism $f = X_1^0$ induced at time $t = 1$ is an irrational rotation. The orbit $\mathcal{O}_f(p)$ is dense for every point $p \in S^1$. To see that f is unstable we approximate f by $g = X_t^0$ with t near one and $t/2\pi$ rational. Every orbit $\mathcal{O}_g(p)$ is periodic and so f is not conjugate to g.

We now show why we defined a conjugacy to be a homeomorphism rather than a diffeomorphism. Let us consider again the diffeomorphism f in Example 9. As we saw f is structurally stable: if g is C^r close to f then there exists a homeomorphism h of S^1 such that $h \circ f = g \circ h$. To construct h we note first that, for each sink a_i of f, we have a sink a_i' of g close to a_i. The same is true for the sources. We put $h(a_i) = a_i'$. It is easy to see that we can choose g close to f such that $a_i' = a_i$ and $g'(a_i) \neq f'(a_i)$. Now suppose that h is, in

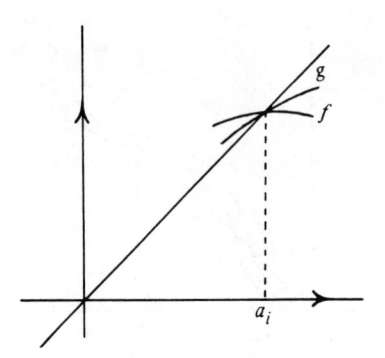

Figure 27

fact, a diffeomorphism. Then we have $h(a_i) = a_i$ and $h'(a_i) \cdot f'(a_i) = g'(a_i) \cdot h'(a_i)$ which implies that $f'(a_i) = g'(a_i)$ contrary to our hypothesis. Thus, if we required the conjugacy to be a diffeomorphism, f would not be stable in $\text{Diff}^r(S^1)$. Similarly we can show that no $f \in \text{Diff}^r(M)$ that has a fixed or periodic point would be stable. This shows that we ought not to impose the condition of being differentiable on a conjugacy. The same idea applies to topological equivalence between vector fields. Although the proof is more complicated it is also true that no vector field with a singularity or a closed orbit would be stable if we required the equivalence to be differentiable. See Exercise 13 of Chapter 2 and also Exercise 5 of Chapter 3.

EXERCISES

1. Show that every C^1 vector field on the sphere S^2 has at least one singularity.

2. Two vector fields $X, Y \in \mathfrak{X}^r(M)$ *commute* if $X_s(Y_t(p)) = Y_t(X_s(p))$ for all $p \in M$ and $s, t \in \mathbb{R}$. Show that if $X, Y \in \mathfrak{X}^1(S^2)$ commute then X and Y have a singularity in common (E. Lima).

3. Let $X = (P, Q)$ be a vector field on $\mathbb{R}^2$ where P and Q are polynomials of degree two. Let γ be a closed orbit of X and $D \subset \mathbb{R}^2$ the disc bounded by γ. Show that X has a unique singularity in D.

4. Let $X \in \mathfrak{X}^1(M^2)$ and let $F \subset M^2$ be a region homeomorphic to the cylinder such that $X_t(F) \subset F$ for all $t \geq 0$. Suppose that X has a finite number of singularities in F. Show that the ω-limit of the orbit of a point $p \in F$ either is a closed orbit or consists of singularities and regular orbits whose ω- and α-limits are singularities.

5. Let $F \subset M^2$ be a region homeomorphic to a Möbius band and let $X \in \mathfrak{X}^1(M^2)$ be a vector field such that $X_t(F) \subset F$ for all $t \geq 0$. If X has a finite number of singularities in F then the ω-limit of the orbit of a point $p \in F$ either is a closed orbit or consists of singularities and regular orbits whose ω- and α-limits are singularities.

Exercises

6. Let γ be an isolated closed orbit of a vector field $X \in \mathfrak{X}^r(M^2)$. Show that there exists a neighbourhood V of γ such that, for $p \in V$, either $\alpha(p) = \gamma$ or $\omega(p) = \gamma$.

7. A closed orbit γ of $X \in \mathfrak{X}^r(M^2)$ is an *attractor* if there exists a neighbourhood V of γ such that $X_t(p) \in V$ for all $t \geq 0$ and $\omega(p) = \gamma$ for all $p \in V$. Show that, if X has a closed orbit that is an attractor, then every vector field Y sufficiently near X also has a closed orbit.

8. Let X be a C^1 vector field on the projective plane. Show that, if X has a finite number of singularities, then the ω-limit of an orbit either is a closed orbit or consists of singularities and regular orbits whose ω- and α-limits are singularities.

9. Let X be a vector field on the torus T^2 which generates an irrational flow X_t. Show that, given $n \in \mathbb{N}$ and $\varepsilon > 0$, there exists a vector field Y with exactly n closed orbits such that $\|Y - X\|_r < \varepsilon$.

10. A *cycle* of a vector field $X \in \mathfrak{X}^r(M)$ is a sequence of singularities $p_1, \ldots, p_j$, $p_{j+1} = p_1$ and regular orbits $\gamma_1, \ldots, \gamma_j$ such that $\alpha(\gamma_i) = p_i$ and $\omega(\gamma_i) = p_{i+1}$. Let $X \in \mathfrak{X}^r(S^2)$, $r \geq 1$, satisfy the following properties:
    (1) X has a finite number of singularities;
    (2) if $p \in S^2$ is a singularity of X then either p is a repelling singularity or the set of orbits γ with $\alpha(\gamma) = p$ is finite.
    Show that, for any orbit γ,
    (a) if $\omega(\gamma)$ contains more than one singularity then $\omega(\gamma)$ contains a cycle;
    (b) if p_1 and p_2 are singularities contained in $\omega(\gamma)$ then there exists a cycle which contains p_1 and p_2.

11.A. Let $G \subset \mathbb{R}^n$ be an additive subgroup. Show that, if G is closed, then it is isomorphic to $\mathbb{R}^k \times \mathbb{Z}^l$ for some k and l with $k + l \leq n$.
    Hint. (a) Show that, if G is discrete, then it is isomorphic to $\mathbb{Z}^l$; that is, there exist vectors $v_1, \ldots, v_l \in \mathbb{R}^n$ such that $G = \{\sum_{i=1}^{l} n_i v_i; n_i \in \mathbb{Z}\}$.
    (b) Show that, if G is not discrete, then it contains a line through the origin.
    (c) Let $E \subset \mathbb{R}^n$ be the subspace of largest dimension contained in G. Let $E^\perp$ be the orthogonal complement of E. Show that $G = E \oplus (E^\perp \cap G)$ and that $E^\perp \cap G$ is a discrete subgroup of $E^\perp$.

11.B. Let $\alpha = (\alpha_1, \ldots, \alpha_n) \in \mathbb{R}^n$. Let $G = \{l\alpha + m; l \in \mathbb{Z}$ and $m \in \mathbb{Z}^n\}$. Suppose that the coordinates of α are independent over the integers; that is, if $\langle m, \alpha \rangle = \sum_{i=1}^{n} m_i \alpha_i = 0$ with $m \in \mathbb{Z}^n$ then $m = 0$. Show that G is dense in $\mathbb{R}^n$.
    Hint. (a) $\pi: \mathbb{R}^n \to \mathbb{R}^{n-1}$ be the projection $\pi(x_1, \ldots, x_n) = (x_1, \ldots, x_{n-1})$. Let $\bar{G}$ be the closure of G. Suppose, by induction, that $\pi(\bar{G}) = \mathbb{R}^{n-1}$.
    (b) Let $E \subset \mathbb{R}^n$ be the subspace of largest dimension contained in $\bar{G}$. Then either the dimension of E is $n - 1$ or it is n and E contains the vector α.
    (c) Let $E_{ij} = \{x; x_k = 0$ if $k \neq i, j\}$. Show that $E_{ij} \cap E$ is a straight line in E_{ij} with rational slope. Deduce that E contains $n - 1$ linearly independent vectors with integer coordinates. The vector product of these vectors is a vector with integer coordinates that is perpendicular to E.

11.C. Find an example of a vector field of class C^∞ on the torus $T^n = S^1 \times \cdots \times S^1$ such that all its orbits are dense in T^n.

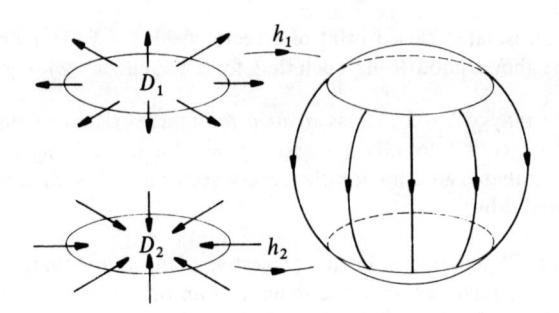

Figure 28

12. Let X^i be a C^∞ vector field defined on a neighbourhood of a disc $D_i \subset \mathbb{R}^2$ for $i = 1, 2$. Suppose that X^i is transversal to the boundary C_i of D_i and that X^1 points out of D_1 while X^2 points into D_2. Show that there exists a C^∞ vector field X on S^2 and embeddings $h_i: D_i \to S^2$ such that:
    (1) $h_1(D_1) \cap h_2(D_2) = \emptyset$;
    (2) $dh_i(p) \cdot X^i(p) = X(h_i(p))$ for all $p \in D_i$;
    (3) if $p \in h_1(C_1)$ then the ω-limit of p is contained in $h_2(D_2)$.

13. Let X^1, X^2 be C^∞ vector fields on manifolds M_1, M_2 of the same dimension. Let $D_i \subset M_i$ be discs such that X^i is transversal to the boundary C_i of D_i, $i = 1, 2$, with X^1 pointing out of D_1 and X^2 pointing into D_2. Show that there exist a C^∞ vector field X on a manifold M and embeddings $h_i: M_i - D_i \to M$ such that:
    (1) $h_1(M_1 - D_1) \cap h_2(M_2 - D_2) = \emptyset$;
    (2) $dh_i(p) \cdot X^i(p) = X(h_i(p))$;
    (3) if $p \in h_1(C_1)$ then the α-limit of p is contained in $h_2(M_2 - D_2)$.

14. Let X be the parallel field $\partial/\partial t$ on the cylinder $S^1 \times [0, 1]$. Let M be the quotient space of $S^1 \times [0, 1]$ by the equivalence relation that identifies $S^1 \times \{0\}$ with $S^1 \times \{1\}$ by an irrational rotation $R: S^1 \times \{0\} \to S^1 \times \{1\}$. Let $\pi: S^1 \times [0, 1] \to M$ be the quotient map. Show that
    (a) there exists a manifold structure on M such that π is a local diffeomorphism and $\pi_* X$ is a C^∞ vector field on M;
    (b) there exists a diffeomorphism $h: M \to T^2$ such that if $Y = h_* X$ then Y_t is an irrational flow.

15. Let M, N and P be manifolds with M and N compact. Show that
    (a) the map comp: $C^r(M, N) \times C^r(N, P) \to C^r(M, P)$, comp$(f, g) = g \circ f$, is continuous;
    (b) the map i: Diff$^r(M) \to$ Diff$^r(M)$, $i(f) = f^{-1}$ is continuous.

16. Let M and N be manifolds, with M compact, and let $S \subset M \times N$ be a submanifold. Consider the set $T_S = \{f \in C^r(M, N);$ graph(f) is transversal to $S\}$, where graph$(f) = \{(p, f(p)); p \in M\}$. Show that T_S is residual in $C^r(M, N)$.

17. For each $f \in C^r(\mathbb{R}^n, \mathbb{R}^m)$ consider the map
$$j^r f: \mathbb{R}^n \to \mathbb{R}^n \times \mathbb{R}^m \times L(\mathbb{R}^n, \mathbb{R}^m) \times \cdots \times L_s^r(\mathbb{R}^n; \mathbb{R}^m)$$
$$x \mapsto (x, f(x), df(x), \ldots, d^r f(x)).$$

Exercises

Let E be the Euclidean space $\mathbb{R}^n \times \mathbb{R}^m \times L(\mathbb{R}^n, \mathbb{R}^m) \times \cdots \times L^r_s(\mathbb{R}^n; \mathbb{R}^m)$. For each open set $U \subset E$ define the subset

$$\mathcal{M}(U) = \{f \in C^r(\mathbb{R}^n, \mathbb{R}^m); j^r f(\mathbb{R}^n) \subset U\}.$$

(a) Show that the sets $\mathcal{M}(U)$ form a base for a topology on $C^r(\mathbb{R}^n, \mathbb{R}^m)$ (the Whitney topology).
(b) Show that $C^r(\mathbb{R}^n, \mathbb{R}^m)$ with the Whitney topology is a Baire space.
(c) Show that the C^∞ maps form a dense subset in $C^r(\mathbb{R}^n, \mathbb{R}^m)$.
(d) Show that, if $k \leq r$, the map

$$C^r(\mathbb{R}^n, \mathbb{R}^m) \to C^{r-k}(\mathbb{R}^n, \mathbb{R}^n \times \mathbb{R}^m \times L(\mathbb{R}^n, \mathbb{R}^m) \times \cdots \times L^k_s(\mathbb{R}^n; \mathbb{R}^m))$$

$$f \mapsto j^k f$$

is continuous.
(e) Let $S \subset \mathbb{R}^n \times \mathbb{R}^m \times L(\mathbb{R}^n, \mathbb{R}^m)$ be a submanifold. Consider the set $T_S = \{f \in C^r(\mathbb{R}^n, \mathbb{R}^m); j^1 f \pitchfork S\}$, where $r \geq 2$. Show that T_S is residual.

18. Let $X^0 \in \mathfrak{X}^r(S^1)$ be a vector field without singularities. Let Σ_1 be the set of vector fields $X = fX^0 \in \mathfrak{X}^r(S^1)$ such that the singularities of X are all nondegenerate except one at which the second derivative of f is nonzero. Let $\Sigma_{1,1}$ be the set of vector fields $X = fX^0$ such that the singularities are all nondegenerate except two at which the second derivative of f is nonzero. Let $\Sigma_{1,2}$ be the set of vector fields $X = fX^0$ such that the singularities of X are nondegenerate except one at which the second derivative of f is zero but the third derivative is nonzero.
(a) Show that Σ_1 is a codimension 1 submanifold of the Banach space $\mathfrak{X}^r(S^1)$ and that Σ_1 is open and dense in $\mathfrak{X}^r(S^1) - G$, where G consists of the structurally stable vector fields as in Section 4.
(b) Show that $\Sigma_2 = \Sigma_{1,1} \cup \Sigma_{1,2}$ is a codimension 2 submanifold of $\mathfrak{X}^r(S^1)$ and that Σ_2 is open and dense in $\mathfrak{X}^r(S^1) - (G \cup \Sigma_1)$.
(c) Describe all the equivalence classes in a neighbourhood of a vector field in Σ_1 and of one in Σ_2.

Remark. Sotomayor [114] considered conditions like these in the context of Bifurcation Theory.

19. (a) Show that, if $g: \mathbb{R} \to \mathbb{R}$ is a C^1 diffeomorphism that commutes with $f: \mathbb{R} \to \mathbb{R}$ given by $f(x) = \lambda x$ where $0 < \lambda < 1$, then g is linear.
(b) Show that, if $g: \mathbb{R}^2 \to \mathbb{R}^2$ is a C^1 diffeomorphism that commutes with a linear contraction whose eigenvalues are complex, then g is linear.
(c) Show, however, that there does exist a nonlinear C^1 diffeomorphism that commutes with a linear contraction.

20. *Poincaré Compactification.* Consider the sphere $S^2 = \{y \in \mathbb{R}^3; \sum_{i=1}^3 y_i^2 = 1\}$ and the plane $P = \{y \in \mathbb{R}^3; y_3 = 1\}$ tangent to the sphere at the north pole. Let $U_i = \{y \in S^2; y_i > 0\}$ and $V_i = \{y \in S^2; y_i < 0\}$. Let $\pi_3: P \to U_3, \tilde{\pi}_3: P \to V_3$ be the central projections, that is, $\pi_3(x)$ and $\tilde{\pi}_3(x)$ are the intersections of the line joining x to the origin with U_3 and V_3. Let L be a linear vector field on $\mathbb{R}^2$. Consider the fields $X^1 = (\pi_3)_* L$ on U_3 and $X^2 = (\tilde{\pi}_3)_* L$ on V_3.
(a) Show that X^1 and X^2 extend to a C^∞ vector field X, which we call $\pi(L)$, on S^2 and that the equator is invariant for X.

(b) Describe the orbits of the fields $\pi(L^i)$, $i = 1, 2, 3, 4$, where the L^i are represented, with respect to the standard basis, by the matrices

$$\begin{pmatrix} \lambda & 0 \\ 0 & \lambda \end{pmatrix}, \quad \begin{pmatrix} \lambda & 0 \\ 1 & \lambda \end{pmatrix}, \quad \begin{pmatrix} \lambda_1 & 0 \\ 0 & \lambda_2 \end{pmatrix}, \quad \begin{pmatrix} \alpha & \beta \\ -\beta & \alpha \end{pmatrix}.$$

(c) Show that, if L^1 and $L^2 = AL^1A^{-1}$ are linear vector fields and A is a linear isomorphism, then $\pi(L^1)$ and $\pi(L^2)$ are topologically equivalent.

(d) Show that, if L is a linear vector field which has two equal eigenvalues or an eigenvalue with real part zero, then $\pi(L)$ is not structurally stable in $\mathfrak{X}^\infty(S^2)$. *Hint.* Use the local charts $\varphi_i \colon U_i \to \mathbb{R}^2$, $\varphi_i \colon V_i \to \mathbb{R}^2$ defined by $\varphi_i(y) = (y_j/y_i, y_k/y_i)$, $\psi_i(y) = (y_j/y_i, y_k/y_i)$ with $j < k$.

Remark. The vector fields that are structurally stable among those induced on the sphere S^n by linear vector fields were characterized by G. Palis in [73].

21. An orbit γ of $X \in \mathfrak{X}^r(M^n)$, $r \geq 1$, is said to be ω-recurrent if $\gamma \subset \omega(\gamma)$. Let M be a compact manifold and γ an ω-recurrent orbit for $X \in \mathfrak{X}^r(M)$. If $f = X_{t=1}$ and $x \in \gamma$ show that x is ω-recurrent, that is $x \in \omega_f(x)$.

Remark. The *Birkhoff centre* $C(X)$ of $X \in \mathfrak{X}^1(M)$ is defined as the closure of the set of those orbits that are both ω- and α-recurrent. The same definition works for $f \in \text{Diff}^r(M)$. This exercise shows that $C(X) = C(f)$ when f is the time 1 diffeomorphism of X.

Chapter 2

Local Stability

In this chapter we shall analyse the local topological behaviour of the orbits of vector fields. We shall show that, for vector fields belonging to an open dense subset of the space $\mathfrak{X}^r(M)$, we can describe the behaviour of the trajectories in a neighbourhood of each point of the manifold. Moreover, the local structure of the orbits does not change for small perturbations of the field. A complete classification via topological conjugacy is then provided.

Such a local question is considered in two parts: near a regular point and near a singularity. The first part, much simpler, is dealt with in Section 1. The second part is developed in Sections 2 through 5. Section 2 is devoted to linear vector fields and isomorphisms for which the notion of hyperbolicity is introduced. In Section 3 this notion is extended to singularities of nonlinear vector fields and fixed points of diffeomorphisms. Local stability for a hyperbolic singularity or a hyperbolic fixed point is proved in Section 4. Finally, in Section 5 we present the local topological classification. Section 6 is dedicated to another important result, the Stable Manifold Theorem. Much related to it is the λ-lemma (Inclination Lemma) that is considered in Section 7, from which we obtain several relevant applications and a new proof of the local stability.

§1 The Tubular Flow Theorem

Definition. Let $X, Y \in \mathfrak{X}^r(M)$ and $p, q \in M$. We say that X and Y are *topologically equivalent at p and q* respectively if there exist neighbourhoods V_p and W_q and a homeomorphism $h: V_p \to W_q$, with $h(p) = q$, which takes orbits of X to orbits of Y preserving their orientation.

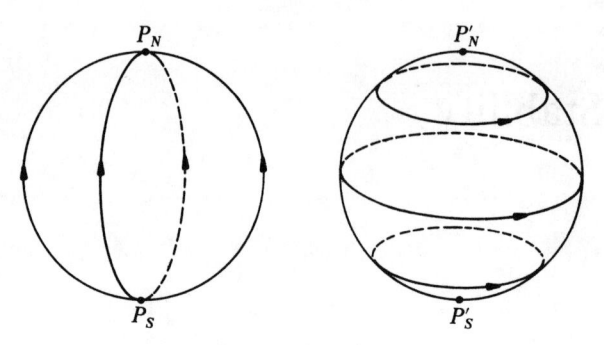

Figure 1

EXAMPLE. Consider the vector fields X and Y on S^2 given in Figure 1. X and Y are not equivalent at P_N, P'_N since each neighbourhood of P'_N contains closed orbits of Y but there are no closed orbits of X near P_N.

Definition. Let $X \in \mathfrak{X}^r(M)$ and $p \in M$. We say that X is *locally stable* at p if for any given neighbourhood $U(p) \subset M$ there exists a neighbourhood $\mathcal{N}_X$ of X in $\mathfrak{X}^r(M)$ such that, for each $Y \in \mathcal{N}_X$, X at p is topologically equivalent to Y at q for some $q \in U$.

The next theorem describes the local behaviour of the orbits in a neighbourhood of a regular point.

1.1 Theorem (Tubular Flow). *Let $X \in \mathfrak{X}^r(M)$ and let $p \in M$ be a regular point of X. Let $C = \{(x^1, \ldots, x^m) \in \mathbb{R}^m; |x^i| < 1\}$ and let X_C be the vector field on C defined by $X_C(x) = (1, 0, \ldots, 0)$. Then there exists a C^r diffeomorphism $h: V_p \to C$, for some neighbourhood V_p of p in M, taking trajectories of X to trajectories of X_C.*

PROOF. Let $x: U \to U_0 \subset \mathbb{R}^m$ be a local chart around p with $x(p) = 0$. Let $x_* X$ be the C^r vector field induced by X on U_0. As $X(p) \neq 0$ we have $x_* X(0) \neq 0$. Let $\varphi: [-\tau, \tau] \times V_0 \to U_0$ be the local flow of $x_* X$ and put $H = \{\omega \in \mathbb{R}^m; \langle \omega, x_* X(0) \rangle = 0\}$, which is a subspace isomorphic to $\mathbb{R}^{m-1}$. Let $\psi: [-\tau, \tau] \times S \to U_0$ be the restriction of φ to $[-\tau, \tau] \times S$ where $S = H \cap V_0$. Take a basis $\{e_1, e_2, \ldots, e_m\}$ of $\mathbb{R} \times H \approx \mathbb{R}^m$ where $e_1 = (1, 0, \ldots, 0)$ and $e_2, \ldots, e_m \subset \{0\} \times H$. It follows that

$$D\psi(0, 0)e_1 = x_* X(0) \quad \text{(by the definition of local flow)}$$

$$D\psi(0, 0)e_j = e_j, \quad j = 2, \ldots, m,$$

since $\psi(0, y) = y$ for all $y \in S$.

Thus, $D\psi(0, 0): \mathbb{R} \times H \to \mathbb{R}^m$ is an isomorphism. By the Inverse Function Theorem, ψ is a diffeomorphism of a neighbourhood of $(0, 0)$ in $[-\tau, \tau] \times S$ onto a neighbourhood of 0 in $\mathbb{R}^m$. Therefore, if $\varepsilon > 0$ is small enough, $C_\varepsilon = \{(t, x) \in \mathbb{R} \times H; |t| < \varepsilon$ and $\|x\| < \varepsilon\}$ and $\tilde{\psi}: C_\varepsilon \to U_0$ is the restriction of

§2 Linear Vector Fields

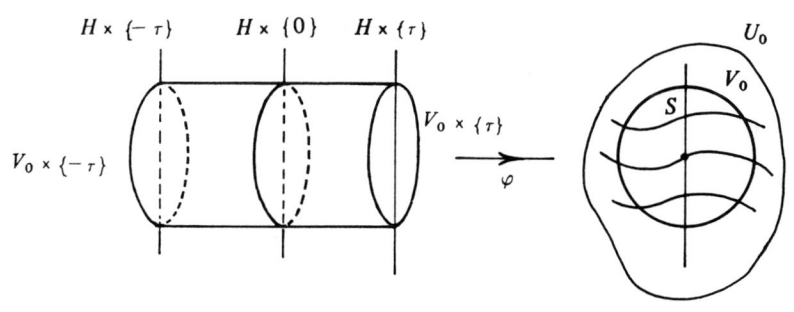

Figure 2

ψ to C_ε, then $\tilde\psi$ is a C^r diffeomorphism onto its image which is open in U_0. Moreover, $\tilde\psi$ takes orbits of the parallel field X_{C_ε} in C_ε to orbits of $x_* X$. Let us consider the C^∞ diffeomorphism $f: C \to C_\varepsilon$, $f(y) = \varepsilon y$ and define $h^{-1} = x^{-1}\tilde\psi f: C \to M$. Then $h: x^{-1}\tilde\psi(C_\varepsilon) \to C$ is a C^r diffeomorphism which satisfies the conditions in the theorem. □

Remark. The diffeomorphism $\tilde h^{-1}: C_\varepsilon \to M$ defined by $\tilde h^{-1} = x^{-1}\tilde\psi$ takes orbits of the unit parallel field X_{C_ε} to orbits of the field X preserving the parameter t.

Corollary 1. *If $X, Y \in \mathfrak{X}^r(M)$ and $p, q \in M$ are regular points of X and Y, respectively, then X is equivalent to Y at p and q.* □

Corollary 2. *If $X \in \mathfrak{X}^r(M)$ and $p \in M$ is a regular point of X then X is locally stable at p.* □

§2 Linear Vector Fields

Let $\mathscr{L}(\mathbb{R}^n)$ be the vector space of linear maps from $\mathbb{R}^n$ to $\mathbb{R}^n$ with the usual norm:

$$\|L\| = \sup\{\|Lv\|; \|v\| = 1\}.$$

First we recall some basic results from linear algebra. If $L \in \mathscr{L}(\mathbb{R}^n)$ and k is a positive integer we write L^k for the linear map $L \circ \cdots \circ L$. It is easy to show, by induction, that $\|L^k\| \le \|L\|^k$. Let us consider the sequence of linear maps $E_m = \sum_{k=0}^m 1/k! \, L^k$ where L^0 means the identity map.

2.1 Lemma. *The sequence E_m converges.*

PROOF. The sequence of real numbers $S_m = \sum_{k=0}^m 1/k! \, \|L\|^k$ is a Cauchy sequence that converges to $e^{\|L\|}$. On the other hand,

$$\|E_{m+m'} - E_m\| = \left\| \sum_{k=m+1}^{m+m'} \frac{1}{k!} L^k \right\| \le \sum_{k=m+1}^{m+m'} \frac{1}{k!} \|L\|^k = \|S_{m+m'} - S_m\|.$$

This shows that $\{E_m\}$ is a Cauchy sequence. As $\mathscr{L}(\mathbb{R}^n)$ is a complete metric space it follows that the sequence $\{E_m\}$ converges. □

Definition. The map $\operatorname{Exp}: \mathscr{L}(\mathbb{R}^n) \to \mathscr{L}(\mathbb{R}^n)$ defined by $\operatorname{Exp}(L) = e^L = \sum_{k=0}^{\infty} 1/k!\, L^k$ is called the *exponential map*.

2.2 Lemma. *Let $\alpha: \mathbb{R} \to \mathscr{L}(\mathbb{R}^n)$ be defined by $\alpha(t) = e^{tL}$. Then α is differentiable and $\alpha'(t) = L\, e^{tL}$.*

PROOF. Let $\alpha_m(t) = I + tL + (t^2/2!)L^2 + \cdots + (t^m/m!)L^m$. It is clear that α_m is differentiable and

$$\alpha_m'(t) = L + tL^2 + \cdots + \frac{t^{m-1}}{(m-1)!} L^m = L\alpha_{m-1}(t).$$

As $\alpha_{m-1}(t)$ converges uniformly to e^{tL} on each bounded subset of $\mathbb{R}$, it follows that $\alpha_m'(t) \to L e^{tL}$ uniformly. Thus, α is differentiable and $\alpha'(t) = L e^{tL}$. □

2.3 Proposition. *Let L be a linear vector field on $\mathbb{R}^n$. Then the map $\varphi: \mathbb{R} \times \mathbb{R}^n \to \mathbb{R}^n$ defined by $\varphi(t, x) = e^{tL}x$ is the flow of the field L.*

PROOF. As the map $\mathscr{L}(\mathbb{R}^n) \times \mathbb{R}^n \to \mathbb{R}^n$, $(L, x) \to Lx$ is bilinear and the map $t \mapsto e^{tL}$ is differentiable it follows by the chain rule that φ is differentiable. Moreover, $\partial/\partial t\, \varphi(t, x) = L\varphi(t, x)$ by Lemma 2.2. As $\varphi(0, x) = x$ for all $x \in \mathbb{R}^n$ the proposition is proved. □

Let $\mathbb{C}^n$ be the set of n-tuples of complex numbers with the usual vector space structure. An element of $\mathbb{C}^n$ can be written in the form $u + iv$ with $u, v \in \mathbb{R}^n$. If $a + ib \in \mathbb{C}$ then $(a + ib)(u + iv) = (au - bv) + i(av + bu)$. Let $\mathscr{L}(\mathbb{C}^n)$ denote the complex vector space of linear maps from $\mathbb{C}^n$ to $\mathbb{C}^n$ with the usual norm; $\|L\| = \sup\{\|Lv\|; v \in \mathbb{C}^n \text{ and } \|v\| = 1\}$. If $L \in \mathscr{L}(\mathbb{R}^n)$ we can define a map $\tilde{L}: \mathbb{C}^n \to \mathbb{C}^n$ by $\tilde{L}(u + iv) = L(u) + iL(v)$. It is easy to see that $\tilde{L}$ is $\mathbb{C}$-linear; that is, $\tilde{L} \in \mathscr{L}(\mathbb{C}^n)$. Let $\operatorname{Exp}: \mathscr{L}(\mathbb{C}^n) \to \mathscr{L}(\mathbb{C}^n)$ be the exponential map, which is defined in the same way as in the real case. Let $\mathscr{C}: \mathscr{L}(\mathbb{R}^n) \to \mathscr{L}(\mathbb{C}^n)$ be the map which associates to each operator L its complexification $\tilde{L}$ defined above. The proposition below follows directly from the definitions.

2.4 Proposition. *The map $\mathscr{C}: \mathscr{L}(\mathbb{R}^n) \to \mathscr{L}(\mathbb{C}^n)$ satisfies the following properties:*

(1) $\mathscr{C}(L + T) = \mathscr{C}(L) + \mathscr{C}(T)$, $\mathscr{C}(\alpha L) = \alpha \mathscr{C}(L)$;
(2) $\mathscr{C}(LT) = \mathscr{C}(L)\mathscr{C}(T)$;
(3) $\mathscr{C}(\operatorname{Exp} L) = \operatorname{Exp} \mathscr{C}(L)$;
(4) $\|\mathscr{C}(L)\| = \|L\|$,

for any $L, T \in \mathscr{L}(\mathbb{R}^n)$ and $\alpha \in \mathbb{R}$. □

§2 Linear Vector Fields

EXAMPLE. Let $L \in \mathscr{L}(\mathbb{R}^2)$ and let $\{e_1, e_2\}$ be a basis for $\mathbb{R}^2$ with respect to which the matrix of L has the form $\begin{pmatrix} \alpha & \beta \\ -\beta & \alpha \end{pmatrix}$. The matrix of $\tilde{L} = \mathscr{C}(L)$ in the basis $\{e_1 + ie_2, e_1 - ie_2\}$ for $\mathbb{C}^2$ is $\begin{pmatrix} \lambda & 0 \\ 0 & \bar{\lambda} \end{pmatrix}$ where $\lambda = \alpha + i\beta$ and $\bar{\lambda} = \alpha - i\beta$. Thus, the matrix of $e^{\tilde{L}}$ in this basis is $\begin{pmatrix} e^\lambda & 0 \\ 0 & e^{\bar{\lambda}} \end{pmatrix}$. On the other hand, $\mathscr{C}(e^L)(e_1 + ie_2) = e^L e_1 + ie^L e_2 = e^{\tilde{L}}(e_1 + ie_2) = e^\lambda(e_1 + ie_2)$. As $e^\lambda = e^\alpha(\cos\beta + i\sin\beta)$ it follows that $e^L e_1 = e^\alpha(\cos\beta \, e_1 - \sin\beta \, e_2)$ and $e^L e_2 = e^\alpha(\sin\beta \, e_1 + \cos\beta \, e_2)$. Therefore, the matrix of e^L in the basis $\{e_1, e_2\}$ is

$$e^\alpha \begin{pmatrix} \cos\beta & \sin\beta \\ -\sin\beta & \cos\beta \end{pmatrix}.$$

2.5 Theorem (Real Canonical Form). *If $L \in \mathscr{L}(\mathbb{R}^n)$ there exists a basis for $\mathbb{R}^n$ with respect to which the matrix of L has the form*

$$\begin{pmatrix} A_1 & & & & & O \\ & \ddots & & & & \\ & & A_r & & & \\ & & & B_1 & & \\ & O & & & \ddots & \\ & & & & & B_s \end{pmatrix},$$

where

$$A_i = \begin{pmatrix} \lambda_i & & & & O \\ 1 & \lambda_i & & & \\ & 1 & \lambda_i & & \\ & & & \ddots & \\ O & & & 1 & \lambda_i \end{pmatrix}, \quad i = 1, \ldots, r, \quad \lambda_i \in \mathbb{R}$$

and

$$B_j = \begin{pmatrix} C_j & & & O \\ I & C_j & & \\ & & \ddots & \\ O & & I & C_j \end{pmatrix}, \quad j = 1, \ldots, s$$

$$C_j = \begin{pmatrix} \alpha_j & \beta_j \\ -\beta_j & \alpha_j \end{pmatrix}, \quad I = \begin{pmatrix} 1 & 0 \\ 0 & 1 \end{pmatrix}, \quad \alpha_j, \beta_j \in \mathbb{R}.$$

The submatrices $A_1, \ldots, A_r, B_1, \ldots, B_s$ are determined uniquely except for their order. □

Corollary. Let $L \in \mathscr{L}(\mathbb{R}^n)$. Given $\varepsilon > 0$ there exists a basis for $\mathbb{R}^n$ with respect to which the matrix of L has the form

$$\begin{pmatrix} A_1 & & & & & \\ & \ddots & & & & \bigcirc \\ & & A_r & & & \\ & & & B_1 & & \\ & \bigcirc & & & \ddots & \\ & & & & & B_s \end{pmatrix}$$

with

$$A_i = \begin{pmatrix} \lambda_i & & \bigcirc \\ \varepsilon & \lambda_i & \\ \bigcirc & \ddots & \ddots \\ & & \varepsilon & \lambda_i \end{pmatrix} \quad B_j = \begin{pmatrix} \alpha_j & \beta_j & & & & & & \\ -\beta_j & \alpha_j & & & & \bigcirc & & \\ \varepsilon & 0 & \ddots & & & & & \\ 0 & \varepsilon & \ddots & & & & & \\ & & \ddots & \ddots & & & & \\ & & & \ddots & \varepsilon & 0 & \alpha_j & \beta_j \\ & & & & 0 & \varepsilon & -\beta_j & \alpha_j \end{pmatrix}. \quad \square$$

2.6 Lemma. *If $A, B \in \mathscr{L}(\mathbb{R}^n)$ satisfy $AB = BA$ then $e^{A+B} = e^A e^B$.*

PROOF. Let $S_m(t) = I + tA + \cdots + (t^m/m!)A^m$. As $AB = BA$ we have $A^k B = BA^k$ and so $S_m(t)B = BS_m(t)$. Since $S_m(t) \to e^{tA}$ we have $e^{tA}B = Be^{tA}$. Let $x \in \mathbb{R}^n$ and consider the curves $\alpha, \beta: \mathbb{R} \to \mathbb{R}^n$, $\alpha(t) = e^{t(A+B)}x$, $\beta(t) = e^{tA}e^{tB}x$. By Lemma 2.2 we have $\alpha'(t) = (A + B)e^{t(A+B)}x = (A + B)\alpha(t)$ and $\beta'(t) = Ae^{tA}e^{tB}x + e^{tA}Be^{tB}x = Ae^{tA}e^{tB}x + Be^{tA}e^{tB}x = (A + B)\beta(t)$ using $e^{tA}B = Be^{tA}$. Therefore α and β are integral curves of the linear vector field $A + B$ and satisfy the same initial condition $\alpha(0) = \beta(0) = x$. By the uniqueness theorem we have $\alpha(t) = \beta(t)$ for all t. In particular, $e^{A+B}x = e^A e^B x$. As this holds for all $x \in \mathbb{R}^n$ it follows that $e^{A+B} = e^A e^B$. $\square$

If $L \in \mathscr{L}(\mathbb{R}^n)$ then the spectrum of $\tilde{L}$, that is, the set of eigenvalues of $\tilde{L}$, is called the *complex spectrum* of L and coincides with the set of roots of the characteristic polynomial of L. The Jordan canonical form of the complexified operator $\tilde{L}$ is represented by

$$\begin{pmatrix} A_1 & \bigcirc \\ & \ddots \\ \bigcirc & A_r \end{pmatrix}, \quad \text{where} \quad A_i = \begin{pmatrix} \lambda_i & & \bigcirc \\ 1 & \lambda_i & \\ & \ddots & \ddots \\ \bigcirc & & 1 & \lambda_i \end{pmatrix}$$

and the λ_i are the eigenvalues of $\tilde{L}$.

We remark that a triangular complex matrix has its diagonal entries as eigenvalues with multiplicity equal to the number of times they appear.

§2 Linear Vector Fields

2.7 Proposition. *If $L \in \mathscr{L}(\mathbb{R}^n)$ and λ is an eigenvalue of $\tilde{L}$ then e^λ is an eigenvalue of e^L with the same multiplicity.*

PROOF. Consider an $m \times m$ matrix

$$\begin{pmatrix} \lambda & & & O \\ 1 & \lambda & & \\ & \ddots & \ddots & \\ O & & 1 & \lambda \end{pmatrix},$$

where $\lambda \in \mathbb{C}$. We have $A = D + N$ where

$$D = \begin{pmatrix} \lambda & & O \\ & \ddots & \\ O & & \lambda \end{pmatrix} \quad \text{and} \quad N = \begin{pmatrix} 0 & & & O \\ 1 & 0 & & \\ & \ddots & \ddots & \\ O & & 1 & 0 \end{pmatrix}.$$

It is easy to see that $N^m = 0$ and that $ND = DN$. By Lemma 2.6 we have $e^A = e^D e^N$. But $e^N = I + N + N^2/2! + \cdots + N^{m-1}/(m-1)!$ since $N^k = 0$ for $k \geq m$. Thus

$$e^N = \begin{pmatrix} 1 & & & & & O \\ 1 & 1 & & & & \\ \frac{1}{2} & \ddots & \ddots & & & \\ \vdots & \ddots & \ddots & \ddots & & \\ 1/(m-1)! & \cdots & \frac{1}{2} & 1 & 1 \end{pmatrix}, \quad e^D = \begin{pmatrix} e^\lambda & & O \\ & \ddots & \\ O & & e^\lambda \end{pmatrix}.$$

Now

$$e^A = e^D e^N = e^\lambda e^N$$

Therefore e^A is a triangular matrix with all its diagonal elements equal to e^λ and so e^λ is an eigenvalue of e^A with multiplicity m.

Now let $L \in \mathscr{L}(\mathbb{R}^n)$. By the Real Canonical Form Theorem the matrix of $\tilde{L}$, with respect to a certain basis of β of $\mathbb{C}^n$, has the form

$$A = \begin{pmatrix} A_1 & & O \\ & \ddots & \\ O & & A_r \end{pmatrix}, \quad \text{with} \quad A_j = \begin{pmatrix} \lambda_j & & & O \\ 1 & \ddots & & \\ & \ddots & \ddots & \\ O & & 1 & \lambda_j \end{pmatrix}$$

It is easy to see that

$$A^k = \begin{pmatrix} A_1^k & & O \\ & \ddots & \\ O & & A_r^k \end{pmatrix},$$

for all $k \in \mathbb{N}$ and, therefore,

$$e^A = \begin{pmatrix} e^{A_1} & & O \\ & \ddots & \\ O & & e^{A_r} \end{pmatrix}.$$

This shows that the eigenvalues of e^A are exactly $e^{\lambda_1}, \ldots, e^{\lambda_r}$ where $\lambda_1, \ldots, \lambda_r$ are the eigenvalues of A. But $e^{\tilde{L}} = \widetilde{e^L}$ is represented with respect to the basis β of $\mathbb{C}^n$ by the matrix e^A which shows that $e^{\lambda_1}, \ldots, e^{\lambda_r}$ are the eigenvalues of the complexification of e^L. $\square$

Definition. A linear vector field $L \in \mathscr{L}(\mathbb{R}^n)$ is *hyperbolic* if the spectrum of L is disjoint from the imaginary axis. The number of eigenvalues of L with negative real part is called the *index* of L.

Note that a hyperbolic linear vector field has only one singularity which is the origin.

2.8 Proposition. *If $L \in \mathscr{L}(\mathbb{R}^n)$ is a hyperbolic vector field then there exists a unique decomposition (called a "splitting") of $\mathbb{R}^n$ as a direct sum $\mathbb{R}^n = E^s \oplus E^u$, where E^s and E^u are invariant subspaces for L and for the flow defined by L such that the eigenvalues of $L^s = L|E^s$ have negative real part and the eigenvalues of $L^u = L|E^u$ have positive real part.*

PROOF. Let $e_1, \ldots, e_n$ be a basis for $\mathbb{R}^n$ for which the matrix of L is in the real canonical form. For an appropriate order of the elements of this basis the matrix of L has the form

$$\begin{pmatrix} A_1 & & & & & & & & \\ & \ddots & & & & & & & \\ & & A_{s'} & & & & & & \\ & & & B_1 & & & & & \\ & & & & \ddots & & & & \\ & & & & & B_{s''} & & & \\ & & & & & & C_1 & & \\ & & & & & & & \ddots & \\ & & & & & & & & C_{u'} \\ & & & & & & & & & D_1 \\ & & & & & & & & & & \ddots \\ & & & & & & & & & & & D_{u''} \end{pmatrix},$$

where

$$A_i = \begin{pmatrix} \lambda_i & & O \\ 1 & \lambda_i & \\ & \ddots & \ddots \\ O & & 1 & \lambda_i \end{pmatrix}, \text{ with } \lambda_i < 0,$$

§2 Linear Vector Fields

$$B_j = \begin{pmatrix} M_j & & & O \\ I & M_j & & \\ & \ddots & \ddots & \\ O & & I & M_j \end{pmatrix}, \quad \text{with} \quad M_j = \begin{pmatrix} \alpha_j & \beta_j \\ -\beta_j & \alpha_j \end{pmatrix}, \quad I = \begin{pmatrix} 1 & 0 \\ 0 & 1 \end{pmatrix}$$

and $\alpha_j < 0$

$$C_k = \begin{pmatrix} \lambda_k & & & O \\ 1 & \lambda_k & & \\ & \ddots & \ddots & \\ O & & 1 & \lambda_k \end{pmatrix}, \quad \text{with} \quad \lambda_k > 0$$

and

$$D_l = \begin{pmatrix} M_l & & & O \\ I & M_l & & \\ & \ddots & \ddots & \\ O & & I & M_l \end{pmatrix}, \quad \text{with} \quad M_l = \begin{pmatrix} \alpha_l & \beta_l \\ -\beta_l & \alpha_l \end{pmatrix} \quad \text{and} \quad \alpha_l > 0.$$

Let E^s be the subspace generated by $e_1, \ldots, e_s$ where $e_1, \ldots, e_s$ correspond to the invariant subspaces associated to $A_1, \ldots, A_{s'}, B_1, \ldots, B_{s''}$. Let E^u be the subspace generated by $e_{s+1}, \ldots, e_n$. It is clear that E^s and E^u are invariant for L and that the matrix of L^s, for the basis $\{e_1, \ldots, e_s\}$, is

$$\begin{pmatrix} A_1 & & & & & O \\ & \ddots & & & & \\ & & A_{s'} & & & \\ & & & B_1 & & \\ & & & & \ddots & \\ O & & & & & B_{s''} \end{pmatrix},$$

while the matrix of L^u, for the basis $\{e_{s+1}, \ldots, e_n\}$, is

$$\begin{pmatrix} C_1 & & & & & O \\ & \ddots & & & & \\ & & C_{u'} & & & \\ & & & D_1 & & \\ & & & & \ddots & \\ O & & & & & D_{u''} \end{pmatrix},$$

which shows the existence of the required decomposition. Uniqueness is immediate. □

Let $L \in \mathscr{L}(\mathbb{R}^n)$ be a hyperbolic vector field. If L_t denotes the flow of L then $L_1 = e^L$ and, as L does not have an eigenvalue on the imaginary axis, it follows from Proposition 2.7 that L_1 does not have an eigenvalue on the unit circle S^1. This suggests the following definition.

Definition. A linear isomorphism $A \in GL(\mathbb{R}^n)$ is *hyperbolic* if the spectrum of A is disjoint from the unit circle $S^1 \subset \mathbb{C}$. In particular, the diffeomorphism induced at time 1 by the flow of a hyperbolic linear vector field is a hyperbolic isomorphism.

2.9 Proposition. *If $A \in GL(\mathbb{R}^n)$ is a hyperbolic isomorphism then there exists a unique decomposition $\mathbb{R}^n = E^s \oplus E^u$ such that E^s and E^u are invariant for A and the eigenvalues of $A^s = A|E^s$ and $A^u = A|E^u$ are the eigenvalues of A of modulus less than 1 and greater than 1 respectively.*

PROOF. Similar to the proof of Proposition 2.8. $\square$

2.10 Proposition. *If $A \in GL(\mathbb{R}^n)$ is a hyperbolic isomorphism then there exists a norm $\|\cdot\|_1$ on $\mathbb{R}^n$ such that $\|A^s\|_1 < 1$ and $\|(A^u)^{-1}\| < 1$, that is A^s is a contraction and A^u is an expansion.*

PROOF. Consider the canonical form for $A^s = A|E^s$,

$$M = \begin{pmatrix} A_1 & & & & & O \\ & \ddots & & & & \\ & & A_{s'} & & & \\ & & & B_1 & & \\ & & & & \ddots & \\ O & & & & & B_{s''} \end{pmatrix},$$

where

$$A_i = \begin{pmatrix} \lambda_i & & & O \\ 1 & \ddots & & \\ & \ddots & \ddots & \\ O & & 1 & \lambda_i \end{pmatrix}, \qquad |\lambda_i| < 1$$

and

$$B_j = \begin{pmatrix} \alpha_j & \beta_j & & & & & O \\ -\beta_j & \alpha_j & & & & & \\ 1 & 0 & \ddots & & & & \\ 0 & 1 & \ddots & \ddots & & & \\ & & \ddots & \ddots & \alpha_j & \beta_j & \\ O & & & 0 & 1 & -\beta_j & \alpha_j \end{pmatrix}, \qquad \alpha_j^2 + \beta_j^2 < 1.$$

§2 Linear Vector Fields

For each $t \in \mathbb{R}$ consider the matrices

$$A_i(t) = \begin{pmatrix} \lambda_i & & O \\ t & \ddots & \\ & \ddots & \ddots \\ O & & t & \lambda_i \end{pmatrix}$$

$$B_j(t) = \begin{pmatrix} \alpha_j & \beta_j & & & & & O \\ -\beta_j & \alpha_j & & & & & \\ t & 0 & \ddots & & & & \\ 0 & t & \ddots & & & & \\ & & & t & 0 & \alpha_j & \beta_j \\ & & & 0 & t & -\beta_j & \alpha_j \\ O & & & & & & \end{pmatrix}$$

and

$$M(t) = \begin{pmatrix} A_1(t) & & & & O \\ & \ddots & & & \\ & & A_{s'}(t) & & \\ & & & B_1(t) & \\ & & & & \ddots \\ O & & & & & B_{s''}(t) \end{pmatrix}$$

We claim that there exists $\delta > 0$ with the following property: if $\varepsilon < \delta$ and $e_1, \ldots, e_s$ is an orthonormal basis of E^s and A is a linear transformation of E^s whose matrix in this basis is $M(\varepsilon)$ then $\|A\| < 1$. In fact, let $A(t)$ be the linear transformation of E^s whose matrix in the basis $e_1, \ldots, e_s$ is $M(t)$. It is easy to see that $\|A(0)\| = \max\{|\lambda_i|, \sqrt{(\alpha_j^2 + \beta_j^2)}\}$. Thus $\|A(0)\| < 1$. As the composition $t \mapsto A(t) \mapsto \|A(t)\|$ is continuous there exists $\delta > 0$ such that $\|A(t)\| < 1$ for $0 < t < \delta$, which proves the claim.

Now let $\delta > 0$ be as above. By the Corollary to Theorem 2.5 there exists a basis $e_1, \ldots, e_s$ of E^s in which the matrix of A^s is $M(\varepsilon)$. We define a new inner product on E^s by $\langle e_i, e_j \rangle_1 = \delta_{ij}$ where $\delta_{ij} = 1$ if $i = j$ and 0 if $i \neq j$. Let $\|\cdot\|_1$ be the norm associated to $\langle \, , \, \rangle_1$. As the basis is orthonormal in the new metric the claim implies $\|A^s\|_1 < 1$. Similarly we change the norm on E^u so that $\|(A^u)^{-1}\|_1 < 1$. We define a norm $\|\cdot\|_1$ on $\mathbb{R}^n$ by $\|v\|_1 = \max\{\|v^s\|_1, \|v^u\|_1\}$, where v^s and v^u are the components of v in E^s and E^u, respectively. It is clear that this norm satisfies the conditions in the proposition. □

Corollary. *If L is a hyperbolic linear vector field with flow L_t and $\mathbb{R}^n = E^s \oplus E^u$ is the splitting of Proposition 2.8 then $L_t(x)$ converges to the origin if $x \in E^s$ and $t \to +\infty$ or if $x \in E^u$ and $t \to -\infty$.*

PROOF. Let $x \in E^s$. It is sufficient to show that $L_n(x) \to 0$ where $n \in \mathbb{N}$ and $n \to \infty$. In fact, if $t \in [0, 1]$ we have, by the continuity of L_t, that, given $\varepsilon > 0$, there exists $\delta_t > 0$ such that $\|L_t(y)\| < \varepsilon$ for $\|y\| < \delta_t$. As $[0, 1]$ is compact there exists $\delta > 0$ such that $\|L_t(y)\| < \varepsilon$ for $\|y\| < \delta$ and all $t \in [0, 1]$. If $L_n(x) \to 0$ as $n \to \infty$ there exists $n_0 \in \mathbb{N}$ such that $\|L_n(x)\| < \delta$ if $n \geq n_0$. If $t > n_0$ then $t = n + s$ for some $n \geq n_0$ and $s \in [0, 1]$. Thus $\|L_t(x)\| = \|L_s L_n(x)\| < \varepsilon$. So it is sufficient to show that $L_n(x) = L_1^n(x)$ tends to 0. By the proposition above there exists a metric on E^s in which L_1 is a contraction, that is $\|L_1\| < 1$. Then $\|L_1^n x\| \leq \|L_1^n\|\|x\| \leq \|L_1\|^n\|x\|$. As $\|L_1\|^n \to 0$ we have $\|L_1^n x\| \to 0$ as required. The second part of the corollary is proved similarly. $\square$

2.11 Proposition. *The set $H(\mathbb{R}^n)$ of hyperbolic isomorphisms of $\mathbb{R}^n$ is open and dense in $GL(\mathbb{R}^n)$.*

PROOF. (a) *Openness.* Let $A \in H(\mathbb{R}^n)$. Let us show that there exists $\delta > 0$ such that, if $\|A - B\| < \delta$, then $B \in H(\mathbb{R}^n)$. Let $\lambda \in S^1$. As λ is not an eigenvalue of $\tilde{A}$, $\det(\tilde{A} - \lambda I) \neq 0$ where I is the identity of $\mathbb{C}^n$. Now, $\det: \mathscr{L}(\mathbb{C}^n) \to \mathbb{C}$ is a continuous map so there exist $\delta_\lambda > 0$ and a neighbourhood V_λ of λ in $\mathbb{C}$ such that, if $\|B - A\| < \delta_\lambda$ and $\mu \in V_\lambda$, then $\det(\tilde{B} - \mu I) \neq 0$. Let $V_{\lambda_1}, \ldots, V_{\lambda_m}$ be a finite subcover of the cover $\{V_\lambda; \lambda \in S^1\}$ of S^1. Put $\delta = \min\{\delta_{\lambda_1}, \ldots, \delta_{\lambda_m}\}$. If $\|B - A\| < \delta$ and $\mu \in S^1$ then $\mu \in V_{\lambda_j}$ for some j, and, therefore, $\det(\tilde{B} - \mu I) \neq 0$. Thus, $B \in H(\mathbb{R}^n)$ as required.

(b) *Density.* Let $A \in GL(\mathbb{R}^n)$ and let $\lambda_1, \ldots, \lambda_n$ be its eigenvalues. It is easy to see that, if $\mu \in \mathbb{R}$, the eigenvalues of $A + \mu I$ are $\lambda_1 + \mu, \ldots, \lambda_n + \mu$. Let $\lambda_{i_1}, \ldots, \lambda_{i_r}$ be the eigenvalues of A which do not belong to S^1. Consider the following numbers:

$\delta_1 = \min\{|\lambda_1|, \ldots, |\lambda_n|\}$

$\delta_2 = \min\{|1 - |\lambda_{i_1}\|, \ldots, |1 - |\lambda_{i_r}\|\}$,

$\delta_3 = \min\{|\alpha|; \alpha + i\beta$ is an eigenvalue of A with $\alpha^2 + \beta^2 = 1$ and $\alpha \neq 0\}$.

It is clear that $\delta_1 > 0$, $\delta_2 > 0$ and $\delta_3 > 0$. If $0 < \mu < \min\{\delta_1, \delta_2, \delta_3\}$ and λ is an eigenvalue of A then $\lambda + \mu \notin S^1$ and so $B = A + \mu I$ is hyperbolic. Given $\varepsilon > 0$ we take $\mu < \varepsilon$ and $\mu < \min\{\delta_1, \delta_2, \delta_3\}$ and then B is hyperbolic and $\|B - A\| = \|\mu I\| < \varepsilon$. This shows that $H(\mathbb{R}^n)$ is dense in $GL(\mathbb{R}^n)$. $\square$

2.12 Proposition. *The set $\mathscr{H}(\mathbb{R}^n)$ of hyperbolic linear vector fields on $\mathbb{R}^n$ is open and dense in $\mathscr{L}(\mathbb{R}^n)$.*

PROOF. (a) *Openness.* The map $\mathrm{Exp}: \mathscr{L}(\mathbb{R}^n) \to GL(\mathbb{R}^n)$ is continuous. By Proposition 2.7 we have $\mathscr{H}(\mathbb{R}^n) = \mathrm{Exp}^{-1}(H(\mathbb{R}^n))$. As $H(\mathbb{R}^n)$ is open it follows that $\mathscr{H}(\mathbb{R}^n)$ is open too.

(b) *Density.* Let $L \in \mathscr{L}(\mathbb{R}^n)$. Let $\delta_1 = \min\{|\alpha|; \alpha + i\beta$ is an eigenvalue of L and $\alpha \neq 0\}$. Given $\varepsilon > 0$ we take $\delta < \min\{\varepsilon, \delta_1\}$. It is easy to see that the vector field $T = L + \delta I$ is hyperbolic and $\|T - L\| < \varepsilon$. $\square$

§2 Linear Vector Fields

Our next aim is to give a necessary and sufficient condition for two hyperbolic linear vector fields to be topologically equivalent.

2.13 Lemma. *Let L be a hyperbolic linear vector field on $\mathbb{R}^n$ with index n. There exists a norm $\|\cdot\|$ on $\mathbb{R}^n$ such that, if $S^{n-1} = \{v \in \mathbb{R}^n; \|v\| = 1\}$, then the vector $L(x)$ at the point x is transversal to S^{n-1} for all $x \in S^{n-1}$.*

PROOF. Let us consider a basis $e_1, \ldots, e_n$ of $\mathbb{R}^n$ for which the matrix of L is

$$A = \begin{pmatrix} A_1(1) & & & & O \\ & \ddots & & & \\ & & A_{s'}(1) & & \\ & & & B_1(1) & \\ & & & & \ddots \\ O & & & & B_{s''}(1) \end{pmatrix},$$

with $A_i(1)$ and $B_j(1)$ as in Proposition 2.8.

Let $\bar{L}$ be a linear vector field on $\mathbb{R}^n$ whose matrix in an orthonormal basis is

$$\bar{A} = \begin{pmatrix} A_1(0) & & & & O \\ & \ddots & & & \\ & & A_{s'}(0) & & \\ & & & B_1(0) & \\ & & & & \ddots \\ O & & & & B_{s''}(0) \end{pmatrix}.$$

It is easy to see that $\bar{L}$ is transversal to S^{n-1}. Since S^{n-1} is compact, if $\varepsilon > 0$ is sufficiently small the field $\bar{\bar{L}}$, whose matrix in this orthonormal basis

$$\bar{\bar{A}} = \begin{pmatrix} A_1(\varepsilon) & & & & O \\ & \ddots & & & \\ & & A_{s'}(\varepsilon) & & \\ & & & B_1(\varepsilon) & \\ & & & & \ddots \\ O & & & & B_{s''}(\varepsilon) \end{pmatrix},$$

is transversal to S^{n-1}. On the other hand, by the corollary to Theorem 2.5, there exists a basis of $\mathbb{R}^n$ in which the matrix of L is $\bar{\bar{A}}$. We define an inner product on $\mathbb{R}^n$ making this basis orthonormal and then, by the argument above, L is transversal to the unit sphere in this norm. □

2.14 Proposition. *If L and T are linear vector fields on $\mathbb{R}^n$ of index n then there exists a homeomorphism $h: \mathbb{R}^n \to \mathbb{R}^n$ such that $hL_t = T_t h$ for all $t \in \mathbb{R}$.*

PROOF. Let $\|\cdot\|_1$ and $\|\cdot\|_2$ be norms on $\mathbb{R}^n$ such that the spheres $S_1^{n-1} = \{v \in \mathbb{R}^n; \|v\|_1 = 1\}$ and $S_2^{n-1} = \{v \in \mathbb{R}^n; \|v\|_2 = 1\}$ are transversal to the vector fields L and T respectively. If $x \in \mathbb{R}^n - \{0\}$ then, by the corollary to

Proposition 2.10, we have $\lim_{t\to\infty} L_t(x) = 0$ and $\lim_{t\to\infty} \|L_{-t}(x)\| = \infty$ so that $\mathcal{O}_L(x)$ does meet S_1^{n-1}. As L is transversal to S_1^{n-1} it follows that $\mathcal{O}_L(x)$ meets S_1^{n-1} in a unique point.

Let $h: S_1^{n-1} \to S_2^{n-1}$ be any homeomorphism (for example, we can put $h(x) = x/\|x\|_2$). We shall extend h to $\mathbb{R}^n$. Define $h(0) = 0$. If $x \in \mathbb{R}^n - \{0\}$ there exists a unique $t_0 \in \mathbb{R}$ such that $L_{-t_0}(x) \in S_1^{n-1}$. Put $h(x) = T_{t_0} h(L_{-t_0}(x))$. It is easy to see that $hL_t = T_t h$ for all $t \in \mathbb{R}$ and that h has an inverse. It remains to show that h is continuous.

Let $x \in \mathbb{R}^n - \{0\}$ and let (x_m) be a sequence converging to x. Take $t_m \in \mathbb{R}$ such that $L_{-t_m}(x_m) \in S_1^{n-1}$ and $t_0 \in \mathbb{R}$ such that $L_{-t_0}(x) \in S_1^{n-1}$. As the flow is continuous it follows that $t_m \to t_0$ and $L_{-t_m}(x_m) \to L_{-t_0}(x)$. Thus $h(x_m) = T_{t_m} h L_{-t_m}(x_m)$ converges to $T_{t_0} h L_{-t_0}(x) = h(x)$ which shows the continuity of h at x. We now show that h is continuous at the origin. From Proposition 2.10 and the compactness of S_2^{n-1} it follows that, given $\varepsilon > 0$, there exists $t_\varepsilon > 0$ such that $\|T_t(y)\| < \varepsilon$ for all $t > t_\varepsilon$ and all $y \in S_2^{n-1}$. On the other hand, as $L(0) = 0$, there exists $\delta > 0$ such that if $\|x\| < \delta$ and $L_{-t}(x) \in S_1^{n-1}$ then $t > t_\varepsilon$. Therefore, $\|h(x)\| < \varepsilon$ if $\|x\| < \delta$, which shows the continuity of h. Similarly we can show that h^{-1} is continuous. □

2.15 Proposition. *Let L and T be hyperbolic linear vector fields. Then L and T are topologically conjugate if and only if they have the same index.*

PROOF. Suppose that L and T have the same index. Let E^s, $E^{s'}$ be the stable subspaces of L and T, respectively. Then $\dim E^s = \dim E^{s'}$. By Proposition 2.14 there exists a homeomorphism $h_s: E^s \to E^{s'}$ conjugating L^s and T^s; that is, $h_s L_t^s = T_t^s h_s$ for all $t \in \mathbb{R}$. Similarly, there exists a homeomorphism $h_u: E^u \to E^{u'}$ conjugating L^u and T^u. We define $h: E^s \oplus E^u \to E^{s'} \oplus E^{u'}$ by $h(x^s + x^u) = h_s(x^s) + h_u(x^u)$. It is easy to see that h is a homeomorphism and conjugates L_t and T_t. Conversely, let h be a topological equivalence between L and T. As 0 is the only singularity of L and T we must have $h(0) = 0$. If $x \in E^s$ we have $\omega(x) = 0$. As a topological equivalence preserves the ω-limit of orbits, we have $\omega(h(x)) = h(\omega(x)) = 0$. Therefore $h(x) \in E^{s'}$ so that $h(E^s) \subset E^{s'}$. Similarly, $h^{-1}(E^{s'}) \subset E^s$. Hence $h|E^s$ is a homeomorphism between E^s and $E^{s'}$. By the Theorem of Invariance of Domain, from Topology, it follows that $\dim E^s = \dim E^{s'}$, which proves the proposition. □

We next intend to show that the eigenvalues of an operator depend continuously on the operator. By that we mean the following. For $L \in \mathscr{L}(\mathbb{R}^n)$, let $\lambda_1, \lambda_2, \ldots, \lambda_k$ be its eigenvalues with multiplicity $m_1, m_2, \ldots, m_k$, respectively. We consider balls $B_\varepsilon(\lambda_i)$ of radius ε and center λ_i, $1 \le i \le k$, so that they are all pairwise disjoint. We want to show that given $\varepsilon > 0$ there exists $\delta > 0$ such that if $T \in \mathscr{L}(\mathbb{R}^n)$ and $\|T - L\| < \delta$, then T has precisely m_i eigenvalues in $B_\varepsilon(\lambda_i)$ counting their multiplicities, for all $1 \le i \le k$.

For $L \in \mathscr{L}(\mathbb{R}^n)$ let $\mathrm{Sp}(L)$ denote the spectrum of L, the set of its eigenvalues. The lemma below shows that $\mathrm{Sp}(L)$ cannot explode for a small perturbation of L.

§2 Linear Vector Fields

2.16 Lemma. *Let $L \in \mathscr{L}(\mathbb{R}^n)$. Given $\varepsilon > 0$ there exists $\delta > 0$ such that, if $T \in \mathscr{L}(\mathbb{R}^n)$ and $\|T - L\| < \delta$, then for each $\lambda' \in \mathrm{Sp}(T)$ there exists $\lambda \in \mathrm{Sp}(L)$ with $|\lambda - \lambda'| < \varepsilon$.*

PROOF. If $\lambda \in \mathrm{Sp}(L)$ then λ is an eigenvalue of the complexified operator $\tilde{L}$, so that $|\lambda| \leq \|\tilde{L}\| = \|L\|$. Thus, if $\|T - L\| < 1$, the spectrum of T is contained in the interior of the disc D with centre at the origin of $\mathbb{C}$ and radius $1 + \|L\|$. Let V_ε be the union of the balls of radius ε with centre the elements of $\mathrm{Sp}(L)$. If $\mu \in D - V_\varepsilon$, then $\det(\tilde{L} - \mu I) \neq 0$. By continuity of the determinant there exist a neighbourhood U_μ of μ in $\mathbb{C}$ and $\delta_\mu > 0$ such that, if $\|T - L\| < \delta_\mu$ and $\mu' \in U_\mu$, then $\det(\tilde{T} - \mu' I) \neq 0$, so that $\mu' \notin \mathrm{Sp}(\tilde{T})$. By the compactness of $D - V_\varepsilon$ we deduce that there exists $\delta > 0$ such that, if $\|T - L\| < \delta$ and $\mu \in D - V_\varepsilon$, then $\det(T - \mu I) \neq 0$. As $\mathrm{Sp}(T) \subset D$ it follows that $\mathrm{Sp}(T) \subset V_\varepsilon$, which proves the lemma. $\square$

If the eigenvalues of L are all distinct it follows from Lemma 2.16 that they change continuously with the operator. Let λ be an eigenvalue of L of multiplicity m and let $E(L, \lambda) \subset \mathbb{C}^n$ be the kernel of $(\tilde{L} - \lambda I)^m$. Then $E(L, \lambda)$ is a subspace of dimension m. Moreover, if $k \geq m$, the kernel of $(\tilde{L} - \lambda I)^k$ is $E(L, \lambda)$.

2.17 Lemma. *If λ is an eigenvalue of $L \in \mathscr{L}(\mathbb{R}^n)$ of multiplicity m then there exist $\varepsilon_0 > 0$ and $\delta > 0$ such that, if $\|T - L\| < \delta$, the sum of the multiplicities of the eigenvalues of T contained in the ball of radius ε_0 and centre λ is at most m.*

PROOF. To get a contradiction suppose for all $\varepsilon > 0$ and $\delta > 0$ there exists $T \in \mathscr{L}(\mathbb{R}^n)$ with $\|T - L\| < \delta$ such that the number of eigenvalues of T, counted with multiplicity, in the ball of radius ε and centre λ is greater than m. Thus there exists an $m' > m$ and a sequence of operators $L_k \to L$ such that $\lambda_{k_1}, \ldots, \lambda_{k_{m'}}$ are eigenvalues of L_k which converge to λ. Let E_k be the kernel of $(\tilde{L}_k - \lambda_{k_1} I) \circ \cdots \circ (\tilde{L}_k - \lambda_{k_{m'}} I)$. We may suppose that the dimension of E_k is m'. Let $e_1^k, \ldots, e_{m'}^k$ be an orthonormal basis of E_k. As $\|e_j^k\| = 1$ and the unit sphere in $\mathbb{C}^n$ is compact we can suppose (by taking a subsequence if necessary) that $e_j^k \to e_j$. The vectors $e_1, \ldots, e_{m'}$ are clearly orthonormal and so span a subspace $\tilde{E}$ of dimension m'. As the operator $(\tilde{L}_k - \lambda_{k_1} I) \circ \cdots \circ (\tilde{L}_k - \lambda_{k_{m'}} I)$ converges to $(\tilde{L} - \lambda I)^{m'}$ we see, by continuity, that the kernel of $(\tilde{L} - \lambda I)^{m'}$ contains $\tilde{E}$ which is absurd since it has dimension $m < m'$. $\square$

2.18 Proposition. *The eigenvalues of an operator $L \in \mathscr{L}(\mathbb{R}^n)$ depend continuously on L.*

PROOF. Let $\lambda_1, \ldots, \lambda_k$ be the distinct eigenvalues of L with multiplicity $n_1, \ldots, n_k$. By Lemma 2.16, given $\varepsilon > 0$, there exists $\delta > 0$ such that, if $\|T - L\| < \delta$, then the eigenvalues of T are contained in balls of radius ε and centre at the points λ_j. It remains to show that the sum of the multiplicities of the eigenvalues of T contained in the ball of centre λ_j is exactly

n_j. By Lemma 2.17, taking $\varepsilon < \varepsilon_0$ if necessary, this sum is less than or equal to n_j. If for some j this sum is strictly less than n_j then the sum of the multiplicities of all the eigenvalues of T would be strictly less than $\sum_{j=1}^{k} n_j = n$, which is absurd. □

Corollary. *If $L \in \mathscr{L}(\mathbb{R}^n)$ is a hyperbolic vector field then there exists a neighbourhood $V \subset \mathscr{L}(\mathbb{R}^n)$ of L such that all $T \in V$ have the same index as L.* □

Another corollary of this proposition is that the roots of a polynomial vary continuously with its coefficients.

2.19 Proposition. *A hyperbolic linear vector field is structurally stable in the space of linear vector fields.*

PROOF. This follows immediately from the corollary above and Proposition 2.15. □

2.20 Proposition. *Let L be a structurally stable linear vector field. Then L is hyperbolic.*

PROOF. Let L be a nonhyperbolic linear vector field and $\delta = \min\{|\alpha|; \alpha + i\beta$ is an eigenvalue of L and $\alpha \neq 0\}$. If $0 < t < \delta$ then $L + tI$ and $L - tI$ are hyperbolic vector fields with different indices and so cannot be topologically equivalent. This shows that, in any neighbourhood of L, there exist two vector fields that are not topologically equivalent. Thus, L is not structurally stable. □

To summarize the results above we may say that a linear vector field is structurally stable in the space of linear fields if and only if it is hyperbolic. Moreover, the structurally stable linear fields form an open dense subset of $\mathscr{L}(\mathbb{R}^n)$.

§3 Singularities and Hyperbolic Fixed Points

In this section we are going to define a subset $\mathscr{G} \subset \mathfrak{X}^r(M)$ such that every $X \in \mathscr{G}$ has a local orbit structure that is stable and simple enough for us to be able to classify it.

The example below shows that a vector field can have an extremely complicated orbit structure near a singularity.

EXAMPLE. If L is a linear vector field on $\mathbb{R}^2$ with a zero eigenvalue there exists a one-dimensional subspace $V \subset \mathbb{R}^2$ consisting of singularities of L. Let L be the vector field whose matrix in the standard basis is $\begin{pmatrix} 0 & 0 \\ 0 & -1 \end{pmatrix}$. In this case $V = \{(t, 0); t \in \mathbb{R}\}$.

§3 Singularities and Hyperbolic Fixed Points

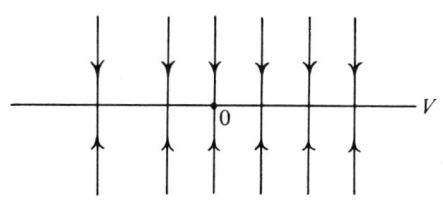

Figure 3

We shall show that arbitrarily close to L there exists an uncountable set $\mathcal{M}$ of vector fields such that, whenever $X, Z \in \mathcal{M}$, X is not locally equivalent to Z at the origin.

Let Y be the constant vector field $Y = (1, 0)$. Let $K \subset \mathbb{R}$ be a compact subset. It is known that there exists a bounded C^∞ map $\rho: \mathbb{R} \to \mathbb{R}$ that vanishes exactly on K. We can also suppose that the first r derivatives of ρ are bounded. Given $\varepsilon > 0$ we choose $n \in \mathbb{N}$ such that $\|(1/n)\rho\|_r < \varepsilon$. Let $Z = L + (1/n)\rho Y$. Then we have $\|Z - L\|_r < \varepsilon$ and the set of singularities of Z is K. Thus, if K and K' are two nonhomeomorphic compact sets, we deduce that the vector fields Z and Z', constructed as above, are not topologically equivalent. Hence there are at least as many equivalence classes of vector fields as there are homeomorphism classes of compact subsets of $\mathbb{R}$.

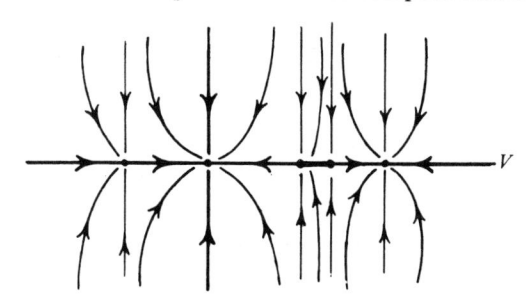

Figure 4

This example motivates the next definition.

Definition. We say that $p \in M$ is a *simple* singularity of a vector field $X \in \mathfrak{X}^r(M)$ if $DX_p: TM_p \to TM_p$ does not have zero as an eigenvalue.

3.1 Proposition. *Let $X \in \mathfrak{X}^r(M)$ and suppose that $p \in M$ is a simple singularity of X. Then there exist neighbourhoods $\mathcal{N}(X) \subset \mathfrak{X}^r(M)$, $U_p \subset M$ of X and p respectively and a continuous function $\rho: \mathcal{N}(X) \to U_p$ which to each vector field $Y \in \mathcal{N}(X)$ associates the unique singularity of Y in U_p. In particular, a simple singularity is isolated.*

PROOF. We shall use the Implicit Function Theorem for Banach spaces. As the problem is a local one we can suppose, by using a local chart, that $M = \mathbb{R}^m$, $p = 0$ and that X is a vector field in $\mathfrak{X}^r(D^m)$, where $D^m = \{x \in \mathbb{R}^m; \|x\| \leq 1\}$. We have that $\mathfrak{X}^r = \mathfrak{X}^r(D^m)$ is a Banach space and the map

$\varphi \colon D^m \times \mathfrak{X}^r \to \mathbb{R}^m$ given by $\varphi(x, Y) = Y(x)$ is of class C^r. We have $\varphi(0, X) = 0$ and, by hypothesis, $D_1\varphi(0, X) = DX(0) \colon \mathbb{R}^m \to \mathbb{R}^m$ is an isomorphism. By the Implicit Function Theorem there exist neighbourhoods U of 0 and $\mathcal{N}$ of X and a unique function $\rho \colon \mathcal{N} \to U$ of class C^r such that $\varphi(\rho(Y), Y) = 0$. Thus, for $x \in U$, $Y(x) = 0$ if and only if $x = \rho(Y)$. Moreover, as $DX(0)$ is an isomorphism and the set of isomorphisms is open, we can suppose, by shrinking $\mathcal{N}$ and U if necessary, that $DY(\rho(Y))$ is an isomorphism so that $\rho(Y)$ is a simple singularity of Y. □

Next we shall characterize a simple singularity of vector field X in M in terms of transversality. For this, let us consider the tangent bundle $TM = \{(p, v); p \in M, v \in TM_p\}$ and let $M_0 = \{(p, 0); p \in M\}$ be the zero section. M_0 is a submanifold of TM diffeomorphic to M and a vector field $X \in \mathfrak{X}^r(M)$ can be thought of as a C^r map from M to TM which we shall denote by the same letter X. Therefore p is a singularity of X if and only if $X(p) \in M_0$.

3.2 Proposition. *Let X be a C^r vector field $(r \geq 1)$ on a manifold M and let $p_0 \in M$ be a singularity of X. Then p_0 is a simple singularity of X if and only if the map $p \mapsto (p, X(p))$ from M to TM is transversal to the zero section M_0 at p_0.*

PROOF. Let $x \colon U \to \mathbb{R}^m$ be a local chart with $x(p_0) = 0$. Let $TU = \{(p, v) \in TM; p \in U\}$. The map $Tx \colon TU \to \mathbb{R}^m \times \mathbb{R}^m$ defined by $Tx(p, v) = (x(p), Dx_p(v))$ is a local chart for TM. Consider the following diagram

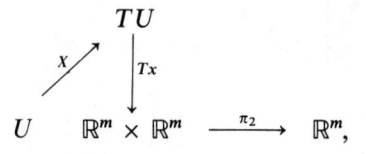

$$\begin{array}{ccc} & TU & \\ {}^X\nearrow & \downarrow {\scriptstyle Tx} & \\ U & \mathbb{R}^m \times \mathbb{R}^m & \xrightarrow{\pi_2} \mathbb{R}^m, \end{array}$$

where π_2 is the projection $\pi_2(x, y) = y$.

Put $h = \pi_2 TxX$. Then X is transversal to M_0 at p_0 if and only if p_0 is a regular point of h, that is, if $dh(p_0) \colon TM_p \to \mathbb{R}^m$ is an isomorphism. On the other hand, $dh(p_0) = Dx(p_0)DX(p_0)$. Thus, $Dh(p_0)$ is an isomorphism if and only if $DX(p_0)$ is an isomorphism, which proves the proposition. □

Let $\mathcal{G}_0 \subset \mathfrak{X}^r(M)$ be the set of vector fields whose singularities are all simple; that is, $\mathcal{G}_0 = \{X \in \mathfrak{X}^r(M); X \colon M \to TM \text{ is transversal to } M_0\}$. As a simple singularity is isolated and M is compact it follows that any $X \in \mathcal{G}_0$ has only a finite number of singularities.

3.3 Proposition. $\mathcal{G}_0$ *is open and dense in $\mathfrak{X}^r(M)$.*

PROOF. (a) *Openness.* As the set of C^r maps from M to TM that are transversal to M_0 is open we conclude that $\mathcal{G}_0$ is open.

(b) *Density.* Let $X \in \mathfrak{X}^r(M)$. By Thom's Transversality Theorem there exist maps $Y \colon M \to TM$ transversal to M_0 arbitrarily close to X. It may happen, however, that Y is not a vector field since we might have $\pi(Y(p)) \neq p$

§3 Singularities and Hyperbolic Fixed Points

for some $p \in M$ where $\pi: TM \to M$ denotes the map $(p, v) \mapsto p$. But $\pi X = id_M$ and, if Y is close enough to X, then $\varphi = \pi Y$ is close to id_M so that it is a diffeomorphism since the set of diffeomorphisms of M is open in $C^r(M, M)$. Now $Z = Y\varphi^{-1}$ is a vector field on M because $\pi Z = \pi Y \varphi^{-1} = \varphi \varphi^{-1} = id_M$. As Y is transversal to M_0 and φ^{-1} is a diffeomorphism it follows that Z is transversal to M_0. If Y is close to X then so is Z. □

EXAMPLE. In this example we shall consider a linear vector field on $\mathbb{R}^2$ with a simple singularity and show that a nonlinear perturbation can give a vector field with an extremely complicated orbit structure. Consider the vector field given by

$$L = \begin{pmatrix} 0 & 1 \\ -1 & 0 \end{pmatrix}.$$

Let $\rho: \mathbb{R} \to \mathbb{R}$ be a C^∞ function such that $\rho(0) = 0$ and $\rho^{(k)}(0) = 0$ for all $k \in \mathbb{N}$. Let X be the vector field on $\mathbb{R}^2$ defined by $X(x, y) = (y + \rho(r^2)x, -x + \rho(r^2)y)$ where $r^2 = x^2 + y^2$. It is easy to see that X is C^∞ and that

$$DX(0, 0) = \begin{pmatrix} 0 & 1 \\ -1 & 0 \end{pmatrix}.$$

Thus, $(0, 0)$ is a simple singularity of X. Let K be a compact subset of $\mathbb{R}^+$ containing 0. We can choose ρ so that $\rho(K) = 0$ and ρ is never zero on $I - K$ where $I = (-\delta, \delta)$ is an interval containing K. Given $\varepsilon > 0$ and $r > 0$ we can choose ρ so that $\|X - L\|_r < \varepsilon$. If $r_0 \in \mathbb{R}^+$ has $\rho(r_0^2) = 0$ then the vector field X is tangent to the circle of radius r_0 so that this circle is a closed orbit for X. On the other hand, if $(a, b) \subset \mathbb{R}^+$ is an interval such that $\rho(t^2) > 0$ for $t \in (a, b)$ and $\rho(a^2) = \rho(b^2) = 0$, then the orbits of X through points of the annulus $D_b - \bar{D}_a = \{z \in \mathbb{R}^2; a < \|z\| < b\}$ are not closed and, in fact, they are spirals whose ω-limit set is the circle of radius b. This follows from the observation that $\langle (x, y), X(x, y) \rangle = r^2 \rho(r^2)$, so that the radial component of X is $r\rho(r^2)$ which is positive for $r \in (a, b)$. Similarly, in an interval (a, b) where $\rho(a^2) = \rho(b^2) = 0$ and $\rho(t^2) < 0$ for $t \in (a, b)$, the orbits in the annulus $D_b - \bar{D}_a$ will have the circle of radius b as α-limit set and the circle of radius a as ω-limit set. See Figure 5.

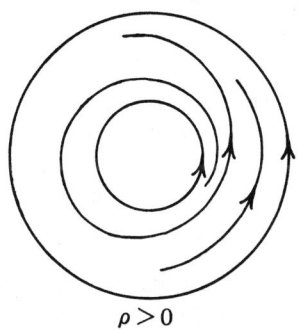

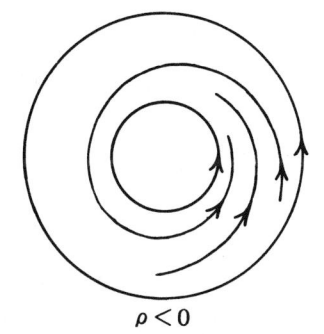

$\rho > 0$ $\rho < 0$

Figure 5

This construction gives a vector field X arbitrarily close to L whose closed orbits intersect the x-axis precisely at the points of the compact set $\{r \in \mathbb{R}^+ ; r^2 \in K\}$ which is homeomorphic to K. If K and K' are two non-homeomorphic compact sets and X and X' are vector fields associated to them by the above construction it follows that X and X' are not topologically equivalent.

Definition. Let $X \in \mathfrak{X}^r(M)$ and let $p \in M$ be a singularity of X. We say that p is a *hyperbolic singularity* if $DX_p : TM_p \to TM_p$ is a hyperbolic linear vector field, that is, DX_p has no eigenvalue on the imaginary axis.

Let $\mathscr{G}_1 \subset \mathfrak{X}^r(M)$ be the set of those vector fields whose singularities are all hyperbolic. It is clear that $\mathscr{G}_1 \subset \mathscr{G}_0$.

3.4 Theorem. *$\mathscr{G}_1$ is open and dense in $\mathfrak{X}^r(M)$.*

PROOF. As $\mathscr{G}_0$ is open and dense in $\mathfrak{X}^r(M)$ and $\mathscr{G}_1 \subset \mathscr{G}_0$, it is sufficient to show that $\mathscr{G}_1$ is open and dense in $\mathscr{G}_0$. Let $X \in \mathscr{G}_0$ and let $p_0, \ldots, p_k \in M$ be the singularities of X. By Proposition 3.1 there exist neighbourhoods $\mathscr{N}(X)$ of X and $U_1, \ldots, U_k$ of $p_1, \ldots, p_k$, respectively, and continuous functions $\rho_j : \mathscr{N} \to U_j, j = 1, \ldots, k$ such that $\rho_j(Y)$ is the only singularity of Y in U_j. We can suppose that these neighbourhoods are pairwise disjoint. If $p \in M - \bigcup_{j=1}^k U_j$ then $X(p) \neq 0$ and, by the compactness of $M - \bigcup_{j=1}^k U_j$, there exists $\delta > 0$ such that $\|X(p)\| > \delta$ for all $p \in M - \bigcup U_j$. Therefore, shrinking $\mathscr{N}$ if necessary, we can suppose that any $Y \in \mathscr{N}$ does not have a singularity in $M - \bigcup U_j$. Let us suppose that $X \in \mathscr{G}_1$. As DX_{p_j} is a hyperbolic linear vector field and such vector fields form an open set we deduce from the continuity of the maps ρ_j, shrinking $\mathscr{N}$ if necessary, that $DY_{\rho_j(Y)}$ will be a hyperbolic linear vector field for all $Y \in \mathscr{N}$. Thus $\mathscr{N} \subset \mathscr{G}_1$, which proves the openness of $\mathscr{G}_1$.

Now suppose that $X \in \mathscr{G}_0$. We shall show that there exists $Y \in \mathscr{G}_1 \cap \mathscr{N}$ arbitrarily close to X. Note that, if $u > 0$ is small enough, then $DX_{p_j} + uI$ is a hyperbolic linear vector field on TM_{p_j} for all $j = 1, \ldots, k$. It will therefore suffice to show that, given a neighbourhood $\mathscr{N}_1 \subset \mathscr{N}$ of X there exists $Y \in \mathscr{N}_1$ such that $Y(p_j) = 0$ and $DY_{p_j} = DX_{p_j} + uI$. Let $V_j \subset U_j$ be a neighbourhood of p_j and $x^j : V_j \to B(3) \subset \mathbb{R}^m$ be a local chart with $x^j(p_j) = 0$ where $B(3)$ is the ball with radius 3 and centre at the origin. Let $\varphi : \mathbb{R}^m \to \mathbb{R}$ be a positive C^∞ function such that $\varphi(B(1)) = 1$ and $\varphi(\mathbb{R}^m - B(2)) = 0$. Let $x_*^j X$ denote the expression of the vector field X in the local chart x^j; that is, $x_*^j X(q) = Dx^j((x^j)^{-1}(q))X((x^j)^{-1}(q))$. Then we define $Y(p) = X(p)$ if $p \in M - \bigcup_j V_j$ and $Y(p) = D(x^j)^{-1}(x^j(p))(x_*^j X(x^j(p)) + u\varphi(x^j(p))x^j(p))$ if $p \in V_j$. It is easy to see that Y is a C^∞ vector field, that $Y(p_j) = 0$ and that $DY_{p_j} = DX_{p_j} + uI$. Moreover, by taking u small enough we have $Y \in \mathscr{N}_1$, which completes the proof. $\square$

Next we shall extend these results to diffeomorphisms of a compact manifold M. We shall omit the proofs of the propositions and the theorem as they are analogous to those just given for vector fields.

Definition. Let $p \in M$ be a fixed point of the diffeomorphism $f \in \text{Diff}^r(M)$. We say that p is an *elementary fixed point* if 1 is not an eigenvalue of $Df_p: TM_p \to TM_p$.

3.5 Proposition. *Let $f \in \text{Diff}^r(M)$ and suppose that p is an elementary fixed point of f. There exist neighbourhoods $\mathcal{N}$ of f in $\text{Diff}^r(M)$ and U of p and a continuous map $\rho: \mathcal{N} \to U$ which, to each $g \in \mathcal{N}$, associates the unique fixed point of g in U and this fixed point is elementary. In particular, an elementary fixed point is isolated.* □

Let Δ denote the diagonal $\{(p, p) \in M \times M; p \in M\}$, which is a submanifold of $M \times M$ of dimension m. If $f \in \text{Diff}^r(M)$ we consider the map $\tilde{f}: M \to M \times M$ given by $\tilde{f}(p) = (p, f(p))$, whose image is the graph of f.

3.6 Proposition. *Let $f \in \text{Diff}^r(M)$ and let $p \in M$ be a fixed point of f. Then p is an elementary fixed point if and only if $\tilde{f}$ is transversal to Δ at p.* □

Let $G_0 \subset \text{Diff}^r(M)$ be the set of diffeomorphisms whose fixed points are all elementary. Thus, $f \in G_0$ if and only if $\tilde{f}$ is transversal to Δ. By using Thom's Transversality Theorem we obtain the following proposition.

3.7 Proposition. *G_0 is open and dense in $\text{Diff}^r(M)$.* □

Definition. Let $p \in M$ be a fixed point of $f \in \text{Diff}^r(M)$. We say that p is a *hyperbolic fixed point* if $Df_p: TM_p \to TM_p$ is a hyperbolic isomorphism, that is, if Df_p has no eigenvalue of modulus 1.

Let $G_1 \subset \text{Diff}^r(M)$ be the set of diffeomorphisms whose fixed points are all hyperbolic.

3.8 Theorem. *G_1 is open and dense in $\text{Diff}^r(M)$.* □

In the next section we shall show that a diffeomorphism $f \in G_1$ is locally stable.

§4 Local Stability

In this section we shall prove a theorem due to Hartman and Grobman according to which a diffeomorphism f is locally conjugate to its linear part at a hyperbolic fixed point. Analogously, a vector field X is locally equivalent to its linear part at a hyperbolic singularity. As a consequence we shall have local stability at a hyperbolic fixed point and at a hyperbolic singularity. The proof we shall present is also valid in Banach spaces [25], [36], [74], [90]. Other generalizations and references can be found in [80].

4.1 Theorem. *Let $f \in \mathrm{Diff}^r(M)$ and let $p \in M$ be a hyperbolic fixed point of f. Let $A = Df_p: TM_p \to TM_p$. Then there exist neighbourhoods $V(p) \subset M$ and $U(0) \subset TM_p$ and a homeomorphism $h: U \to V$ such that*

$$hA = fh.$$

Remark. As this is a local problem we can, by using a local chart, suppose that $f: \mathbb{R}^m \to \mathbb{R}^m$ is a diffeomorphism with 0 as a hyperbolic fixed point. Before proving Theorem 4.1 we shall need a few lemmas.

4.2 Lemma. *Let E be a Banach space, suppose that $L \in \mathscr{L}(E, E)$ satisfies $\|L\| \leq a < 1$ and that $G \in \mathscr{L}(E, E)$ is an isomorphism with $\|G^{-1}\| \leq a < 1$. Then*

(a) *$I + L$ is an isomorphism and $\|(I + L)^{-1}\| \leq 1/(1 - a)$,*
(b) *$I + G$ is an isomorphism and $\|(I + G)^{-1}\| \leq a/(1 - a)$.*

PROOF OF LEMMA 4.2. (a) Given $y \in E$, define $u: E \to E$ by $u(x) = y - L(x)$. Then $u(x_1) - u(x_2) = L(x_2 - x_1)$. Thus, $\|u(x_1) - u(x_2)\| \leq a\|x_1 - x_2\|$ so that u is a contraction. Hence, u has a unique fixed point $x \in E$; that is, $x = u(x) = y - Lx$. Therefore, there exists a unique $x \in E$ such that $(L + I)x = y$; that is, $I + L$ is a bijection. Let $y \in E$ have $\|y\| = 1$ and take $x \in E$ such that $(I + L)^{-1}y = x$. As $x + Lx = y$, we have $\|x\| - a\|x\| \leq 1$ so that $\|x\| \leq 1/(1 - a)$. Thus, $\|(I + L)^{-1}\| \leq 1/(1 - a)$.

(b) First note that $I + G = G(I + G^{-1})$. As $\|G^{-1}\| \leq a < 1$, the first part of the lemma says that $I + G^{-1}$ is invertible. Thus $(I + G)^{-1} = (I + G^{-1})^{-1}G^{-1}$ and, therefore,

$$\|(I + G)^{-1}\| \leq \|(I + G^{-1})^{-1}\|\|G^{-1}\| \leq \frac{1}{1 - a} \cdot a = \frac{a}{1 - a},$$

which proves the lemma. □

As $A = Df_0$ is a hyperbolic isomorphism, there exists an invariant splitting $\mathbb{R}^m = E^s \oplus E^u$ and a norm $\|\cdot\|$ on $\mathbb{R}^m$ in which

$$\|A^s\| \leq a < 1, \quad \text{where} \quad A^s = A|E^s: E^s \to E^s,$$

$$\|(A^u)^{-1}\| \leq a < 1, \quad \text{where} \quad A^u = A|E^u: E^u \to E^u.$$

Let $C_b^0(\mathbb{R}^m)$ be the Banach space of bounded continuous maps from $\mathbb{R}^m$ to $\mathbb{R}^m$ with the uniform norm: $\|u\| = \sup\{\|u(x)\|; x \in \mathbb{R}^m\}$. As $\mathbb{R}^m = E^s \oplus E^u$ we have a decomposition $C_b^0(\mathbb{R}^m) = C_b^0(\mathbb{R}^m, E^s) \oplus C_b^0(\mathbb{R}^m, E^u)$ where $u = u^s + u^u$ with $u^s = \pi_s \circ u$ and $u^u = \pi_u \circ u$ obtained from the natural projections $\pi_s: E^s \oplus E^u \to E^s$ and $\pi_u: E^s \oplus E^u \to E^u$.

4.3 Lemma. *There exists $\varepsilon > 0$ such that, if $\varphi_1, \varphi_2 \in C_b^0(\mathbb{R}^m)$ have Lipschitz constant less than or equal to ε, then $A + \varphi_1$ and $A + \varphi_2$ are conjugate.*

§4 Local Stability

PROOF. We must find a homeomorphism $h: \mathbb{R}^m \to \mathbb{R}^m$ that satisfies the equation

$$h(A + \varphi_1) = (A + \varphi_2)h. \tag{1}$$

Let us try a solution of the form $h = I + u$ with $u \in C_b^0(\mathbb{R}^m)$. Then we need

$$(I + u)(A + \varphi_1) = (A + \varphi_2)(I + u) \tag{2}$$

or

$$A + \varphi_1 + u(A + \varphi_1) = A + Au + \varphi_2(I + u)$$

or equivalently,

$$Au - u(A + \varphi_1) = \varphi_1 - \varphi_2(I + u). \tag{3}$$

We shall show that there exists a unique $u \in C_b^0(\mathbb{R}^m)$ that satisfies equation (3). Consider the linear operator

$$\mathscr{L}: C_b^0(\mathbb{R}^m) \to C_b^0(\mathbb{R}^m),$$

$$\mathscr{L}(u) = Au - u(A + \varphi_1).$$

We claim that $\mathscr{L}$ is invertible and that $\|\mathscr{L}^{-1}\| \leq \|A^{-1}\|/(1 - a)$. In fact, $\mathscr{L} = \bar{A}\mathscr{L}^*$ where $\mathscr{L}^*: C_b^0(\mathbb{R}^m) \to C_b^0(\mathbb{R}^m)$ is given by $\mathscr{L}^*(u) = u - A^{-1}u(A + \varphi_1)$ and $\bar{A}: C_b^0(\mathbb{R}^m) \to C_b^0(\mathbb{R}^m)$ by $\bar{A}(u) = A \circ u$. As $\bar{A}$ is invertible, we must show that $\mathscr{L}^*$ is invertible for then $\mathscr{L}^{-1} = \mathscr{L}^{*-1}\bar{A}^{-1}$. We remark that $C_b^0(\mathbb{R}^m, E^s)$ and $C_b^0(\mathbb{R}^m, E^u)$ are invariant under $\mathscr{L}^*$ since E^s and E^u are invariant under A^{-1}. Thus we can write $\mathscr{L}^* = \mathscr{L}^{*s} \oplus \mathscr{L}^{*u}$ where $\mathscr{L}^{*s} = \mathscr{L}^*|C_b^0(\mathbb{R}^m, E^s)$ and $\mathscr{L}^{*u} = \mathscr{L}^*|C_b^0(\mathbb{R}^m, E^u)$. It is easy to see that if ε is small enough then $A + \varphi_1$ is a homeomorphism and so the operator $u^s \mapsto A^{-1}u^s(A + \varphi_1)$ is invertible and its inverse $u^s \mapsto A^s u^s (A + \varphi_1)^{-1}$ is a contraction with norm bounded by the number $a < 1$ involved in the hyperbolicity of A. By part (b) of Lemma 4.2, $\mathscr{L}^{*s}$ is invertible and $\|(\mathscr{L}^{*s})^{-1}\| \leq a/(1 - a)$. From part (a) of Lemma 4.2 we also conclude that $\mathscr{L}^{*u}$ is invertible and $\|(\mathscr{L}^{*u})^{-1}\| \leq 1/(1 - a) = \max\{1/(1 - a), a/(1 - a)\}$. Therefore, $\mathscr{L}$ is invertible and

$$\|\mathscr{L}^{-1}\| = \|\mathscr{L}^{*-1}\bar{A}^{-1}\| \leq \|A^{-1}\|/(1 - a),$$

which proves our claim.

Now consider the map

$$\mu: C_b^0(\mathbb{R}^m) \to C_b^0(\mathbb{R}^m), \qquad \mu(u) = \mathscr{L}^{-1}[\varphi_1 - \varphi_2(I + u)].$$

We have

$$\|\mu(u_1) - \mu(u_2)\| = \|\mathscr{L}^{-1}[\varphi_2(I + u_2) - \varphi_2(I + u_1)]\|$$
$$\leq \|A^{-1}\|(1 - a)^{-1}\varepsilon\|u_2 - u_1\|.$$

If ε is small enough to make $\varepsilon\|A^{-1}\|(1 - a)^{-1} < 1$ then μ is a contraction and so has a unique fixed point u, say, in $C_b^0(\mathbb{R}^m)$. As $u \in C_b^0(\mathbb{R}^m)$ is a solution of (3) if and only if it is a fixed point of μ we conclude that (3) has a unique solution in $C_b^0(\mathbb{R}^m)$. It remains to prove that $I + u$ is a homeomorphism. For this we first note that the method used above also shows that the equation

$$(A + \varphi_1)(I + v) = (I + v)(A + \varphi_2)$$

also has a unique solution $v \in C_b^0(\mathbb{R}^m)$. We claim that

$$(I + u)(I + v) = (I + v)(I + u) = I.$$

In fact,
$$(I + u)(I + v)(A + \varphi_2) = (I + u)(A + \varphi_1)(I + v)$$
$$= (A + \varphi_2)(I + u)(I + v).$$

On the other hand, as $(I + u)(I + v) = I + v + u(I + v)$ with $w = v + u(I + v) \in C_b^0(\mathbb{R}^m)$ and $I(A + \varphi_2) = (A + \varphi_2)I$, the uniqueness of the solution of the equation $(I + w)(A + \varphi_2) = (A + \varphi_2)(I + w)$ shows that

$$(I + u)(I + v) = I.$$

Similarly we have $(I + v)(I + u) = I$. This shows that $I + u$ is a homeomorphism which conjugates $A + \varphi_1$ and $A + \varphi_2$ and that proves the lemma. □

4.4 Lemma. *Given $\varepsilon > 0$ there exists a neighbourhood U of 0 and an extension of $f | U$ to $\mathbb{R}^m$ of the form $A + \varphi$ where $\varphi \in C_b^0(\mathbb{R}^m)$ has Lipschitz constant at most ε.*

PROOF. Let $\alpha: \mathbb{R} \to \mathbb{R}$ be a C^∞ function with the following properties:

$$\alpha(t) = 0, \quad \text{if } t \geq 1;$$
$$\alpha(t) = 1, \quad \text{if } t \leq \tfrac{1}{2};$$
$$|\alpha'(t)| < K, \quad \forall\, t \in \mathbb{R}, K > 2.$$

Let $f = A + \psi$ with $\psi(0) = 0$ and $D\psi_0 = 0$. Let B_e be a ball with centre at the origin and radius e such that $\|D\psi_x\| < \varepsilon/2K$ for $x \in B_e$. Define $\varphi: \mathbb{R}^m \to \mathbb{R}^m$ by $\varphi(x) = \alpha(\|x\|/e)\psi(x)$. It is clear that $\varphi(x) = 0$ if $\|x\| \geq e$. Let us show that φ satisfies the conditions in the lemma. In fact, since $\varphi(x) = \psi(x)$ for $\|x\| \leq e/2$ we see that $A + \varphi$ is an extension of $f | B_{e/2}$. On the other hand, if $x_1, x_2 \in B_e$ we have

$$\|\varphi(x_1) - \varphi(x_2)\| = \|[\alpha(\|x_1\|/e)\psi(x_1) - \alpha(\|x_2\|/e)\psi(x_2)]\|$$
$$= \|[\alpha(\|x_1\|/e) - \alpha(\|x_2\|/e)]\psi(x_1)$$
$$+ \alpha(\|x_2\|/e)[\psi(x_1) - \psi(x_2)]\|$$
$$\leq (K\|x_1 - x_2\|/e)(\varepsilon/2K)\|x_1\|$$
$$+ (\varepsilon/2K)\|x_1 - x_2\| \leq \varepsilon\|x_1 - x_2\|.$$

If $x_1 \in B_e$ and $x_2 \notin B_e$ we have $\|\varphi(x_1) - \varphi(x_2)\| \leq \varepsilon/2\|x_1 - x_2\| \leq \varepsilon\|x_1 - x_2\|$ and, if $x_1, x_2 \notin B_e$, we have $\|\varphi(x_1) - \varphi(x_2)\| = 0 \leq \varepsilon\|x_1 - x_2\|$. Thus, φ has Lipschitz constant at most ε, which proves the lemma. □

§4 Local Stability

PROOF OF THEOREM 4.1. Let $\varepsilon > 0$ be as in Lemma 4.3. Let $A + \varphi$ be an extension of $f \mid U(0)$ where $U(0)$ is a neighbourhood of 0 and φ has Lipschitz constant at most ε. By Lemma 4.3 there exists a homeomorphism $h \colon \mathbb{R}^m \to \mathbb{R}^m$ such that $hA = (A + \varphi)h$. Thus $hA = fh$ on a neighbourhood of 0 as required. □

Remarks. (1) The conjugacy between $A + \varphi_1$ and $A + \varphi_2$ in Lemma 4.3 is unique among maps having a finite distance from the identity. However, there are infinitely many conjugacies that do not satisfy this last condition. For example, consider a contraction $\varphi \colon \mathbb{R} \to \mathbb{R}$ with $\varphi(0) = 0$. Suppose that $\varphi(1) = a > 0$ and $\varphi(-1) = b < 0$. We pick any homeomorphism $h \colon [-1, b] \cup [a, 1] \to [-1, b] \cup [a, 1]$ with $h(-1) = -1, h(b) = b, h(a) = a$, $h(1) = 1$. We extend h to the whole real line putting $h(0) = 0$ and $h(x) = \varphi^n h \varphi^{-n}(x)$ where $n \in \mathbb{Z}$ is chosen so that $\varphi^{-n}(x) \in [-1, b] \cup [a, 1]$. In this way we find as many solutions of the equation $h\varphi = \varphi h$ as there are homeomorphisms from $[-1, b] \cup [a, 1]$ to itself.

(2) Even if we require the conjugacy between A and f to be near the identity it is not unique because it depends on the extension $A + \varphi$ of f to the whole space $\mathbb{R}^m$.

Next we shall show from Theorem 4.1 that a hyperbolic linear isomorphism is locally stable in $\mathscr{L}(\mathbb{R}^m)$. In fact, as we shall see in the next section, a hyperbolic isomorphism is globally stable in $\mathscr{L}(\mathbb{R}^m)$.

4.5 Proposition. *Let A be a hyperbolic isomorphism. There exists $\delta > 0$ such that, if $B \in \mathscr{L}(\mathbb{R}^m)$ and $\|B - A\| < \delta$. then B is locally conjugate to A.*

PROOF. We shall look for a homeomorphism $h \colon \mathbb{R}^m \to \mathbb{R}^m$ that conjugates A and an extension of $B \mid U$, where U is a neighbourhood of the origin.

Let $\alpha \colon \mathbb{R} \to \mathbb{R}$ be a C^∞ map with the properties:

$$\alpha(t) = 1, \quad \text{if } |t| \leq 1;$$

$$\alpha(t) = 0, \quad \text{if } |t| \geq 2;$$

$$\alpha(\mathbb{R}) \subset [0, 1].$$

We can write B as $A + (B - A)$. Let $\varphi \colon \mathbb{R}^m \to \mathbb{R}^m$ be defined by $\varphi(x) = \alpha(\|x\|) \cdot (B - A)(x)$. Then we have $\varphi \mid B_1 = (B - A) \mid B_1$, where B_1 is the ball with centre 0 and radius 1, and $\varphi(x) = 0$ if $\|x\| \geq 2$. For all $x \in \mathbb{R}^m$, $\|D\varphi_x\| \leq K \|B - A\| + \|B - A\|$, where $K = \sup\{|\alpha'(t)|; t \in \mathbb{R}\}$ is greater than 1. Given $\varepsilon > 0$ we take $\delta < \varepsilon/2K$ so that $\|D\varphi_x\| < \varepsilon$. It follows that φ has Lipschitz constant less than ε. By Lemma 4.3 there exists a homeomorphism $h \colon \mathbb{R}^m \to \mathbb{R}^m$ such that

$$hA = (A + \varphi)h.$$

Since $A + \varphi$ is an extension of B, h is a local conjugacy between A and B. □

4.6 Theorem. *Let $f \in \text{Diff}^r(M)$, $r \geq 1$, and let $p \in M$ be a hyperbolic fixed point of f. Then f is locally stable at p.*

PROOF. By Proposition 3.5 there exist neighbourhoods $\tilde{\mathcal{N}}(f)$ and $W(p)$ and a continuous map $\rho\colon \tilde{\mathcal{N}}(f) \to W$ which associates to each $g \in \tilde{\mathcal{N}}(f)$ the unique fixed point of g in W, $\rho(g)$, and this fixed point is hyperbolic. If we take a small enough neighbourhood $\mathcal{N}(f) \subset \tilde{\mathcal{N}}(f)$ we shall have Df_p near $Dg_{\rho(g)}$ and so, by Proposition 4.5, these linear isomorphisms are locally conjugate. As f is locally conjugate to Df_p and g is locally conjugate to $Dg_{\rho(g)}$ it follows, by transitivity, that f is locally conjugate to g. $\square$

We shall now extend these results to vector fields. Let V be a neighbourhood of 0 in $\mathbb{R}^m$ and let $X\colon V \to \mathbb{R}^m$ be a C^r vector field, $r \geq 1$. We recall that 0 is a hyperbolic singularity of X if $L = DX_0$ is a hyperbolic linear vector field. We shall show that, if 0 is a hyperbolic singularity of X then the orbits of X in a neighbourhood of 0 have the same topological behaviour as the orbits of the linear vector field L. For this we shall need some lemmas.

4.7 Lemma (Gronwall's Inequality). *Let $u, v\colon [a, b] \to \mathbb{R}$ be continuous nonnegative functions that, for some $\alpha \geq 0$, satisfy*

$$u(t) \leq \alpha + \int_a^t u(s)v(s)\, ds \quad \forall\, t \in [a, b].$$

Then

$$u(t) \leq \alpha \exp\left[\int_a^t v(s)\, ds\right].$$

PROOF. Let $\omega\colon [a, b] \to \mathbb{R}$ be the map $\omega(t) = \alpha + \int_a^t u(s)v(s)\, ds$. Suppose first that $\alpha > 0$. We have $\omega(a) = \alpha$ and $\omega(t) \geq \alpha > 0$ for all $t \in [a, b]$. As $\omega'(t) = v(t)u(t) \leq v(t)\omega(t)$, we have $\omega'(t)/\omega(t) \leq v(t)$. Integrating from a to t we obtain

$$\omega(t)/\alpha \leq \exp\left[\int_a^t v(s)\, ds\right].$$

Thus

$$u(t) \leq \omega(t) \leq \alpha \exp\left[\int_a^t v(s)\, ds\right].$$

If $\alpha = 0$ the previous case implies that, for all $\alpha_1 > 0$, $u(t) \leq \alpha_1 \exp[\int_a^t v(s)\, ds]$. Thus, $u(t) = 0$ and the inequality remains true. $\square$

4.8 Lemma. *Let $Y\colon \mathbb{R}^m \to \mathbb{R}^m$ be a C^r vector field with $Y(0) = 0$ that satisfies a Lipschitz condition with constant K. Then the flow of Y is defined on $\mathbb{R} \times \mathbb{R}^m$ and $\|Y_t(x) - Y_t(y)\| \leq e^{K|t|}\|x - y\|$ for all $x, y \in \mathbb{R}^m$.*

§4 Local Stability

PROOF. Let $x \in \mathbb{R}^m$. To get a contradiction suppose the maximal interval of the integral curve of Y through the point x is (a, b) with $b < \infty$. Let $\varphi: (a, b) \to \mathbb{R}^m$ be the integral curve through the point x. We have

$$\varphi(t) = x + \int_0^t Y(\varphi(s))\, ds.$$

Therefore, if $t \geq 0$,

$$\|\varphi(t)\| \leq \|x\| + \int_0^t \|Y(\varphi(s))\|\, ds \leq \|x\| + \int_0^t K\|\varphi(s)\|\, ds.$$

From Gronwall's inequality we obtain

$$\|\varphi(t)\| \leq e^{K|t|}\|x\| \leq e^{Kb}\|x\| \quad \text{if } t \geq 0.$$

Let $t_n \to b$ and consider the sequence $\{\varphi(t_n)\}$ whose terms all belong to the closed ball with centre 0 and radius $M = e^{Kb}\|x\|$. Since

$$\varphi(t_n) - \varphi(t_m) = \int_{t_m}^{t_n} Y(\varphi(s))\, ds,$$

we have that

$$\|\varphi(t_n) - \varphi(t_m)\| \leq KM|t_n - t_m|.$$

Thus, $\varphi(t_n)$ is a Cauchy sequence and so it converges to a point $y \in \mathbb{R}^n$. The local flow of Y around y enables us to extend the integral curve φ to the right of b contrary to our initial hypothesis. Thus, the flow of Y is defined on $\mathbb{R} \times \mathbb{R}^m$. Since

$$Y_t(x) - Y_t(y) = x - y + \int_0^t [Y(Y_s(x)) - Y(Y_s(y))]\, ds,$$

we deduce that, for $t \geq 0$,

$$\|Y_t(x) - Y_t(y)\| \leq \|x - y\| + \int_0^t K\|Y_s(x) - Y_s(y)\|\, ds.$$

From Gronwall's inequality we have

$$\|Y_t(x) - Y_t(y)\| \leq e^{K|t|}\|x - y\| \quad \text{if } t \geq 0.$$

For $t \leq 0$ we obtain the same expression by applying this argument to the vector field $-Y$. □

4.9 Lemma. *Let $X: V \to \mathbb{R}^m$ be a C^r vector field with $X(0) = 0$. Let $L = DX_0$. Given $\varepsilon > 0$ there exists a C^r vector field $Y: \mathbb{R}^m \to \mathbb{R}^m$ with the following properties:*

(1) *the field Y is Lipschitz with some Lipschitz constant K so that the flow induced by Y is defined on $\mathbb{R} \times \mathbb{R}^m$;*
(2) *$Y = L$ outside a ball B_l;*

(3) *there exists an open set $U \subset V$ containing 0 such that $Y = X$ on U;*
(4) *if $Y_t = L_t + \varphi_t$ there exists $M > 0$ such that $\|\varphi_t\| \le M$ for all $t \in [-2, 2]$ and φ_1 has Lipschitz constant at most ε. Moreover, $D(\varphi_1)_0 = 0$ or, equivalently, $D(Y_1)_0 = e^L = L_1$.*

PROOF. As $L = DX_0$ we have $X = L + \psi$ where $\psi: V \to \mathbb{R}^m$ is a C^r map that satisfies $\psi(0) = 0$ and $D\psi_0 = 0$. Let $\alpha: \mathbb{R} \to \mathbb{R}$ be a C^∞ map such that $\alpha(\mathbb{R}) \subset [0, 1]$, $\alpha(t) = 1$ if $|t| \le l/2$ and $\alpha(t) = 0$ if $t \ge l$.

Let $\tilde{\varphi}: \mathbb{R}^m \to \mathbb{R}^m$ be defined by $\tilde{\varphi}(x) = \alpha(\|x\|) \cdot \psi(x)$ if $x \in V$ and $\tilde{\varphi}(x) = 0$ if $x \in \mathbb{R}^m - V$. Given $\delta > 0$ we can choose $l > 0$ so that the map $\tilde{\varphi}$ is C^r and has Lipschitz constant at most δ. It is clear that $\tilde{\varphi} = \psi$ on $B_{l/2}$ and $\tilde{\varphi} = 0$ outside B_l. Let $Y: \mathbb{R}^m \to \mathbb{R}^m$ be the vector field defined by $Y = L + \tilde{\varphi}$. Again it is clear that $Y = X$ on $B_{l/2}$, $Y = L$ outside B_l and Y satisfies (1). It remains to show that condition (4) holds. In fact, $\|Y_t(x) - Y_t(y)\| \le e^{2K}\|x - y\|$ for $t \in [-2, 2]$ by Lemma 4.8. Put $\varphi_t = Y_t - L_t$. That there exists M such that $\|\varphi_t\| < M$ for $t \in [-2, 2]$ follows from the fact that Y_t and L_t are bounded on $B(0, l)$ for $t \in [-2, 2]$ and $Y = L$ outside $B(0, l)$. We also have

$$\varphi_t(x) - \varphi_t(y) = \int_0^t [\tilde{\varphi}(Y_s(x)) - \tilde{\varphi}(Y_s(y))] \, ds + \int_0^t L(\varphi_s(x) - \varphi_s(y)) \, ds.$$

From Gronwall's inequality we now conclude that

$$\|\varphi_t(x) - \varphi_t(y)\| \le 2e^{2K}\delta e^{2\|L\|}\|x - y\| \quad \forall t \in [-2, 2].$$

Provided δ is small enough, condition (4) will be satisfied. Finally, let us see that $D(\varphi_1)_0 = 0$. We want to show that given $\rho > 0$ there exists $r > 0$ such that $\|\varphi_1(x)\| \le \rho\|x\|$ for $\|x\| \le r$. Since $(D\tilde{\varphi})_0 = 0$, we can choose r so that $\|\tilde{\varphi}(z)\| \le \eta\|z\|$ for $\|z\| < r$ with $\eta < \rho e^{-K} e^{-\|L\|}$. As in the expression above, we have

$$\varphi_1(x) = \int_0^1 \tilde{\varphi}(Y_s(x)) \, ds + \int_0^1 L(\varphi_s(x)) \, ds.$$

Using the fact that $\|Y_s(x)\| \le e^K\|x\|$ for $0 \le s \le 1$, from Gronwall's inequality we get

$$\|\varphi_1(x)\| \le \eta e^K \|x\| e^{\|L\|} \le \rho\|x\|.$$

This completes the proof of Lemma 4.9. $\square$

4.10 Theorem. *Let $X: V \to \mathbb{R}^m$ be a C^r vector field and 0 a hyperbolic singularity of X. Put $L = DX_0$. Then X is locally equivalent to L at 0.*

PROOF. Let $Y: \mathbb{R}^m \to \mathbb{R}^m$ be a C^r vector field as in Lemma 4.9. As $Y = X$ on the neighbourhood U of 0, the identity map on U takes X-orbits to Y-orbits. Therefore, Y is locally equivalent to X. It remains now to prove that Y is locally conjugate to L. In fact we shall show that there exists a homeomorphism of $\mathbb{R}^m$ that conjugates the flows Y_t and L_t.

§4 Local Stability

By Lemma 4.9, $Y_t = L_t + \varphi_t$ and, since $DY_0 = L$, we have $D(Y_1)_0 = e^L = L_1$. Thus, the diffeomorphism $Y_1 = L_1 + \varphi_1$ has the origin as a hyperbolic fixed point and φ_1 has Lipschitz constant at most ε. By Lemma 4.3 there exists a unique homeomorphism $h: \mathbb{R}^m \to \mathbb{R}^m$ at a finite distance from the identity (that is, $h = I + u$ with $u \in C_b^0(\mathbb{R}^m)$) that satisfies $hY_1 = L_1 h$.

We define $H: \mathbb{R}^m \to \mathbb{R}^m$ by

$$H = \int_0^1 L_{-t} h Y_t \, dt.$$

It is clear that H is a continuous map and condition (4) of Lemma 4.9 ensures that H is at a finite distance from the identity. We now show that

$$L_s H = H Y_s \quad \text{for all } s \in \mathbb{R}.$$

For this it is sufficient to consider $s \in [0, 1]$. We have

$$L_{-s} H Y_s = L_{-s}\left(\int_0^1 L_{-t} h Y_t \, dt\right) Y_s = \int_0^1 L_{-(s+t)} h Y_{t+s} \, dt.$$

By putting $u = t + s - 1$ we obtain

$$\int_0^1 L_{-(s+t)} h Y_{t+s} \, dt = \int_{-1+s}^s L_{-u-1} h Y_{u+1} \, du$$

$$= \int_{-1+s}^0 L_{-u} L_{-1} h Y_1 Y_u \, du + \int_0^s L_{-(u+1)} h Y_{u+1} \, du.$$

We put $v = u + 1$ in the first term and $v = u$ in the second, use $L_{-1} h Y_1 = h$ and deduce that

$$L_{-s} h Y_s = \int_0^s L_{-v} h Y_v \, dv + \int_s^1 L_{-v} h Y_v \, dv = H.$$

This shows that H is a continuous map at a finite distance from the identity and it conjugates the flows L_t and Y_t. It remains to show that H is a homeomorphism. In fact, as $L_1 H = H Y_1$ and $L_1 h = h Y_1$, the uniqueness of the solution of this equation shows that $H = h$, which proves the theorem. □

4.11 Theorem. *Let $X \in \mathfrak{X}^r(M)$ and let $p \in M$ be a hyperbolic singularity of X. Then X is locally stable at p.*

PROOF. We have seen that there exist neighbourhoods $\mathcal{N}(X)$ in $\mathfrak{X}^r(M)$ and U of p in M such that every vector field $Y \in \mathcal{N}(X)$ has a unique hyperbolic singularity p_Y in U. Moreover, for $\mathcal{N}(X)$ small enough, p_Y has the same index as p. Thus, the linear vector fields DX_p and DY_{p_Y} are topologically equivalent. As X is locally equivalent to DX_p and Y is locally equivalent to DY_{p_Y} this proves the theorem. □

§5 Local Classification

We have seen that the subset $\mathscr{G}_1(M) \subset \mathfrak{X}^r(M)$ consisting of those vector fields whose singularities are all hyperbolic is open and dense in $\mathfrak{X}^r(M)$. Moreover, these vector fields are locally stable at each point of M. We shall now describe the possible types of local topological behaviour for vector fields in $\mathscr{G}_1$.

Let us consider the linear fields $L_0, L_1, \ldots, L_m$ on $\mathbb{R}^m$ which are represented, with respect to the standard basis, by the matrices:

$$L_0 = \begin{pmatrix} 1 & & \bigcirc \\ & 1 & \\ & & \ddots \\ \bigcirc & & 1 \end{pmatrix}, \quad L_1 = \begin{pmatrix} -1 & & \bigcirc \\ & 1 & \\ & & \ddots \\ \bigcirc & & 1 \end{pmatrix}, \ldots, \quad L_m = \begin{pmatrix} -1 & & \bigcirc \\ & -1 & \\ & & \ddots \\ \bigcirc & & -1 \end{pmatrix}.$$

If L is a hyperbolic linear vector field on $\mathbb{R}^m$ then L is conjugate to L_i, where i is the index of L.

Let $C: \mathbb{R}^m \to \mathbb{R}^m$ be the constant vector field given by $C(x) = (1, 0, \ldots, 0)$. We may collect the results proven in previous sections into the following theorem.

5.1 Theorem. *Let $X \in \mathscr{G}_1(M)$ and choose $p \in M$.*

(a) *If p is a regular point of X then X is locally equivalent at p to the constant vector field C at 0.*
(b) *If p is a singularity of X then X is locally equivalent to L_i where i is the index of p.*

PROOF. (a) This comes from the Tubular Flow Theorem.
 (b) Apply the Grobman–Hartman Theorem and Proposition 2.15. □

Now let us consider the space $\text{Diff}^r(M)$ of C^r diffeomorphisms of M. We saw that the set $G_1(M) \subset \text{Diff}^r(M)$ consisting of those diffeomorphisms whose fixed points are all hyperbolic is open and dense in $\text{Diff}^r(M)$. We shall now describe the possible types of topological behaviour of the orbits of a diffeomorphism in G_1 near its fixed points. By the theorem of Hartman and Grobman it is sufficient to classify the hyperbolic linear isomorphisms.

5.2 Proposition. *Let $A \in \mathscr{L}(\mathbb{R}^m)$ be a hyperbolic linear isomorphism. There exists $\varepsilon > 0$ such that, if $B \in \mathscr{L}(\mathbb{R}^m)$ satisfies $\|A - B\| < \varepsilon$, then B is conjugate to A.*

PROOF. By Proposition 4.5, B is locally conjugate to A; that is, there exists a homeomorphism $h: V(0) \to U(0)$ such that $hA = Bh$. Let E^s and E^u be the stable and unstable subspaces of A and let $E^{s'}$ and $E^{u'}$ be the corresponding subspaces for B. Let $V^s = V(0) \cap E^s$, $V^u = V(0) \cap E^u$, $U^{s'} = U(0) \cap E^{s'}$ and $U^{u'} = U(0) \cap E^{u'}$. By the continuity of h we have $h(V^s) = U^{s'}$ and $h(V^u) = U^{u'}$. We shall define a homeomorphism $h^s: E^s \to E^{s'}$ conjugating $A^s = A|E^s$

§5 Local Classification

and $B^{s'} = B|E^{s'}$. If $x \in V^s$, which is a neighbourhood of the origin in E^s, we put $h^s(x) = h(x) \in E^{s'}$. If $x \in E^s - V^s$ then, as $A^n(x) \to 0$ as $n \to \infty$, there exists $r \in \mathbb{N}$ such that $A^r(x) \in V^s$. We put $h^s(x) = B^{-r}hA^r(x)$. As h conjugates A and B in V^s we see immediately that h^s does not depend on the choice of r. It is also easy to check that h^s is a homeomorphism and conjugates A^s and $B^{s'}$. Similarly we define a homeomorphism $h^u: E^u \to E^{u'}$ conjugating A and B. Then we can define $\tilde{h}: E^s \oplus E^u \to E^{s'} \oplus E^{u'}$ by $\tilde{h}(x^s + x^u) = h^s(x^s) + h^u(x^u)$. It is clear that $\tilde{h}$ is a homeomorphism that conjugates A and B. □

We leave it to the reader to prove the next proposition.

5.3 Proposition. *Let A and B be hyperbolic isomorphisms of $\mathbb{R}^m$. Let E^s and E^u be the stable and unstable subspaces of A and $E^{s'}$, $E^{u'}$ the corresponding subspaces for B. Then A and B are conjugate if and only if $A^s = A|E^s$ is conjugate to $B^{s'} = B|E^{s'}$ and $A^u = A|E^u$ is conjugate to $B^{u'} = B|E^{u'}$.* □

We want to give a necessary and sufficient condition for two hyperbolic isomorphisms to be conjugate.

5.4 Proposition. *Let A_1 and A_2 be the isomorphisms of $\mathbb{R}^m$ which are represented, with respect to the standard basis, by the matrices:*

$$\tilde{A}_1 = \begin{pmatrix} \frac{1}{2} & & \bigcirc \\ & \frac{1}{2} & \\ & & \ddots \\ \bigcirc & & \frac{1}{2} \end{pmatrix}, \quad \tilde{A}_2 = \begin{pmatrix} -\frac{1}{2} & & \bigcirc \\ & \frac{1}{2} & \\ & & \ddots \\ \bigcirc & & \frac{1}{2} \end{pmatrix}.$$

Let A be a hyperbolic isomorphism of index m. If A preserves the orientation, i.e. $\det(A) > 0$, then A is conjugate to A_1. If A reverses the orientation, i.e. $\det(A) < 0$, then A is conjugate to A_2.

PROOF. By local stability A is conjugate to each isomorphism in some neighbourhood of A. Thus, we can simplify the argument by assuming that A is diagonalizable.

Let $\{v_1, \ldots, v_m\}$ be a basis for $\mathbb{R}^m$ in which the matrix $\tilde{A}$, that represents A, is in real canonical form:

$$\tilde{A} = \begin{pmatrix} \lambda_1 & & & & & & & \bigcirc \\ & \ddots & & & & & & \\ & & \lambda_{s'} & & & & & \\ & & & \mu_1 & & & & \\ & & & & \ddots & & & \\ & & & & & \mu_{s''} & & \\ & & & & & & B_1 & \\ & & & & & & & \ddots \\ \bigcirc & & & & & & & B_{s'''} \end{pmatrix},$$

where $-1 < \lambda_i < 0, 0 < \mu_i < 1$ and
$$B_i = \begin{pmatrix} \alpha_i & \beta_i \\ -\beta_i & \alpha_i \end{pmatrix}$$
with $\beta_i \neq 0$ and $\alpha_i^2 + \beta_i^2 < 1$.

We remark that two hyperbolic isomorphisms, B_0 and B_1, are conjugate if they are in the same connected component of the space of hyperbolic isomorphisms, which is open in $\mathscr{L}(\mathbb{R}^m)$. In fact let $\alpha\colon [0, 1] \to GL(\mathbb{R}^m)$ be a continuous curve such that $\alpha(t)$ is hyperbolic for all $t \in [0, 1]$ and $\alpha(0) = B_0$, $\alpha(1) = B_1$. By local stability, for each $t \in [0, 1]$, $\alpha(t)$ has an open neighbourhood $V_t \subset GL(\mathbb{R}^m)$ such that every isomorphism in V_t is conjugate to $\alpha(t)$. As $\alpha([0, 1])$ is compact, there exist $t_1 = 0 < t_2 < \cdots < t_{k-1} < t_k = 1$ such that $V_{t_1} \cup \cdots \cup V_{t_k} \supset \alpha([0, 1])$. As $V_{t_i} \cap V_{t_{i+1}} \neq \varnothing$ we see that $\alpha(t_i)$ is conjugate to $\alpha(t_{i+1})$; thus, B_0 is conjugate to B_1.

By this remark it is sufficient to find a continuous path $\alpha\colon [0, 1] \to GL(\mathbb{R}^m)$ through hyperbolic isomorphisms such that the matrix of $\alpha(0)$, in the basis $v_1, \ldots, v_m$, is $\tilde{A}$ and the matrix of $\alpha(1)$ is $\tilde{A}_1$ or $\tilde{A}_2$. This is because the isomorphisms A_1 and A_2 are similar (and therefore conjugate) to the isomorphisms represented, respectively, by $\tilde{A}_1$ and $\tilde{A}_2$ in the basis $\{v_1, \ldots, v_m\}$. First we construct a continuous path $\alpha_1\colon [0, 1] \to GL(\mathbb{R}^m)$ such that $\alpha_1(0) = A$, $\alpha_1(t)$ is hyperbolic for all $t \in [0, 1]$ and

$$\alpha_1(1) = \begin{pmatrix} -\frac{1}{2} & & & & & & \\ & \ddots & {\scriptstyle s'} & & & & O \\ & & -\frac{1}{2} & & & & \\ & & & \frac{1}{2} & & & \\ & & & & \ddots & {\scriptstyle s''+s'''} & \\ O & & & & & \frac{1}{2} \end{pmatrix}$$

Afterwards we shall construct a continuous path $\alpha_2\colon [0, 1] \to GL(\mathbb{R}^m)$ such that $\alpha_2(0) = \alpha_1(1)$ and $\alpha_2(1) = A_1$ or A_2. We put

$$\alpha_1(t) = \begin{pmatrix} \lambda_1(t) & & & & & & & O \\ & \ddots & & & & & & \\ & & \lambda_{s'}(t) & & & & & \\ & & & \mu_1(t) & & & & \\ & & & & \ddots & & & \\ & & & & & \mu_{s''}(t) & & \\ & & & & & & B_1(t) & \\ & & & & & & & \ddots \\ O & & & & & & & B_{s'''}(t) \end{pmatrix}$$

$$\lambda_i(t) = (1 - t)\lambda_i - (\tfrac{1}{2})t,$$
$$\mu_i(t) = (1 - t)\mu_i + (\tfrac{1}{2})t,$$
$$B_i(t) = \begin{pmatrix} \cos(\omega_i t) & \sin(\omega_i t) \\ -\sin(\omega_i t) & \cos(\omega_i t) \end{pmatrix} \begin{pmatrix} \alpha_i & \beta_i \\ -\beta_i & \alpha_i \end{pmatrix} \quad \text{if } t \in [0, \tfrac{1}{2}]$$

§5 Local Classification

and
$$B_i(t) = \begin{pmatrix} (\tfrac{1}{2})(2t-1) + 2(1-t)\sqrt{(\alpha_i^2 + \beta_i^2)} & O \\ O & (\tfrac{1}{2})(2t-1) + 2(1-t)\sqrt{(\alpha_i^2 + \beta_i^2)} \end{pmatrix}$$

if $t \in [\tfrac{1}{2}, 1]$, where

$$\cos(\tfrac{1}{2}\omega_i) = \alpha_i/\sqrt{(\alpha_i^2 + \beta_i^2)} \quad \text{and} \quad \sin(\tfrac{1}{2}\omega_i) = -\beta_i/\sqrt{(\alpha_i^2 + \beta_i^2)}.$$

It is easy to see that $-1 < \lambda_i(t) < 0$, $0 < \mu_i(t) < 1$ for all $t \in [0, 1]$ and that the eigenvalues of $B_i(t)$ have modulus less than 1 for $t \in [0, 1]$. This implies that $\alpha_1(t)$ is hyperbolic for all $t \in [0, 1]$. As α_1 is continuous we deduce that $\alpha_1(0) = A$ is conjugate to $\alpha_1(1)$ whose matrix is

$$\begin{pmatrix} -\tfrac{1}{2} & & & & & & \\ & \ddots & s' & & & & \\ & & -\tfrac{1}{2} & & & O & \\ & & & \tfrac{1}{2} & & & \\ & & & & \ddots & s''+s''' & \\ & O & & & & \tfrac{1}{2} & \\ & & & & & & \end{pmatrix}$$

Now we construct the curve α_2. Suppose that A reverses the orientation, that is, $\det(A) < 0$. As α_1 is continuous and $\alpha_1(t)$ is an isomorphism for all $t \in [0, 1]$, we have $\det(\alpha_1(t)) < 0$ for all $t \in [0, 1]$ and, in particular, $\det(\alpha_1(1)) < 0$. Thus, in this case the number s' of negative entries on the diagonal of $\alpha_1(1)$ is odd. We put

$$\alpha_2(t) = \begin{pmatrix} -\tfrac{1}{2} & & & & & O \\ & C_1(t) & & & & \\ & & \ddots & & & \\ & & & C_{(s'-1)/2}(t) & & \\ & & & & \tfrac{1}{2} & \\ & O & & & & \ddots \\ & & & & & & \tfrac{1}{2} \end{pmatrix},$$

where

$$C_j(t) = \begin{pmatrix} \cos(\pi t) & \sin(\pi t) \\ -\sin(\pi t) & \cos(\pi t) \end{pmatrix} \begin{pmatrix} -\tfrac{1}{2} & 0 \\ 0 & -\tfrac{1}{2} \end{pmatrix}.$$

Then we have

$$C_j(0) = \begin{pmatrix} -\tfrac{1}{2} & 0 \\ 0 & -\tfrac{1}{2} \end{pmatrix}, \quad C_j(1) = \begin{pmatrix} \tfrac{1}{2} & 0 \\ 0 & \tfrac{1}{2} \end{pmatrix}.$$

Moreover, the eigenvalues of $C_j(t)$ are of the form $-\tfrac{1}{2}e^{\pi it}$ and so have modulus less than 1 for all $t \in [0, 1]$. As $\alpha_2(0) = \alpha_1(1)$ and $\alpha_2(1)$ is represented by the matrix $\tilde{A}_2$, we have A conjugate to A_2. Similarly, if $\det(A) > 0$ then A is conjugate to A_1. Using degree theory, as presented in [64] or [38], we can prove that A_1 is not conjugate to A_2. In fact, if h is a homeomorphism conjugating A_1 and A_2 we have: $\deg h = (\deg h)(\deg A_1) = \deg(hA_1) = \deg(A_2 h) = (\deg A_2)(\deg h) = -\deg h$ and this is a contradiction because $\deg h = \pm 1$. $\square$

Remark. If A is a hyperbolic isomorphism of index 0 then A is conjugate to one of the following isomorphisms:

$$A_1 = \begin{pmatrix} 2 & & O \\ & 2 & \\ & & \ddots \\ O & & & 2 \end{pmatrix} \quad \text{or} \quad A_2 = \begin{pmatrix} -2 & & O \\ & 2 & \\ & & \ddots \\ O & & & 2 \end{pmatrix}$$

according as $\det(A) > 0$ or $\det(A) < 0$, respectively. The proof of this is entirely analogous to that of Proposition 5.4.

We shall now classify hyperbolic fixed points of diffeomorphisms using the equivalence relation of local conjugacy.

5.5 Theorem. *Let $f \in \text{Diff}^r(M)$ and suppose that $p \in M$ is a hyperbolic fixed point of f. Then f is locally conjugate at p to one of the following linear isomorphisms A_i^j of $\mathbb{R}^m$, where $m = \dim M$. The isomorphisms A_i^j, $0 \le i \le m$ and $1 \le j \le 4$, have index i and are represented in the standard basis of $\mathbb{R}^m$ by the matrices*

$$A_0^1 = A_0^2 = \begin{pmatrix} 2 & & O \\ & 2 & \\ & & \ddots \\ O & & & 2 \end{pmatrix} \quad A_0^3 = A_0^4 = \begin{pmatrix} -2 & & O \\ & 2 & \\ & & \ddots \\ O & & & 2 \end{pmatrix},$$

$$A_i^1 = \begin{pmatrix} \frac{1}{2} & & & & O \\ & \frac{1}{2} & & & \\ & & \ddots & {}^i & \\ & & & \frac{1}{2} & \\ & & & & 2 \\ & & & & & \ddots \\ O & & & & & & 2 \end{pmatrix}, \quad A_i^2 = \begin{pmatrix} -\frac{1}{2} & & & & O \\ & \frac{1}{2} & & & \\ & & \ddots & {}^i & \\ & & & \frac{1}{2} & \\ & & & & 2 \\ & & & & & \ddots \\ O & & & & & & 2 \end{pmatrix},$$

$$A_i^3 = \begin{pmatrix} \frac{1}{2} & & & & O \\ & \ddots & {}^i & & \\ & & \frac{1}{2} & & \\ & & & -2 & \\ & & & & 2 \\ & & & & & \ddots \\ O & & & & & & 2 \end{pmatrix}, \quad A_i^4 = \begin{pmatrix} -\frac{1}{2} & & & & O \\ & \frac{1}{2} & & & \\ & & \ddots & {}^i & \\ & & & \frac{1}{2} & \\ & & & & -2 \\ & & & & & \ddots \\ O & & & & & & 2 \end{pmatrix},$$

$$A_m^1 = A_m^3 = \begin{pmatrix} \frac{1}{2} & & O \\ & \frac{1}{2} & \\ & & \ddots \\ O & & & \frac{1}{2} \end{pmatrix}, \quad A_m^2 = A_m^4 = \begin{pmatrix} -\frac{1}{2} & & O \\ & \frac{1}{2} & \\ & & \ddots \\ O & & & \frac{1}{2} \end{pmatrix}.$$

PROOF. This follows immediately from the Theorem of Hartman and Grobman, Proposition 5.3 and Proposition 5.4. □

§6 Invariant Manifolds

Let $f \in \text{Diff}^r(M)$ and suppose that $p \in M$ is a hyperbolic fixed point of f. The set $W^s(p)$ of points in M that have p as ω-limit is called the stable manifold of p and the set $W^u(p)$ of points that have p as α-limit is called the unstable manifold of p. It is clear that $W^s(p)$ and $W^u(p)$ are invariant by f. Using the hyperbolicity of p we shall describe in this section the structure of these sets and we shall analyse the way they change under perturbations of the diffeomorphism f. Analogous definitions and results are valid for singularities of vector fields.

EXAMPLE 1. If $A \in GL(\mathbb{R}^m)$ is a hyperbolic isomorphism there is an invariant splitting $\mathbb{R}^m = E^s \oplus E^u$ such that, for $q \in E^s$, $A^n(q) \to 0$ as $n \to \infty$ and, for $q \in E^u$, $A^{-n}(q) \to 0$ as $n \to \infty$. Moreover, for any other q, $\|A^n(q)\| \to \infty$ both as $n \to \infty$ and as $n \to -\infty$. Thus, $W^s(0) = E^s$ and $W^u(0) = E^u$.

Let us suppose that $M \subset \mathbb{R}^k$ and let d be the metric induced on M from $\mathbb{R}^k$. For $\beta > 0$ we shall write $B_\beta \subset M$ for the ball with centre p and radius β.

Definition. The sets
$$W^s_\beta(p) = \{q \in B_\beta \,;\, f^n(q) \in B_\beta, \forall n \geq 0\},$$
$$W^u_\beta(p) = \{q \in B_\beta \,;\, f^{-n}(q) \in B_\beta, \forall n \geq 0\}$$
are called the *local stable* and *unstable manifolds*, of size β, of the point p.

We recall that a topological immersion of $\mathbb{R}^s$ in M is a continuous map $F: \mathbb{R}^s \to M$ such that every point $x \in \mathbb{R}^s$ has a neighbourhood V with the following property: the restriction of F to V, $F|V$, is a homeomorphism onto its image. In this case we say that $F(\mathbb{R}^s) \subset M$ is an immersed topological submanifold of dimension s. A topological embedding of $\mathbb{R}^s$ in M is an injective topological immersion which is a homeomorphism onto its image.

6.1 Proposition. *If $\beta > 0$ is sufficiently small we have*:

(1) $W^s_\beta(p) \subset W^s(p)$ and $W^u_\beta(p) \subset W^u(p)$; that is, those points in a neighbourhood of p whose positive (respectively negative) orbit remains in the neighbourhood have p as ω-limit (respectively α-limit);
(2) $W^s_\beta(p)$ (respectively $W^u_\beta(p)$) is an embedded topological disc in M whose dimension is that of the stable (respectively unstable) subspace of $A = Df_p$;
(3) $W^s(p) = \bigcup_{n \geq 0} f^{-n}(W^s_\beta(p))$ and $W^u(p) = \bigcup_{n \geq 0} f^n(W^u_\beta(p))$. Hence there exists an injective topological immersion $\varphi_s: E^s \to M$ ($\varphi_u: E^u \to M$) whose image is $W^s(p)$ (respectively $W^u(p)$), where E^s and E^u are the stable and unstable subspaces of $A = Df_p$.

PROOF. (1) and (2): By the Grobman–Hartman Theorem there exists a neighbourhood U of 0 in TM_p and a homeomorphism $h: B_\beta \to U$ which conjugates f and the isomorphism A. As A is a hyperbolic isomorphism it follows that if $x \in U$ has $A^n(x) \in U$ for all $n \geq 0$ then $x \in E^s$ and so $A^n(x) \to 0$ when $n \to \infty$. Let $q \in W^s_\beta(p)$. As $f^n(q) \in B_\beta$, for $n \geq 0$, and $hf^n(q) = A^n h(q)$ we have $A^n h(q) \in U$ for $n \geq 0$ so that $A^n h(q) \to 0$. Thus, $f^n(q) = h^{-1} A^n h(q)$ converges to $p = h^{-1}(0)$ which shows that $W^s_\beta(p) \subset W^s(p)$. Moreover, $h^{-1}(E^s \cap U) = W^s_\beta(p)$ which proves part (2). Similarly, $W^u_\beta(p) \subset W^u(p)$ and $W^u_\beta(p) = h^{-1}(U \cap E^u)$.

(3) As $W^s(p)$ is invariant by f and $W^s_\beta(p) \subset W^s(p)$, we have $f^{-n}(W^s_\beta(p)) \subset W^s(p)$ for all n so that $\bigcup_{n \geq 0} f^{-n}(W^s_\beta(p)) \subset W^s(p)$. On the other hand, if $q \in W^s(p)$ then $\lim_{n \to \infty} f^n(q) = p$ so there exists $n_0 \in \mathbb{N}$ such that $f^n(q) \in B_\beta$ for all $n \geq n_0$. Thus, $f^{n_0}(q) \in W^s_\beta(p)$ and so $q \in f^{-n_0} W^s_\beta(p)$. Similarly, we may show that $W^u(p) = \bigcup_{n \geq 0} f^n W^u_\beta(p)$. We shall now define a map $\varphi_s: E^s \to M$ whose image is $W^s(p)$. If $x \in E^s$ there exists $n_0 \in \mathbb{N}$ such that $A^{n_0}(x) \in U$ where U is the neighbourhood of 0 considered above. We define $\varphi_s(x) = f^{-n_0} h^{-1} A^{n_0}(x)$. As h^{-1} conjugates A and f, it follows that φ_s is well defined, that is, it does not depend on the choice of n_0. It is easy to see that φ_s is an injective topological immersion and that $\varphi_s(E^s) = W^s(p)$. Similarly, we may construct an injective topological immersion $\varphi_u: E^u \to M$ whose image is $W^u(p)$. □

Remarks. (1) If $p \in M$ is a fixed point of f then the stable manifold of p for f coincides with the unstable manifold of p for f^{-1}. This duality permits us to translate each property of the stable manifold into a property of the unstable manifold.

(2) Although the local stable manifold is an *embedded* topological disc, the global stable manifold may not be an embedded submanifold of M as Example 2 below shows.

(3) It is important to stress that the Grobman–Hartman Theorem only provides $W^s(p)$ with the structure of a topological submanifold, as we saw in Proposition 6.1. However, the next theorem is independent of the Grobman–Hartman Theorem and shows that $W^s(p)$ is in fact a differentiable immersed submanifold of the same class as the diffeomorphism. We presented Proposition 6.1 as motivation for the main result of this section.

EXAMPLE 2. Let $f: S^2 \to S^2$ be the diffeomorphism induced at time 1 by the flow of the vector field X whose orbit structure is as follows: the north pole p_N is the only singularity in the northern hemisphere; the south pole p_S is a saddle whose stable and unstable manifolds form a "figure eight" that encircles two other singularities. See Figure 6. In this example the stable manifold of p_S is not an embedded submanifold of S^2.

EXAMPLE 3. Let $f = Y_1$ where Y is the vector field on S^2 whose orbit structure is shown in Figure 7. In this example $W^s(p_S)$ and $W^u(p_S)$ are embedded submanifolds of S^2.

§6 Invariant Manifolds

Northern hemisphere Southern hemisphere

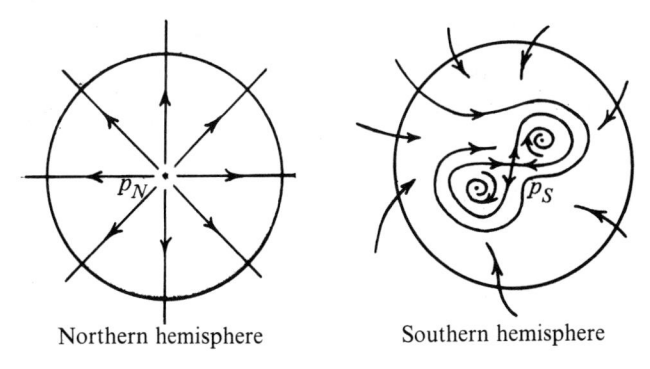

Figure 6

Definition. Let S and S' be C^r submanifolds of M and let $\varepsilon > 0$. We say that S and S' are ε C^r-close if there exists a C^r diffeomorphism $h: S \to S' \subset M$ such that $i'h$ is ε-close to i in the C^r topology. Here $i: S \to M$ and $i': S' \to M$ denote the inclusions.

6.2 Theorem (The Stable Manifold Theorem). *Let $f \in \text{Diff}^r(M)$, let p be a hyperbolic fixed point of f and E^s the stable subspace of $A = Df_p$. Then:*

(1) $W^s(p)$ *is a C^r injectively immersed manifold in M and the tangent space to $W^s(p)$ at the point p is E^s;*
(2) *Let $D \subset W^s(p)$ be an embedded disc containing the point p. Consider a neighbourhood $\mathcal{N} \subset \text{Diff}^r(M)$ such that each $g \in \mathcal{N}$ has a unique hyperbolic fixed point p_g contained in a certain neighbourhood U of p. Then, given $\varepsilon > 0$, there exists a neighbourhood $\tilde{\mathcal{N}} \subset \mathcal{N}$ of f such that, for each $g \in \tilde{\mathcal{N}}$, there exists a disc $D_g \subset W^s(p_g)$ that is ε C^r-close to D.*

We are going to present a proof of this theorem using the implicit function theorem in Banach spaces. The proof is due to M. Irwin [43]. We base our presentation on a set of notes by J. Franks.

Northern hemisphere Southern hemisphere

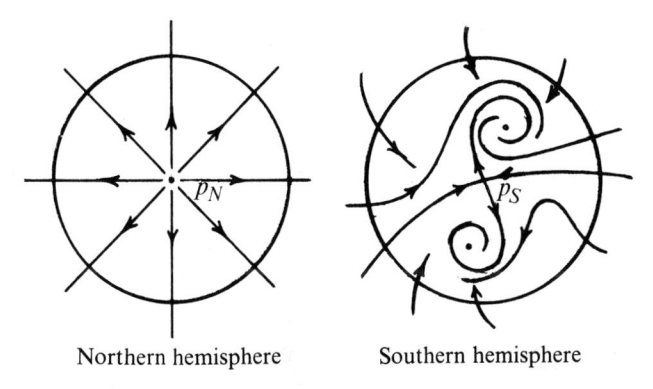

Figure 7

We shall prove that the local stable manifold, $W^s_\beta(p)$, is the graph of a C^r map and that the points of $W^s_\beta(p)$ have p as their ω-limit. Thus, the global stable manifold is of class C^r, since $W^s(p) = \bigcup_{n\geq 0} f^{-n} W^s_\beta(p)$. We can therefore restrict ourselves to the case where f is a diffeomorphism defined on a neighbourhood V of 0 in $\mathbb{R}^m$, with 0 as a hyperbolic fixed point.

Let $A = Df(0)$. Let us consider the A-invariant splitting $\mathbb{R}^m = E^s \oplus E^u$ and norms $\|\cdot\|_s, \|\cdot\|_u$ on E^s, E^u such that $\|A^s\|_s < a < 1$ and $\|(A^u)^{-1}\|_u < a < 1$. On $\mathbb{R}^m$ we use the norm $\|x_s \oplus x_u\| = \max\{\|x_s\|_s, \|x_u\|_u\}$. For $\beta > 0$ we write B_β for the open ball with centre 0 and radius β and we put $B^s_\beta = B_\beta \cap E^s$, $B^u_\beta = B_\beta \cap E^u$. We choose β so that, in B_β, we can write

$$f = A + \Phi, \qquad \Phi(0) = 0, \qquad \|D\Phi\| < \varepsilon$$

for some ε, $0 < \varepsilon < \tfrac{1}{2}(a^{-1} - 1)$. We shall also use the notation $f = (f^s, f^u)$, $A = (A^s, A^u)$ and $\Phi = (\Phi^s, \Phi^u)$. To prove Theorem 6.2 we need the following lemma.

6.3 Lemma. *If $z = (x_s, x_u)$ and $z' = (x_s, x'_u)$ satisfy $f^n(z) \in B_\beta$ and $f^n(z') \in B_\beta$ for all $n \geq 0$ then $z = z'$.*

PROOF. Consider two points $y = (y_s, y_u)$ and $y' = (y'_s, y'_u)$ in B_β such that $\|y_s - y'_s\| \leq \|y_u - y'_u\|$. We claim that

$$\|f^u(y) - f^u(y')\| \geq (a^{-1} - \varepsilon)\|y_u - y'_u\|$$

and

$$\|f^s(y) - f^s(y')\| \leq \|f^u(y) - f^u(y')\|.$$

In fact, $f^u(y) - f^u(y') = A^u(y) - A^u(y') + \Phi^u(y) - \Phi^u(y')$. By the Mean Value Theorem we have $\|f^u(y) - f^u(y')\| \geq a^{-1}\|y_u - y'_u\| - \varepsilon\|y - y'\|$. As $\|y - y'\| = \|y_u - y'_u\|$ we deduce that

$$\|f^u(y) - f^u(y')\| \geq (a^{-1} - \varepsilon)\|y_u - y'_u\|.$$

Similarly

$$\|f^s(y) - f^s(y')\| \leq a\|y_s - y'_s\| + \varepsilon\|y - y'\|.$$

As $\|y - y'\| = \|y_u - y'_u\| \geq \|y_s - y'_s\|$ we obtain

$$\|f^s(y) - f^s(y')\| \leq (a + \varepsilon)\|y_u - y'_u\| \leq \|f^u(y) - f^u(y')\|.$$

From these inequalities we conclude that

$$\|f(y) - f(y')\| \geq (a^{-1} - \varepsilon)\|y - y'\|.$$

Now consider the original points z, z' and put $y = f^n(z)$, $y' = f^n(z')$ for $n > 0$. From the above argument we obtain

$$\|f^n(z) - f^n(z')\| \geq (a^{-1} - \varepsilon)^n \|z - z'\|.$$

§6 Invariant Manifolds

As $a^{-1} - \varepsilon > 1$ we conclude that $z = z'$. If not, the distance between $f^n(z)$ and $f^n(z')$ would tend to infinity with n contradicting the fact that $f^n(z)$ and $f^n(z')$ belong to B_β for all $n \geq 0$. □

PROOF OF THEOREM 6.2. We want to show that the set of points whose positive orbits remain in a neighbourhood of 0 form a C^r submanifold. We shall also show that this set coincides with the set of points whose positive orbits converge to 0. First let us motivate the proof. Let K be the space of sequences $\gamma(n)$, $n \geq 0$, in $\mathbb{R}^m$ such that $\gamma(n) \to 0$, with the norm $\|\gamma\| = \sup_n \|\gamma(n)\|$. Let G be the subset of K defined by $G = \{\gamma \in K;\ \gamma(n) \in B_\beta$ for $n \geq 0\}$. Suppose, for some $z \in B_\beta$, that $\gamma \in G$ and $\gamma(n) = f^n(z)$. Then

$$\gamma(n) = A(A + \Phi)^{n-1}(z) + \Phi(\gamma(n-1))$$
$$= A^2(A + \Phi)^{n-2}(z) + A\Phi(\gamma(n-2)) + \Phi(\gamma(n-1))$$
$$= A^n(z) + \sum_{i=0}^{n-1} A^{n-1-i} \Phi(\gamma(i)).$$

The second component of $\gamma(n)$ has the expression

$$(A^u)^n \left[z_u + \sum_{i=0}^{n-1} (A^u)^{-1-i} \Phi^u(\gamma(i)) \right].$$

As A^u is an expansion we conclude that $\sum_{i=0}^{n-1} (A^u)^{-1-i} \Phi^u(\gamma(i))$ converges to $-z_u$.

This motivates us to define a map $F: B_\beta^s \times G \to K$ by

$$F(x, \gamma)(n) = \gamma(n) - \left((A^s)^n(x) + \sum_{i=0}^{n-1} (A^s)^{n-1-i} \Phi^s(\gamma(i)), \right.$$
$$\left. - \sum_{i=n}^{\infty} (A^u)^{n-1-i} \Phi^u(\gamma(i)) \right).$$

We shall show that $F(x, \gamma) \in K$ and then that for each x there exists $\gamma \in G$ such that $F(x, \gamma) = 0$.

Choose $b > 0$ such that $\|\Phi(z)\| < b$ for $z \in B_\beta$. As $0 < a < 1$, $\sum_0^\infty a^j$ is bounded and converges to $(1 - a)^{-1}$. Firstly, notice that $\gamma(n) \to 0$ and $(A^s)^n(x) \to 0$ since $\|(A^s)^n\| < a^n$. Also

$$\left\| \sum_{i=n}^{\infty} (A^u)^{n-1-i} \Phi^u(\gamma(i)) \right\| \leq (1-a)^{-1} \sup_{i \geq n} \|\Phi^u(\gamma(i))\|.$$

Thus, given $\varepsilon > 0$, we choose n large enough so that $\|\Phi^u(\gamma(i))\| < (1-a)\varepsilon$ for $i \geq n$. Now the second component of $F(x, \gamma)$ tends to 0. Look at the first component. For $0 \leq m \leq n$ we have

$$\left\| \sum_{i=0}^{n-1} (A^s)^{n-1-i} \Phi^s(\gamma(i)) \right\| \leq (1-a)^{-1} a^{n-m} b + (1-a)^{-1} \sup_{i \geq m} \|\Phi^s(\gamma(i))\|.$$

Given $\varepsilon > 0$ we make m large enough for the second term to be less than $\varepsilon/2$. Then we can choose n large enough for the first term to be less than $\varepsilon/2$. Thus, $F(x, \gamma) \in K$.

Now we shall use the Implicit Function Theorem. If we fix $\gamma \in G$ the map $x \to F(x, \gamma)$ from B^s_β to K is affine and continuous. In particular it is of class C^{r-1}. We shall show that, for $u \in K$,

$$D_2 F(x, \gamma)(u)(n) = u(n) - \left(\sum_{i=0}^{n-1} (A^s)^{n-1-i} D\Phi^s(\gamma(i))(u(i)), \right.$$
$$\left. - \sum_{i=n}^{\infty} (A^u)^{n-1-i} D\Phi^u(\gamma(i))(u(i)) \right).$$

To simplify the expressions below we write λ for the right-hand side of this equation. We want to prove that, given $\delta > 0$,

$$\|F(x, \gamma + u)(n) - F(x, \gamma)(n) - \lambda\| \leq \delta \|u\|$$

for small $\|u\|$. The left-hand side of this is less than or equal to

$$\left\| \sum_{i=0}^{n-1} (A^s)^{n-1-i} [\Phi^s(\gamma(i) + u(i)) - \Phi^s(\gamma(i)) - D\Phi^s(\gamma(i))(u(i))] \right\|$$
$$+ \left\| \sum_{i=n}^{\infty} (A^u)^{n-1-i} [\Phi^u(\gamma(i) + u(i)) - \Phi^u(\gamma(i)) - D\Phi^u(\gamma(i))(u(i))] \right\|.$$

Since $D\Phi$ is continuous on the closure of B_δ, it is uniformly continuous on B_δ. Thus, given $\delta' > 0$, there exists $\rho > 0$ such that $\|D\Phi(z + u) - D\Phi(z)\| < \delta'$ for $z \in B_\delta$ and $\|u\|$ so small that $z + u \in B_\delta$ and $\|u\| < \rho$. Applying the Mean Value Theorem to $\Phi(z + u) - \Phi(z)$, we get

$$\|\Phi(z + u) - \Phi(z) - D\Phi(z)u\| \leq \delta' \|u\|.$$

We conclude that each term in the above expression is smaller than $(1 - a)^{-1} \delta' \|u\|$. So it is enough to consider $2(1 - a)^{-1} \delta' < \delta$. This shows that $\lambda = D_2 F(x, \gamma)(u)(n)$. It is easy to check directly that $D_2 F$ is continuous.

We conclude that F is of class C^1 since $D_1 F$ and $D_2 F$ are continuous. Moreover we can check directly that $D_2 F(0, 0)$ is the identity I on K. By the Implicit Function Theorem there exists a ball B^s_r in E^s and a C^1 map $\varphi: B^s_r \to G$ such that $\varphi(0) = 0$ and $F(x, \varphi(x)) = 0$ for all $x \in B^s_r$. We have

$$\varphi(x)(0) = \left(x, - \sum_{i=0}^{\infty} (A^u)^{-1-i} \Phi^u(\varphi(x)(0)) \right).$$

Then we define $h: B^s_r \to E^u$ as the second component of this expression, so that $\varphi(x)(0) = (x, h(x))$. The map h is of class C^1 since φ is C^1 and the map $\theta: G \to \mathbb{R}^m$ taking γ to $\gamma(0)$ is continuous linear. We remark that $h = \pi_2 \circ \theta \circ \varphi$ where $\pi_2: \mathbb{R}^m \to E^u$ is the natural projection. As $F(x, \varphi(x))(n) = 0$ for all $n \geq 0$ we have

$$\varphi(x)(n) = \left((A^s)^n(x) + \sum_{i=0}^{n-1} (A^s)^{n-1-i} \Phi^s(\varphi(x)(i)), - \sum_{i=n}^{\infty} (A^u)^{n-1-i} \Phi^u(\varphi(x)(i)) \right).$$

§6 Invariant Manifolds

From this it follows that $\varphi(x)(n+1) = (A + \Phi)\varphi(x)(n) = f(\varphi(x)(n))$. By induction we have $\varphi(x)(n) = f^n(x, h(x))$ for all $n > 0$. This shows that if $z \in \text{graph}(h)$ then $f^n(z)$ tends to 0 as $n \to \infty$. On the other hand, if z and its positive orbit remain in a small neighbourhood of 0 then Lemma 6.3 shows that $z \in \text{graph}(h)$. Thus $f(\text{graph}(h)) \subset \text{graph}(h)$ and $\text{graph}(h)$ represents the local stable manifold of 0 for f. It remains to prove that $Dh(0) = 0$ to conclude that the local stable manifold is tangent to E^s at 0. As $F(x, \varphi(x)) = 0$ we have $D_1 F(x, \varphi(x)) + D_2 F(x, \varphi(x)) D\varphi(x) = 0$. From this it follows that $D\varphi(0) = -D_1 F(0, 0)$ since $D_2 F(0, 0) = I$. If $v \in E^s$ then $D\varphi(0)(v) - D_1 F(0, 0)(v) = -u$ where $u(n) = (-(A^s)^n v, 0)$ for $n \geq 0$. As $h = \pi_2 \circ \theta \circ \varphi$ we have $Dh(0)(v) = \pi_2 \circ \theta \circ D\varphi(0)v$ since π_2 and θ are linear. Thus $Dh(0)v = \pi_2((A^s)^0 v, 0) = 0$ as we wanted to prove.

We proved above that the stable manifold is of class C^1. In fact it is C^r if f is C^r. This comes from the fact that F above is C^r; we saw that $D_1 F$ is C^{r-1} and it can be checked directly that $D_2 F$ is C^{r-1}. The continuous dependence of the local stable manifold on the perturbation Φ is proved in the same way as above. It is enough to make the map F depend also on the bounded C^r perturbation $F = F(x, y, \Phi)$. Then one checks that $D_3 F$ is C^{r-1} so that F is C^r. The rest of the argument is as before. We just notice that the fixed point for $A + \Phi$ may vary (continuously) with Φ. But this is not very relevant for the proof above: with a C^r diffeomorphism we can translate the fixed point back into the origin. □

It is interesting to remark that the proof works as well for maps: A^s may have some eigenvalues equal to zero. However, in this way we can only show the existence of the stable manifold. The unstable manifold also exists as we can show using another proof, the so called graph transform, explained below.

Another interesting remark is that the proof above is valid in Banach spaces: we only add as an assumption that the local map f and its derivatives are uniformly continuous in B_δ.

Now let us consider a vector field $X \in \mathfrak{X}^r(M)$ and let $p \in M$ be a hyperbolic singularity of X. The stable manifold of p for the vector field X, $W^s(p, X)$, is the set of points of M whose ω-limit is p. Let $f = X_1$ be the diffeomorphism induced at time $t = 1$. As we have seen, p is a hyperbolic fixed point of f. If $W^s(p, f)$ denotes the stable manifold of p for f then $W^s(p, f) = W^s(p, X)$. In fact, if $x \in W^s(p, X)$, that is, if $X_t(x) \to p$ as $t \to \infty$, then clearly $X_n(x) = f^n(x) \to p$ as $n \to \infty$. Thus, $W^s(p, X) \subset W^s(p, f)$. On the other hand, let U be any neighbourhood of p. As $X(p) = 0$ there exists a neighbourhood V of p such that $X_t(V) \subset U$ for $0 \leq t \leq 1$. If $x \in W^s(p, f)$ then there exists $n_0 \in \mathbb{N}$ such that $f^n(x) \in V$ for $n \geq n_0$. Thus $X_t(x) \in U$ for $t \geq n_0$. This shows that $X_t(x) \to p$ as $t \to \infty$. Thus $W^s(p, f) \subset W^s(p, X)$.

We shall also present a sketch of another more geometrical proof of Theorem 6.2 that can be found in [39] and [77]. We shall omit the details of this second proof since they are technically more complicated. We emphasize,

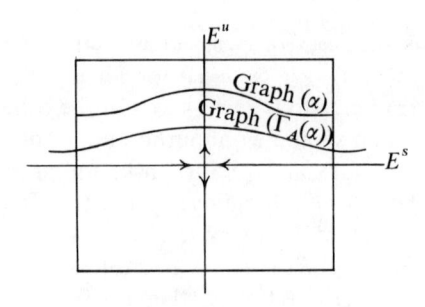

Figure 8

however, that this proof lends itself to important generalizations as in [40].

Let $\mathscr{F} \subset C^r(B_\beta^s, B_\beta^u)$ be the set of C^r maps whose Lipschitz constant is at most 1. Then $\mathscr{F}$ is a closed subset of $C^r(B_\beta^s, B_\beta^u)$ and hence a Baire space. If $\alpha \in \mathscr{F}$ then the graph of α is a C^r submanifold of $\mathbb{R}^m$. As A contracts vectors in E^s and expands vectors in E^u it follows that $A^{-1}(\text{graph}(\alpha)) \cap B_\beta$ is the graph of a C^r map from B_β^s to B_β^u which we denote by $\Gamma_A(\alpha)$. In fact, $\Gamma_A(\alpha)(x_s) = (A^u)^{-1}\alpha(A^s x_s)$. As $\Gamma_A(\alpha)$ also has Lipschitz constant at most 1 we have defined a map $\Gamma_A : \mathscr{F} \to \mathscr{F}$ called the *graph transform* associated to A. It is easy to see that, if $\alpha \in \mathscr{F}$, the sequence $\{\Gamma_A^n(\alpha)\}$ converges to the zero map whose graph is the local stable manifold of A. As A is the derivative of f at the point 0 it is reasonable to expect that we can, by taking β small enough, define a graph transform $\Gamma_f : \mathscr{F} \to \mathscr{F}$ associated to f. If Γ_f has an attracting fixed point $\alpha_f \in C^r(B_\beta^s, B_\beta^u)$, that is, if $\Gamma_f(\alpha_f) = \alpha_f$ and $\Gamma_f^n(\alpha) \to \alpha_f$ for any $\alpha \in \mathscr{F}$, then $f(\text{graph}(\alpha_f)) = \text{graph}(\alpha_f)$ and $\text{graph}(\alpha_f)$ is the set of points of B_β whose positive orbits for f remain in B_β. Thus $\text{graph}(\alpha_f) = W_\beta^s(0)$. In other words the proof of the theorem can be done in the following steps:

(1) Γ_f is well defined and has an attracting fixed point $\alpha_f \in C^r(B_\beta^s, B_\beta^u)$;
(2) if g is near f in the C^r topology then Γ_g is well defined and has an attracting fixed point, which varies continuously with g.

§7 The λ-lemma (Inclination Lemma). Geometrical Proof of Local Stability

In this section we shall discuss a local fact which is relevant to several results in Dynamical Systems and explain some of them [75]. In particular, we shall give another more geometrical proof of the Grobman–Hartman Theorem. Again we observe that the same result (and the proof) is valid in Banach spaces.

Let f be a C^r diffeomorphism of a neighbourhood V in $\mathbb{R}^m$ with 0 as a hyperbolic fixed point. Consider the hyperbolic isomorphism $A = Df(0)$

§7 The λ-lemma (Inclination Lemma). Geometrical Proof of Local Stability

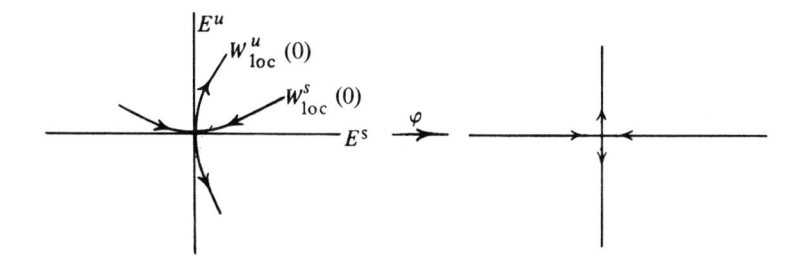

Figure 9

and the invariant splitting $\mathbb{R}^m = E^s \oplus E^u$. In Section 6 we saw that the local stable manifold of the fixed point 0, $W^s_{\text{loc}}(0)$, is the graph of a C^r map $\varphi_s: B^s_\beta \to E^u$ with $\varphi_s(0) = 0$ and $D\varphi_s(0) = 0$. Here $B^s_\beta(0) \subset E^s$ denotes the ball with centre 0 and radius β. In the same way the local unstable manifold, $W^u_{\text{loc}}(0)$, is the graph of a C^r map $\varphi_u: B^u_\beta \to E^s$ with $\varphi_u(0) = 0$ and $D\varphi_u(0) = 0$. Let us consider the map

$$\varphi: B^s_\beta \oplus B^u_\beta \to E^s \oplus E^u, \qquad \varphi(x_s, x_u) = (x_s - \varphi_u(x_u), \ x_u - \varphi_s(x_s)).$$

It is clear that φ is C^r and that $D\varphi(0)$ is the identity. Thus φ is a diffeomorphism when restricted to some neighbourhood of 0 in $\mathbb{R}^m$. Let us consider the diffeomorphism $\tilde{f} = \varphi_* f = \varphi f \varphi^{-1}$. Then $\tilde{f}$ is a diffeomorphism of a neighbourhood of the origin with $\tilde{f}(0) = 0$ and $D\tilde{f}(0) = A$. Moreover, the local stable manifold of $\tilde{f}$ is a neighbourhood of the origin in E^s while the local unstable manifold is a neighbourhood of the origin in E^u. In other words, we can always assume that the local stable (respectively unstable) manifold of a hyperbolic fixed point of a diffeomorphism f is a neighbourhood of the fixed point in the stable (respectively unstable) subspace of the linear part of f.

Let $\|\cdot\|$ be a norm on $\mathbb{R}^m$ such that $\|A^s\| \leq a < 1$ and $\|(A^u)^{-1}\| \leq a < 1$ where A^s and A^u are the restrictions of A to E^s and E^u respectively. If $f^s: B^s_\beta \to E^s$ is the restriction of f to $W^s_\beta(0) \subset B^s_\beta \subset E^s$ then $Df^s(0) = A^s$. We have that f^s is a contraction for small enough β since A^s is a contraction. Therefore, if $B^s \subset B^s_\beta$ is an open ball with centre at the origin then $f(\partial B^s) \subset B^s$ where $\partial B^s = \bar{B}^s - B^s$ is the boundary of B^s. The annulus $G^s(0) = \overline{B^s - f(B_s)}$ is called a *fundamental domain* for the stable manifold of 0. It is clear that $\partial G^s(0) = \partial B^s \cup f(\partial B^s)$.

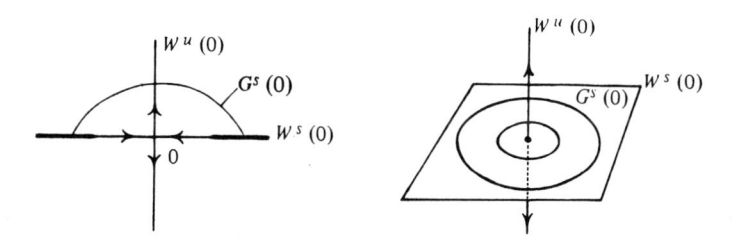

Figure 10

If $x \in W^s(0) - \{0\}$ then the orbit of x has at least one and at most two points in the fundamental domain $G^s(0)$, that is, $\bigcup_{n \in \mathbb{Z}} f^n(G^s(0)) = W^s(0) - \{0\}$ and, if $x \in \text{int } G^s(0)$, $f^n(x) \notin G^s(0)$ for all $n \in \mathbb{Z} - \{0\}$.

Any neighbourhood $N^s(0)$ of $G^s(0)$ that is disjoint from $W^u_{\text{loc}}(0)$ is called a *fundamental neighbourhood* for the stable manifold of 0. Similarly, we define a fundamental domain $G^u(0)$ and a fundamental neighbourhood $N^u(0)$ for the unstable manifold of 0.

Let $B^s \subset E^s$ be a ball contained in $W^s_{\text{loc}}(0)$, $B^u \subset E^u$ a ball contained in $W^u_{\text{loc}}(0)$ and $V = B^s \times B^u$. Consider a point $q \in W^s_{\text{loc}}(0)$ and a disc D^u of dimension $u = \dim E^u$ transversal to $W^s_{\text{loc}}(0)$ at q.

7.1 Lemma (The λ-lemma). *Let $V = B^s \times B^u$, let $q \in W^s(0) - \{0\}$ and let D^u be as above. Let D^u_n be the connected component of $f^n(D^u) \cap V$ to which $f^n(q)$ belongs. Given $\varepsilon > 0$, there exists $n_0 \in \mathbb{N}$ such that, if $n > n_0$, then D^u_n is ε C^1-close to B^u.*

PROOF. The expression for f in the neighbourhood V of 0 is given by

$$f(x_s, x_u) = (A^s x_s + \varphi_s(x_s, x_u), A^u x_u + \varphi_u(x_s, x_u))$$

where

$$(Df)_0 = (A^s, A^u), \qquad x_s \in B^s, \qquad x_u \in B^u$$

$$\|A^s\| \leq a < 1, \qquad \|(A^u)^{-1}\| \leq a < 1,$$

$$\left.\frac{\partial \varphi_s}{\partial x_u}\right|_{B^u} = \left.\frac{\partial \varphi_u}{\partial x_s}\right|_{B^s} = 0.$$

As $\partial \varphi_i / \partial x_j(0, 0) = 0$ for $i, j = s, u$, the continuity of these partial derivatives implies that there exists k with $a_1 = a + k < 1, 0 < k < 1, b = (a^{-1} - k) > 1$, $k < (b-1)^2/4$ and there exists $V' \subset V$ such that

$$k \geq \max_{V'} \left\|\frac{\partial \varphi_i}{\partial x_j}\right\|, \qquad i, j = s, u.$$

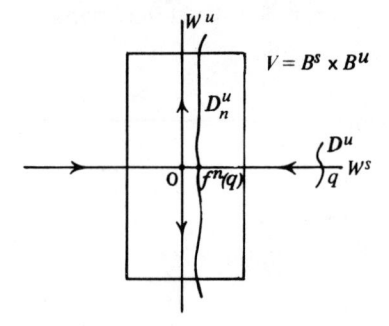

Figure 11

§7 The λ-lemma (Inclination Lemma). Geometrical Proof of Local Stability

We can suppose that $q \in V'$ and $B^u \subset V'$. Let v_0 be any unit vector in $(TD^u)_q$. We can write $v_0 = (v_0^s, v_0^u)$ in the product $V = B^s \times B^u$. Let λ_0 be the slope of v_0, $\lambda_0 = \|v_0^s\|/\|v_0^u\|$, with $\|v_0^u\| \neq 0$ since D^u is transversal to B^s at q. Consider

$$q_1 = f(q), \qquad v_1 = Df_q(v_0)$$
$$q_2 = f^2(q), \qquad v_2 = Df_{q_1}(v_1)$$
$$\vdots \qquad\qquad \vdots$$
$$q_n = f^n(q), \qquad v_n = Df_{q_{n-1}}(v_{n-1}).$$

For $q \in \partial B^s$,

$$Df_q(v_0) = \begin{pmatrix} A^s + \partial\varphi_s/\partial x_s(q) & \partial\varphi_s/\partial x_u(q) \\ O & A^u + \partial\varphi_u/\partial x_u(q) \end{pmatrix} \begin{pmatrix} v_0^s \\ v_0^u \end{pmatrix}$$

$$= \begin{pmatrix} A^s v_0^s + \partial\varphi_s/\partial x_s(q) v_0^s + \partial\varphi_s/\partial x_u(q) v_0^u \\ A^u v_0^u + \partial\varphi_u/\partial x_u(q) v_0^u \end{pmatrix}.$$

Thus

$$\lambda_1 = \frac{\|v_1^s\|}{\|v_1^u\|} = \frac{\|A^s v_0^s + \partial\varphi_s/\partial x_s(q) v_0^s + \partial\varphi_s/\partial x_u(q) v_0^u\|}{\|A^u v_0^u + \partial\varphi_u/\partial x_u(q) v_0^u\|}.$$

The numerator is bounded above by

$$\|A^s v_0^s\| + \|\partial\varphi_s/\partial x_s(q) v_0^s\| + \|\partial\varphi_s/\partial x_u(q) v_0^u\| \leq a\|v_0^s\| + k\|v_0^s\| + k\|v_0^u\|.$$

The denominator is bounded below by

$$\|A^u v_0^u\| - \|\partial\varphi_u/\partial x_u(q) v_0^u\| \geq a^{-1}\|v_0^u\| - k\|v_0^u\|.$$

Hence

$$\lambda_1 \leq \frac{a\lambda_0 + k\lambda_0 + k}{a^{-1} - k} \leq \frac{\lambda_0 + k}{b} = \frac{\lambda_0}{b} + \frac{k}{b},$$

$$\lambda_2 = \frac{\|v_2^s\|}{\|v_2^u\|} \leq \frac{\lambda_1 + k}{b} \leq \frac{\lambda_0}{b^2} + k \sum_{i=1}^{2} \frac{1}{b^i},$$

$$\lambda_n = \frac{\|v_n^s\|}{\|v_n^u\|} \leq \frac{\lambda_0}{b^n} + k \sum_{i=1}^{n} \frac{1}{b^i} \leq \frac{\lambda_0}{b^n} + \frac{k}{b-1}.$$

As $\lambda_0/b^n \to 0$ as $n \to \infty$ and $k/(b-1) < (b-1)/4$, there exists $n_0 \in \mathbb{Z}^+$ such that, for any $n > n_0$ we have $\lambda_n \leq (b-1)/4$. Let $0 < k_1 < \min(\varepsilon, k)$. As $\partial\varphi_s/\partial x_u|_{B^u} = 0$ and B^u is compact, there exists $\delta < \varepsilon$ such that, for $V_1 = \delta B^s \times B^u \subset V$, we have

$$\max_{V_1} \left\|\frac{\partial\varphi_s}{\partial x_u}\right\| \leq k_1.$$

Here δB^s means the ball whose radius is δ times the radius of B^s.

As v_0 can be chosen so that λ_0 is the maximum possible slope of unit vectors in $(TD^u)_q$, there exists n_0 such that, for $n \geq n_0$, all the nonzero vectors of $(TD^u_n)_{q_n}$ have slope $\lambda_n \leq (b-1)/4$ and $q_{n_0} \in V_1$. Then, by the continuity of the tangent plane of $D^u_{n_0}$, there exists a disc $\tilde{D}^u$ embedded in $D^u_{n_0}$ with centre q_{n_0} and such that the slope λ of any unit vector in $(T\tilde{D}^u)_p$, $p \in \tilde{D}^u$ satisfies $\lambda \leq (b-1)/2$.

Let $v \in (T\tilde{D}^u)_p$ for $p \in \tilde{D}^u$. In the decomposition $v = (v^s, v^u)$, v has slope $\lambda_{n_0} = \|v^s\|/\|v^u\|$. Let us calculate the slope of the iterates of v. We have

$$Df_p(v) = \begin{pmatrix} A^s v^s + \partial\varphi_s/\partial x_s(p)v^s + \partial\varphi_s/\partial x_u(p)v^u \\ \partial\varphi_u/\partial x_s(p)v^s + A^u v^u + \partial\varphi_u/\partial x_u(p)v^u \end{pmatrix}$$

so

$$\lambda_{n_0+1} = \frac{\|A^s v^s + \partial\varphi_s/\partial x_s(p)v^s + \partial\varphi_u/\partial x_u(p)v^u\|}{\|\partial\varphi_u/\partial x_s(p)v^s + A^u v^u + \partial\varphi_u/\partial x_u(p)v^u\|}$$

whose numerator is less than $a\|v^s\| + k\|v^s\| + k_1\|v^u\|$ and whose denominator is greater than

$$\|A^u v^u\| - \|\partial\varphi_u/\partial x_u(p)v^u\| - \|\partial\varphi_u/\partial x_s(p)v^s\| \geq a^{-1}\|v^u\| - k\|v^u\| - k\|v^s\|.$$

Hence

$$\lambda_{n_0+1} \leq \frac{a\lambda_{n_0} + k\lambda_{n_0} + k_1}{a^{-1} - k - k\lambda_{n_0}} \leq \frac{\lambda_{n_0} + k_1}{b - k\lambda_{n_0}}$$

$$\leq \frac{\lambda_{n_0} + k_1}{b - \frac{1}{2}k(b-1)} \leq \frac{\lambda_{n_0} + k_1}{b - \frac{1}{2}(b-1)} = \frac{\lambda_{n_0} + k_1}{\frac{1}{2}(b+1)}.$$

Let $b_1 = \frac{1}{2}(b+1)$, $b_1 > 1$. Then $\lambda_{n+n_0} \leq \lambda_{n_0}/b_1^n + k_1/(b_1 - 1)$. There exists $\tilde{n}$ such that, for $n \geq \tilde{n}$,

$$\lambda_{n+n_0} \leq \varepsilon\left(1 + \frac{1}{b_1 - 1}\right).$$

As we could have considered v such that λ_{n_0} was the maximum possible slope of unit tangent vectors to $\tilde{D}^u$, we see that, for $n \geq \tilde{n}$, any nonzero vector tangent to $f^n(\tilde{D}^u) \cap V_1$ has slope less than $\varepsilon[1 + (b_1 - 1)^{-1}]$. Thus, given $\varepsilon > 0$, there exists $\bar{n}$ such that, for $n \geq \bar{n}$, all the nonzero tangent vectors to $f^n(\tilde{D}^u) \cap V_1$ have slope less than ε.

Let us compare the norm of a vector tangent to $f^n(\tilde{D}^u) \cap V_1$ with that of its image by Df:

$$(v_n^s, v_n^u) \to Df(v_n^s, v_n^u) = (v_{n+1}^s, v_{n+1}^u)$$

$$\frac{\sqrt{\|v_{n+1}^s\|^2 + \|v_{n+1}^u\|^2}}{\sqrt{\|v_n^s\|^2 + \|v_n^u\|^2}} = \frac{\|v_{n+1}^u\|}{\|v_n^u\|}\sqrt{\frac{1 + \lambda_{n+1}^2}{1 + \lambda_n^2}}.$$

§7 The λ-lemma (Inclination Lemma). Geometrical Proof of Local Stability

From the expressions for v^u_{n+1} and v^u_n we conclude that

$$\frac{\|v^u_{n+1}\|}{\|v^u_n\|} \geq a^{-1} - k - k\lambda_n.$$

As the slopes λ_{n+1} and λ_n are arbitrarily small, we see that the norms of the iterates of nonzero vectors tangent to $f^n(\tilde{D}^u) \cap V_1$ are growing by a ratio that approaches $b = a^{-1} - k > 1$. Hence the diameter of $f^n(\tilde{D}^u) \cap V_1$ increases, and this, together with the fact that its tangent spaces have uniformly small slope, implies that there exists $\bar{n}$ such that, for $n > \bar{n}$, $f^n(\tilde{D}^u) \cap V_1$ is C^1 close to B^u via the canonical projection onto B^u. This completes the proof of the λ-lemma. □

Remarks. (1) The λ-lemma can be stated for a family of discs transversal to $W^s(0)$ provided that this family is continuous in the C^1 topology. Thus, let $F: G^s(0) \to C^1(B^u, M)$ be a continuous map which associates to each point q of the fundamental domain $G^s(0)$ a disc $D^u_q = F(q)B^u$ transversal to B^s. Let $U = B^s \times B^u$ as above. Then, given $\varepsilon > 0$, there exists $n_0 \in \mathbb{N}$ such that $f^n(D^u_q) \cap U$ is a disc ε C^1-close to B^u for any $q \in G^s(0)$ and $n \geq n_0$.

(2) Although this is not necessary for the majority of applications, these discs can be proven to be C^r close, $r \geq 1$, if F is a continuous family of C^r discs; that is, if we have a continuous map $F: G^s(0) \to C^r(B^u, M)$.

(3) The following fact is an immediate consequence of the λ-lemma. Suppose D^s is a small s-dimensional disc transversal to $W^u(0)$. Then, there exists $n_0 > 0$ and a sequence of points $z_n \in D^u$, $n \geq n_0$, such that $f^n(z_n) \in D^s$.

Corollary 1. *Let $p_1, p_2, p_3 \in M$ be hyperbolic fixed points of $f \in \text{Diff}^r(M)$. If $W^u(p_1)$ has a point of transversal intersection with $W^s(p_2)$ and $W^u(p_2)$ has a point of transversal intersection with $W^s(p_3)$ then $W^u(p_1)$ has a point of transversal intersection with $W^s(p_3)$.*

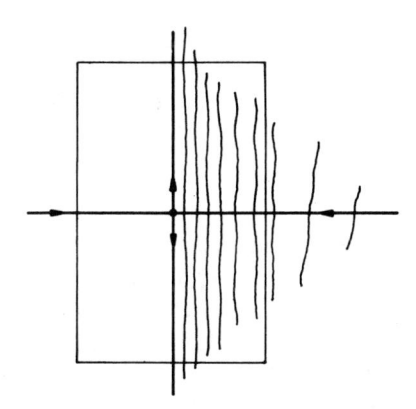

Figure 12

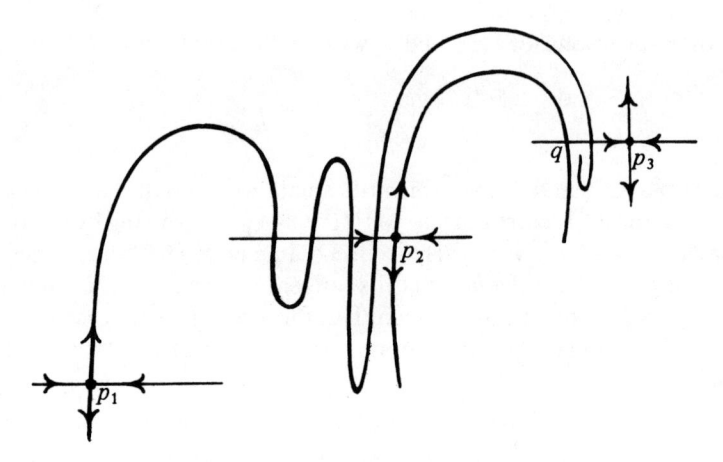

Figure 13

PROOF. Let q be a point of transversal intersection of $W^u(p_2)$ and $W^s(p_3)$. We consider a closed disc $D \subset W^u(p_2)$ containing p_2 and q. As D has a point of transversal intersection with $W^s(p_3)$ it follows that there exists $\varepsilon > 0$ such that if $\tilde{D}$ is a disc ε C^1-close to D then $\tilde{D}$ also has a point of transversal intersection with $W^s(p_3)$. Now let q_2 be a point of transversal intersection of $W^u(p_1)$ with $W^s(p_2)$ and $D^u \subset W^u(p_1)$ a disc containing q_2 of the same dimension as $W^u(p_2)$. By the λ-lemma there exists an integer n_0 such that $f^{n_0}(D^u)$ contains a disc $\tilde{D}$ that is ε C^1-close to D. Thus there exists a point $\tilde{q} \in \tilde{D} \cap W^s(p_3)$. As $W^u(p_1)$ is invariant we have $f^{n_0}(D^u) \subset W^u(p_1)$ so that $\tilde{q}$ is a point of transversal intersection of $W^u(p_1)$ and $W^s(p_3)$. $\square$

Corollary 2. *Let $p \in M$ be a hyperbolic fixed point of $f \in \text{Diff}^r(M)$ and let $N^s(p)$ be a fundamental neighbourhood of $W^s(p)$. Then $\bigcup_{n \geq 0} f^n(N^s(p)) \supset U - W^u_{\text{loc}}(p)$ for some neighbourhood U of p.*

PROOF. First, notice that the iterates by f of a fundamental domain $G^s(p) \subset N^s(p)$ cover $W^s(p) - \{p\}$; that is, $\bigcup_{n \in \mathbb{Z}} f^n(G^s(p)) = W^s(p) - \{p\}$. Moreover, by the λ-lemma, every point in a certain neighbourhood U of p that is not in $W^u_{\text{loc}}(p)$ belongs to some iterate of a section that is transversal to $G^s(p)$ and contained in $N^s(p)$. $\square$

We shall now prove the λ-lemma for vector fields. Let $p \in M$ be a hyperbolic singularity for $X \in \mathfrak{X}^r(M)$. Let $W^s_{\text{loc}}(p)$ and $W^u_{\text{loc}}(p)$ be the local stable and unstable manifolds of the point p. Let B^s be a disc embedded in $W^s_{\text{loc}}(p)$ such that ∂B^s is transversal to the field X in $W^s(p)$. The sphere $\mathcal{G}^s(p) = \partial B^s$ is called a *fundamental domain* for $W^s(p)$. It is easy to see that, if $x \in W^s(p) - \{p\}$, the orbit of x intersects $\mathcal{G}^s(p)$ in only one point. Similarly, we can define a fundamental domain $\mathcal{G}^u(p)$ for $W^u(p)$.

§7 The λ-lemma (Inclination Lemma). Geometrical Proof of Local Stability

Let D^u be a disc transversal to $W^s_{\text{loc}}(p)$ that contains a point $q \in W^s_{\text{loc}}(p)$ and has dim D^u = dim $W^u_{\text{loc}}(p)$. Let K be the compact set $K = \{X_t(q); t \in [0, 1]\}$ and, for each point $X_t(q) \in K$, consider the disc $D^u(X_t(q)) = X_t(D^u)$ which contains the point $X_t(q)$ and is transversal to $W^s_{\text{loc}}(p)$ since X_t is a diffeomorphism and $W^s_{\text{loc}}(p)$ is invariant for X_t. Let $f = X_1$ be the diffeomorphism induced at time 1. Then p is a hyperbolic fixed point for f and the stable and unstable manifolds of p for the diffeomorphism f coincide with the stable and unstable manifolds of p for the vector field X. Let B^s be a disc embedded in $W^s_{\text{loc}}(p)$ containing p, B^u a disc embedded in $W^s_{\text{loc}}(p)$ containing p and $V = B^s \times B^u$ a neighbourhood of p. By the λ-lemma for diffeomorphisms, given $\varepsilon > 0$, there exists $n_0 \in \mathbb{N}$ such that, if $n > n_0$, $D^u_n(x)$ is ε C^1-close to B^u, where $D^u_n(x)$ is the connected component of $f^n(D^u(x)) \cap V$ that contains $f^n(x)$ and $x \in K$. This proves the following lemma.

7.2 Lemma. *Given $\varepsilon > 0$ there exists $t_0 > 0$ such that, if $t > t_0$ and D^u_t is the connected component of $X_t(D^u) \cap V$ that contains $X_t(q)$, then D^u_t is ε C^1-close to B^u.* □

Now we present another, more geometric, proof of the Grobman–Hartman Theorem. We shall use the λ-lemma and the Stable Manifold Theorem, whose proof is independent of the Grobman–Hartman Theorem as we have already remarked. The proof we present is for flows but similar arguments work as well for diffeomorphisms.

7.3 Lemma. *Let $p \in M$ be a hyperbolic singularity of a vector field $X \in \mathfrak{X}^r(M)$. There exists a neighbourhood U of p and a continuous map $\pi_s: U \to B_s$, where $B_s = U \cap W^s_{\text{loc}}(p)$ is a disc containing p, with the following properties:*

(1) $\pi_s^{-1}(p) = B_u = U \cap W^u_{\text{loc}}(p)$ *is a disc containing p;*
(2) *for each $x \in B_s$, $\pi_s^{-1}(x)$ is a C^r submanifold of M transversal to $W^s_{\text{loc}}(p)$ at the point x;*
(3) π_s *is of class C^r except possibly at the points of B^u;*
(4) *the fibration defined by π_s is invariant for the flow of X, that is, if $t \geq 0$ then $X_t(\pi_s^{-1}(x)) \supset \pi_s^{-1}(X_t(x))$.*

PROOF. We can suppose, by using a local chart, that X is a vector field on a neighbourhood V of the origin in $\mathbb{R}^m = E^s \oplus E^u$ with 0 as a hyperbolic singularity. We can also suppose that $W^s_{\text{loc}}(0)$ is an open subset of E^s containing 0 and that $W^u_{\text{loc}}(0)$ is an open subset of E^u containing 0. Let $\mathscr{S}^s(0)$ be a fundamental domain for $W^s(0)$. $\mathscr{S}^s(0)$ is a sphere contained in E^s and transversal to the vector field X on E^s. Let $B^u \subset E^u$ be a disc containing 0. If we take B^u small enough the cylinder $\mathscr{S}^s(0) \times B^u$ is transversal to the vector field. In $\mathscr{S}^s(0) \times B^u$ we have a C^r map $\pi_s: \mathscr{S}^s(0) \times B^u \to W^s_{\text{loc}}(0)$ which is the projection on the first factor. The fibres $\pi_s^{-1}(x)$ through points $x \in \mathscr{S}^s(0)$ are

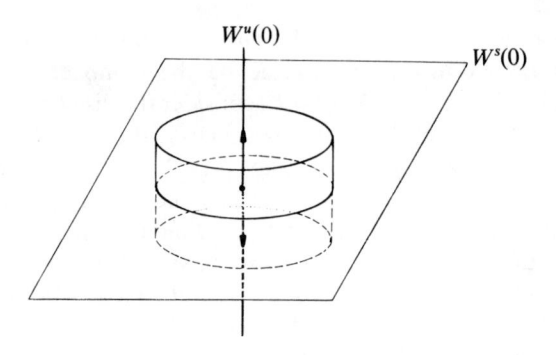

Figure 14

discs transversal to $W^s_{\text{loc}}(0)$. By Corollary 2 of the λ-lemma $\bigcup_{t \geq 0} X_t(\mathscr{G}^s \times B^u) \supset U - E^u$ where U is a neighbourhood of p. If $x \in U \cap E^u$ we define $\pi_s(x) = p$. If $x \in U - E^u$ then there exists $t > 0$ such that $X_{-t}(x) \in \mathscr{G}^s \times B^u$. Then we define $\pi_s(x) = X_t \pi_s X_{-t}(x)$. It is clear that π_s is of class C^r in $U - E^u$. The continuity of π_s at points of E^u follows from the λ-lemma. □

Using the fibres constructed above we can prove the local stability of a hyperbolic singularity. In fact, let $p \in M$ be a hyperbolic singularity of $X \in \mathfrak{X}^r(M)$. Let N be a neighbourhood of X such that any $Y \in N$ has a singularity p_Y near to p and of the same index. Define homeomorphisms $h^s \colon W^s_{\text{loc}}(p) \to W^s_{\text{loc}}(p_Y)$, $h^u \colon W^u_{\text{loc}}(p) \to W^u_{\text{loc}}(p_Y)$ that conjugate the flows of X and Y by first defining them on fundamental domains and then extending them as in Proposition 2.14 to $W^s_{\text{loc}}(p)$ and $W^u_{\text{loc}}(p)$ using the flows of X and Y. Consider fibrations $\pi_s^X \colon U_p \to W^s_{\text{loc}}(p)$, $\pi_u^X \colon U_p \to W^u_{\text{loc}}(p)$, $\pi_s^Y \colon V_{p_Y} \to W^s_{\text{loc}}(p_Y)$, $\pi_u^Y \colon V_{p_Y} \to W^u_{\text{loc}}(p_Y)$ as in Lemma 7.3.

If $q \in U_p$ define $h(q) = \tilde{q}$ where $\tilde{q}$ is such that $\pi_s^Y(\tilde{q}) = h^s(\pi_s^X q)$ and $\pi_u^Y(\tilde{q}) = h^u(\pi_u^X q)$. It is easy to see that h is a homeomorphism that conjugates the flows of X and Y. We remark that the fibrations considered above define continuous coordinate systems in which the flows are expressed as products and consequently the homeomorphism h is the product of h^s and h^u.

A proof of the Grobman–Hartman Theorem for diffeomorphisms using the λ-lemma is in [75]. The constructions are similar but more elaborate.

EXERCISES

1. Show that a linear vector field L is hyperbolic if and only if the ω-limit of each orbit is either the origin or empty.

2. Show that there exists a linear vector field L on $\mathbb{R}^4$ and an orbit γ of L such that the ω-limit of γ contains γ but γ is neither singular nor a closed orbit.

3. We say that a linear isomorphism $A\colon \mathbb{R}^n \to \mathbb{R}^n$ embeds in a *flow* if there exists a vector field X generating a flow X_t with $A = X_1$. Characterize, by their canonical forms, the hyperbolic linear isomorphisms that embed in flows.

4. Let $f\colon M \to \mathbb{R}$ be of class C^r, $r \geq 2$, and let $X = \operatorname{grad} f$. Show that $p \in M$ is a hyperbolic singularity of X if and only if $df(p) = 0$ and $d^2 f(p)$ is a nondegenerate bilinear form.

5. Let $X = \operatorname{grad} f$ where $f\colon M \to \mathbb{R}$ is of class C^{r+1}. Show that if $p \in M$ is a singularity of X then the eigenvalues of dX_p are real.

6. Give an example of a vector field $X \in \mathfrak{X}^r(S^2)$ such that $X \in \mathscr{G}_1$ and $X_{t=1} \notin G_1$, that is, $X_{t=1}$ has a nonhyperbolic fixed point.

7. Let $X = \operatorname{grad} f$ where $f\colon M \to \mathbb{R}$ is of class C^{r+1}, $r \geq 1$. Show that $X \in \mathscr{G}_1$ if and only if $X_{t=1} \in G_1$.

8. We say that a C^r function $f\colon M \to \mathbb{R}$, $r \geq 2$, is a *Morse function* if $\operatorname{grad} f \in \mathscr{G}_1$, that is, if the singularities of $\operatorname{grad} f$ are all hyperbolic. Show that the set of Morse functions is open and dense in $C^r(M)$.

   Hint. Let $\varphi\colon \mathbb{R}^n \to \mathbb{R}$ be of class C^∞. Show that 0 is a regular value of the map $\Phi\colon \mathbb{R}^n \times L(\mathbb{R}^n, \mathbb{R}) \to L(\mathbb{R}^n, \mathbb{R})$ defined by $\Phi(x, A) = d\varphi(x) + A$.

9. Let X and Y be C^1 vector fields on $\mathbb{R}^m$. Suppose that 0 is an attracting hyperbolic singularity for X and Y. Show that there exists a homeomorphism h of a neighbourhood of the origin which conjugates the diffeomorphisms $X_{t=1}$ and $Y_{t=1}$ but does not take orbits of X to orbits of Y.

10. Let $p \in M$ be a hyperbolic fixed point of a diffeomorphism f. Let $\{p_n\}$ be a sequence of periodic points of f with $p_n \neq p$ and $p_n \to p$. Show that there exists a sequence of periodic points of f that converge to a point other than p on the unstable manifold of p.

11. Show that if $f \in \operatorname{Diff}^r(M)$, $r \geq 1$, is structurally stable then all the fixed points of f are hyperbolic.

12. Let $p \in M$ be a hyperbolic periodic point of $f \in \operatorname{Diff}^r(M)$. Show that given $n \in \mathbb{N}$ there exists a neighbourhood V of p such that any periodic point of f in $V - \{p\}$ has period greater than n.

13. Let $0 \in \mathbb{R}^n$ be a hyperbolic singularity for the vector fields X and $Y \in \mathfrak{X}^r(\mathbb{R}^n)$. Show that if there exists a C^1 diffeomorphism, $f\colon \mathbb{R}^n \to \mathbb{R}^n$, taking orbits of X to orbits of Y then the eigenvalues of $L = DX(0)$ are proportional to those of $\tilde{L} = DY(0)$.

    Hint. Under the above hypothesis there exists a function $\lambda\colon \mathbb{R}^n \to \mathbb{R}$ such that $Df(x) \cdot X(x) = \lambda(x) Y(f(x))$. Show that, for each $v \neq 0$, there exists $\tilde{\lambda}(v) = \lim_{t \to 0} \lambda(tv)$ and that $Df(0) \cdot Lv = \tilde{\lambda}(v) \tilde{L} Df(0) \cdot v$.

14. Let $f\colon \mathbb{R}^n \to \mathbb{R}^n$ be a C^1 diffeomorphism with $f(0) = 0$. Let E^u and E^0 be invariant subspaces for f such that $\mathbb{R}^n = E^0 \oplus E^u$ and write $f^u\colon E^u \to E^u$ and $f^0\colon E^0 \to E^0$ for the restrictions of f to E^u and E^0, respectively. Suppose that the eigenvalues of $df^u(0)$ have absolute value > 1 and that the eigenvalues of $df^0(0)$ have absolute

value ≤ 1. Let $D^u \subset E^u$ be a disc containing 0 and D a disc transversal to E^0 containing a point $q \in E^0$ such that $f^n(q) \to 0$ as $n \to \infty$. Show that there exists a neighbourhood V of 0 in $\mathbb{R}^n$ such that, for all $\varepsilon > 0$, there exists $n_0 \in \mathbb{N}$ satisfying the following property: if $n \geq n_0$ then $f^n(D)$ contains a disc ε C^1-close to $V \cap D^u$.

15. Let $f: \mathbb{R}^2 \to \mathbb{R}^2$ be a C^3 diffeomorphism, $f(x, y) = (f_1(x, y), f_2(x, y))$, with the following properties:
    (1) $f_1(0, y) = 0$ for all $y \in \mathbb{R}$;
    (2) $f_2(x, 0) = 0$ for all $x \in \mathbb{R}$;
    (3) $\partial f_2/\partial y(0, 0) > 1$;
    (4) if $\alpha(x) = f_1(x, 0)$ then $\alpha'(0) = 1$, $\alpha''(0) = 0$ and $\alpha'''(0) < 0$.
    Show that there exists $a > 0$ and a neighbourhood V of $(0, 0)$ such that, given $\varepsilon > 0$ and a segment D transversal to the axis $x = 0$ through a point $(0, y) \in V$, there exists $n_0 \in \mathbb{N}$ such that if $n > n_0$ then $f^{-n}(D)$ contains a disc ε C^1-close to the interval $\{(x, 0); -a \leq x \leq a\}$.

16. Show that the diffeomorphism f in the previous exercise is locally conjugate to the diffeomorphism
    $$g(x, y) = (x - x^3, 2y).$$

17. Let $f: \mathbb{R}^2 \to \mathbb{R}^2$ be a C^2 diffeomorphism, $f(x, y) = (f_1(x, y), f_2(x, y))$ with the following properties:
    (1) $f_1(0, y) = 0$ for all $y \in \mathbb{R}$;
    (2) $f_2(x, 0) = 0$ for all $x \in \mathbb{R}$;
    (3) $\partial f_2/\partial y(0, 0) > 1$;
    (4) if $\alpha(x) = f_1(x, 0)$ then $\alpha'(0) = 1$ and $\alpha''(0) \neq 0$. Show that f is locally conjugate to the diffeomorphism
    $$g(x, y) = (x + x^2, 2y).$$

Chapter 3

The Kupka–Smale Theorem

Let M be a compact manifold of dimension m and $\mathfrak{X}^r(M)$ the space of C^r vector fields on M, $r \geq 1$, with a C^r norm. In Chapter 2 we showed that the set $\mathscr{G}_1 \subset \mathfrak{X}^r(M)$, consisting of fields whose singularities are hyperbolic, is open and dense in $\mathfrak{X}^r(M)$. This is an example of a generic property, i.e. a property that is satisfied by almost all vector fields. In this chapter we shall analyse other generic properties in $\mathfrak{X}^r(M)$. The original proof of the results dealt with here can be found in [44], [82] and [107].

First we introduce the concept of hyperbolicity for closed orbits. As in the case of singularities a hyperbolic closed orbit γ persists under small perturbations of the original vector field. Moreover, the structure of the trajectories of the field is very simple and is stable under small perturbations. In particular, the set of points which has γ as ω-limit (α-limit) is a differentiable manifold called the stable (unstable) manifold of γ. In a sense that will be made precise in the text, compact parts of these manifolds change only a little when we change the field a little.

Let us consider two hyperbolic singularities σ_1 and σ_2. If the stable manifold of σ_1 intersects the unstable manifold of σ_2 then σ_1 and σ_2 are related by the existence of orbits which are born in σ_2 and die in σ_1. If the intersection is transversal then a small perturbation of the field will have hyperbolic singularities that are related in the same manner. Analogous concepts and properties are valid for closed orbits as we shall see later.

We shall show here that all these properties hold for the fields in a residual subset of $\mathfrak{X}^r(M)$. At the end of the chapter we shall establish similar properties for $\mathrm{Diff}^r(M)$.

§1 The Poincaré Map

In the previous chapter we described the topological behaviour of the orbits of a vector field in the neighbourhood of a hyperbolic singularity. Now we are going to make an analogous study for closed orbits. As in the case of singularities we need to restrict ourselves to a subset of the space of vector fields in order to obtain a simple description of the orbit structure in neighbourhoods of the closed orbits.

Let γ be a closed orbit of a vector field $X \in \mathfrak{X}^r(M)$. Through a point $x_0 \in \gamma$ we consider a section Σ transversal to the field X.

The orbit through x_0 returns to intersect Σ at time τ, where τ is the period of γ. By the continuity of the flow of X the orbit through a point $x \in \Sigma$ sufficiently close to x_0 also returns to intersect Σ at a time near to τ. Thus if $V \subset \Sigma$ is a sufficiently small neighbourhood of x_0 we can define a map $P: V \to \Sigma$ which to each point $x \in V$ associates $P(x)$, the first point where the orbit of x returns to intersect Σ. This map is called the Poincaré map associated to the orbit γ (and the section Σ). Knowledge of this map permits us to give a description of the orbits in a neighbourhood of γ. Thus, if $x \in V$ is a fixed point of P then the orbit of x is closed and its period is approximately equal to the period of γ if x is near to x_0. In the same way if x is a periodic point of P of period k, i.e. $P(x) \in V, P^2(x) \in V, \ldots, P^k(x) = x$, then the orbit through x is periodic with period approximately equal to $k\tau$.

If $P^k(x)$ is defined for all $k > 0$ the positive orbit through x will be contained in a neighbourhood of γ and if, in addition, $P^k(x) \to x_0$ as $k \to \infty$, then the ω-limit of the orbit of x is γ. We can also detect the orbits which have γ as α-limit using the inverse of P, which is the Poincaré map associated to the field $-X$.

From the continuity of the flows of X and $-X$ it follows that P is a homeomorphism from a neighbourhood of x_0 in Σ into Σ. Later we shall show, using the differentiability of the flow via the Tubular Flow Theorem, that P is in fact a local diffeomorphism of the same class as the field. We shall then be able to use the derivative of P at x_0 to describe the orbit structure in the neighbourhood of γ. For that we shall need some preliminary results.

A *tubular flow* for $X \in \mathfrak{X}^r(M)$ is a pair (F, f) where F is an open set in M and f is a C^r diffeomorphism of F onto the cube $I^m = I \times I^{m-1} = \{(x, y) \in$

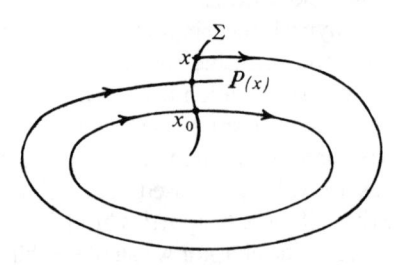

Figure 1

§1 The Poincaré Map

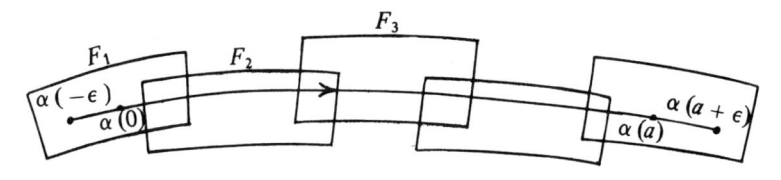

Figure 2

$\mathbb{R} \times \mathbb{R}^{m-1}$; $|x| < 1$ and $|y^i| < 1$, $i = 1, \ldots, m - 1\}$ which takes the trajectories of X in F to the straight lines $I \times \{y\} \subset I \times I^{m-1}$. If f_*X denotes the field in I^m induced by f and X, i.e. $f_*X(x, y) = Df_{f^{-1}(x, y)} \cdot X(f^{-1}(x, y))$, then f_*X is parallel to the constant field $(x, y) \to (1, 0)$.

The open set F is called a flow box for the field X. In the previous chapter we saw that, if $p \in M$ is a regular point of X then there exists a flow box containing p (Tubular Flow Theorem).

1.1 Proposition (Long Tubular Flow). *Let $\gamma \subset M$ be an arc of a trajectory of X that is compact and not closed. Then there exists a tubular flow (F, f) of X such that $F \supset \gamma$.*

PROOF. Let $\alpha: [-\varepsilon, a + \varepsilon] \to M$ be an integral curve of X such that $\alpha([0, a]) = \gamma$ and $\alpha(t) \neq \alpha(t')$ if $t \neq t'$. Let us consider the compact set $\tilde{\gamma} = \alpha([-\varepsilon, a + \varepsilon])$. As the points of $\tilde{\gamma}$ are regular there exists, by the Tubular Flow Theorem, a cover of $\tilde{\gamma}$ by flow boxes. Let δ be the Lebesgue number of this cover. We take a finite cover $\{F_1, \ldots, F_k\}$ of $\tilde{\gamma}$ by flow boxes of diameter less than $\delta/2$. By construction it follows that, if $F_i \cap F_j \neq \emptyset$, then $F_i \cup F_j$ is contained in some flow box of X. Using this property we can reorder the F_i, reducing them in size if necessary, so that each F_i intersects only F_{i-1} and F_{i+1}.

Let $-\varepsilon = t_1 < t_2 < \cdots < t_n = a + \varepsilon$ be such that $p_i = \alpha(t_i) \in F_i \cap \tilde{\gamma}$ and let us write I_d^{m-1} for $\{(0, y) \in I \times I^{m-1}; |y_j| < d, j = 1, \ldots, m - 1\}$. Let (F_i, f_i) be the tubular flows corresponding to the flow boxes above. It is clear that $\Sigma_1 = f_1^{-1}(I_d^{m-1})$ is a section transversal to X because f_1 is a local diffeomorphism and $p_0 \in \Sigma_1$. If $\Sigma_i = X_{t_i - t_1}(\Sigma_1)$ it follows that Σ_i is a section transversal to X which contains the point p_i. If d is sufficiently small $\Sigma_i \subset F_i$.

For each $p \in \tilde{\gamma}$ we take $t \in [0, a + 2\varepsilon]$ such that $p = X_t(p_1)$ and consider the section $\Sigma_p = X_t(\Sigma_1)$. Using the Tubular Flow Theorem we have $\Sigma_p \cap \Sigma_q = \emptyset$ if $p \neq q$ and also that $F = \bigcup_{p \in \tilde{\gamma}} \Sigma_p$ is a neighbourhood of γ.

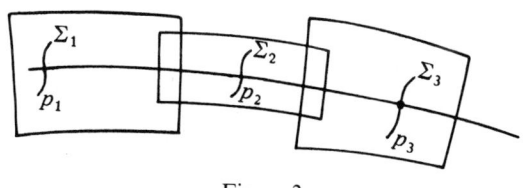

Figure 3

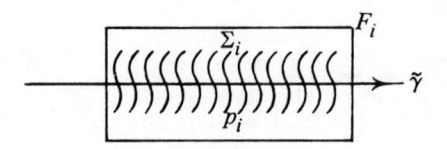

Figure 4

In this neighbourhood we have a C^r fibration whose fibre over the point p is Σ_p, i.e. the projection $\pi_1: F \to \tilde{\gamma}$, which associates to each $z \in F$ the point p such that $z \in \Sigma_p$, is a C^r map. We have another C^r projection defined on F, $\pi_2: F \to \Sigma_1$, which associates to each point $z \in F$ the intersection of the orbit of z with Σ_1. More precisely, if $z \in \Sigma_p$ and $p = X_t(p_1)$ then $\pi_2(z) = X_{-t}(z)$. Let us consider two diffeomorphisms $g_1: \tilde{\gamma} \to [-1, 1]$ and $g_2: \Sigma_1 \to I^{m-1}$. Then we define $f: F \to I \times I^{m-1}$ by $f(z) = (g_1\pi_1(z), g_2\pi_2(z))$. It is clear that (F, f) is a tubular flow which contains γ. □

Remark. The diffeomorphism f obtained above takes orbits of X in F to orbits of the constant field $C: I \times I^{m-1} \to \mathbb{R}^m$, $C(x, y) = (1, 0)$. In general f does not preserve the parameter t, i.e. f_*X is not equal to the field C. However, we can find a neighbourhood of γ, $\tilde{F} \subset F$, and a diffeomorphism $\tilde{f}: \tilde{F} \to (-b, b) \times I^{m-1}$, where $b > 0$, such that $\tilde{f}_*X$ is the constant field. In fact, take $p \in \gamma$ and $b > 0$ such that $\gamma \subset \bigcup_{t \in (-b, b)} X_t(p) \subset F$. Let $\Sigma_p \subset F$ be a section transversal to X through the point p small enough for $\tilde{F} = \bigcup_{t \in (-b, b)} X_t(\Sigma_p)$ to be contained in F. If $z \in \tilde{F}$ and $X_{-t}(z) \in \Sigma_p$ then put $\tilde{f}(z) = (t, hX_{-t}(z))$ where $h: \Sigma_p \to I^{m-1}$ is a diffeomorphism. It is easy to see that $\tilde{f}$ is a C^r diffeomorphism and that $\tilde{f}_*X$ is the constant field.

1.2 Proposition. *Let γ be a closed orbit of a vector field $X \in \mathfrak{X}^r(M)$ and let Σ be a section transversal to X through a point $p \in \gamma$. If $P_\Sigma: U \subset \Sigma \to \Sigma$ is the Poincaré map then P_Σ is a C^r diffeomorphism from a neighbourhood V of p in Σ onto an open set in Σ.*

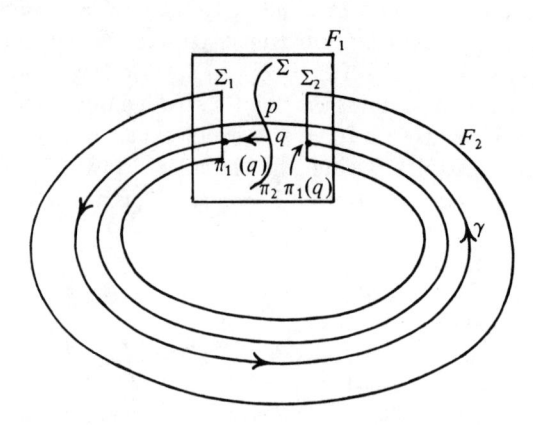

Figure 5

§1 The Poincaré Map

PROOF. Let (F_1, f_1) be a tubular flow containing p and let (F_2, f_2) be a long tubular flow such that $\gamma \subset F_1 \cup F_2$ as in Figure 5.

Let Σ_1 and Σ_2 be the components of the boundary of F_2 which are transversal to X, i.e. $\Sigma_1 = f_2^{-1}(\{-1\} \times I^{m-1})$ and $\Sigma_2 = f_2^{-1}(\{1\} \times I^{m-1})$. Write $\pi_1: V \subset \Sigma \to \Sigma_1$, $\pi_2: \Sigma_1 \to \Sigma_2$ and $\pi_3: \Sigma_2 \to \Sigma$ for the projections along the trajectories of X, where V is a small neighbourhood of p in Σ. $P_\Sigma = \pi_3 \circ \pi_2 \circ \pi_1$. It is easy to see from the Tubular Flow Theorem that π_1, π_2 and π_3 are maps of class C^r. Thus P_Σ is C^r. As P has an inverse of class C^r, which is the Poincaré map corresponding to the field $-X$, it follows that P_Σ is a C^r diffeomorphism from V onto an open set in Σ, which finishes the proof. □

Let Σ_1 and Σ_2 be sections transversal to X through points p_1 and p_2 of a closed orbit γ as in Figure 6.

Let $h: \Sigma_1 \to \Sigma_2$ be the map which associates to each $q \in \Sigma_1$ the first point in which the orbit of q intersects Σ_2. By the Tubular Flow Theorem h is a C^r diffeomorphism. If P_{Σ_1} and P_{Σ_2} are the Poincaré maps for the sections Σ_1 and Σ_2 respectively we have $P_{\Sigma_2} = h \circ P_{\Sigma_1} \circ h^{-1}$. So $DP_{\Sigma_2}(p_2) = Dh(p_1) \circ DP_{\Sigma_1}(p_1) \circ Dh^{-1}(p_2)$ and therefore $DP_{\Sigma_2}(p_2)$ has the same eigenvalues as $DP_{\Sigma_1}(p_1)$. This shows that the next definition depends only on the field and not on the section Σ.

Definition. Let $p \in \gamma$ where γ is a closed orbit of X. Let Σ be a section transversal to X through the point p. We say that γ is a *hyperbolic closed orbit* of X if p is a hyperbolic fixed point of the Poincaré map $P: V \subset \Sigma \to \Sigma$.

Remark. As the flow of a vector field depends continuously on the field the Poincaré map also depends continuously on the field. More precisely, let $P_X: V \subset \Sigma \to \Sigma$ be the Poincaré map of X. Let $\mathscr{V}$ be a neighbourhood of X in $\mathfrak{X}^r(M)$ such that, for all Y in $\mathscr{V}$, Σ is still a section transversal to Y and the orbit of Y through each point of V still returns to intersect Σ. Then the map $\mathscr{V} \to C^r(V, \Sigma)$ which associates to each $Y \in \mathscr{V}$ its Poincaré map P_Y is continuous.

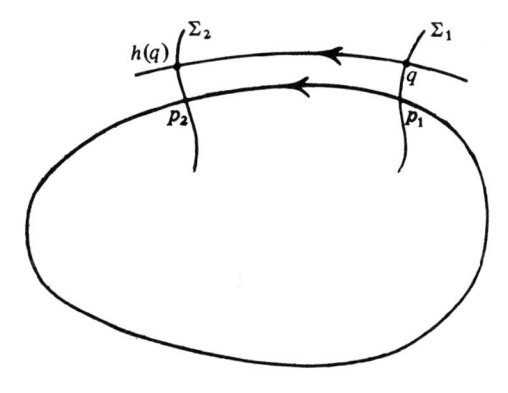

Figure 6

From this remark we conclude that, if γ is a hyperbolic closed orbit of the vector field X, there exists a neighbourhood $\mathscr{V}$ of X in $\mathfrak{X}^r(M)$ such that every $Y \in \mathscr{V}$ has a hyperbolic closed orbit γ_Y close to γ. This is because an analogous property holds for hyperbolic fixed points of diffeomorphisms as we saw in the last chapter.

In what follows we shall show that, if γ is a hyperbolic closed orbit of a vector field $X \in \mathfrak{X}^r(M)$, then X is locally stable at γ. That is, for each vector field Y belonging to a neighbourhood $\mathscr{V}$ of X, there exists a homeomorphism $h: V \to V'$ where V is a neighbourhood of γ, taking orbits of X to orbits of Y. As we have already remarked, we cannot require the homeomorphism h to conjugate the flows X_t and Y_t because this would imply, among other things, that the closed orbit $\gamma_Y \subset V'$ has the same period as γ. It is clear that there exist vector fields Y arbitrarily close to X such that γ_Y has period different from γ: it is enough to take $Y = (1 + n^{-1})X$ with n large enough.

Let Σ be a section transversal to the field X through the point $p \in \gamma$. We say that Σ is an *invariant section* if there exists a neighbourhood $U \subset \Sigma$ of p such that $X_\omega(U) \subset \Sigma$ where ω is the period of γ. The next lemma shows that we can reparametrize the vector field X in such a way as to make a given transversal section invariant. The proof is quite technical although the result is intuitively clear.

1.3 Lemma. *Let $X \in \mathfrak{X}^r(M)$ and let γ be a hyperbolic closed orbit of X with period ω. Let Σ be a section transversal to X through a point $p \in \gamma$. Then there exists a continuous map $\mu: \mathscr{V} \to \mathfrak{X}^r(M)$, where $\mathscr{V}$ is a neighbourhood of X, such that:*

(a) *$\mu(Y) = \rho_Y \cdot Y$, where $\rho_Y: M \to \mathbb{R}$ is a positive differentiable function that takes the value 1 outside a neighbourhood of a point of γ;*
(b) *there exists a neighbourhood $U \subset \Sigma$ of p such that $Y^*_\omega(U) \subset \Sigma$ where $Y^* = \mu(Y)$. That is, Σ is an invariant section of $Y^* = \mu(Y)$ for all $Y \in \mathscr{V}$.*

PROOF. Let $\Sigma' = X_{-(\omega - t_0)}(\Sigma)$ with $0 < t_0 < \omega$. We have $X_{t_0}(p) = p' \in \Sigma'$. Let us consider the C^r map, $\alpha: \Sigma \to \mathbb{R}$, that associates to each $y \in \Sigma$ the least positive time $\alpha(y)$ for which $X_{\alpha(y)}(y) \in \Sigma'$. Notice that if $\alpha(y) = t_0$, for all y in a neighbourhood of p in Σ, then Σ is invariant.

Let $U \subset \tilde{U} \subset \Sigma$ be neighbourhoods of p with $\bar{U} \subset \tilde{U}$. Using a bump function that takes the value 1 on $\bar{U}$ and 0 outside $\tilde{U}$, we define a C^r map,

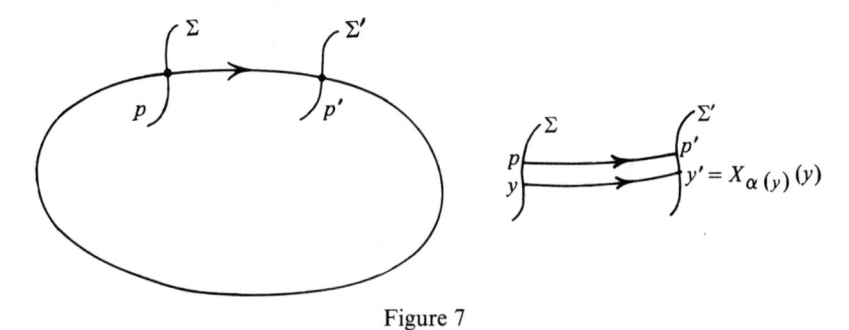

Figure 7

§1 The Poincaré Map

$\beta: \Sigma \to \mathbb{R}$, that is equal to α on U and is constant at the value t_0 outside $\tilde{U}$. With the same bump function we define for each vector field Y in a neighbourhood of X a C^r function, $\beta_Y: \Sigma \to \mathbb{R}$, which coincides with the function α_Y on U and is equal to t_0 outside $\tilde{U}$. Here $\alpha_Y: \Sigma \to \mathbb{R}$ is the function that associates to each $y \in \Sigma$ the least positive time $\alpha_Y(y)$ for which $Y_{\alpha_Y(y)}(y) \in \Sigma'_Y$ where $\Sigma'_Y = Y_{-(\omega - t_0)}(\Sigma)$.

Next we construct the desired reparametrization. Let $G: \mathscr{V} \times \Sigma \times \mathbb{R} \to \mathbb{R}$ be a C^r map satisfying the following condition: for each $Y \in \mathscr{V}$ and each $y \in \Sigma$, $G_{Y,y}(t) = G(Y, y, t)$ is a polynomial in t of degree $2r + 3$ whose coefficients are determined by

$$G_{Y,y}(0) = 0, \qquad G_{Y,y}(t_0) = \beta_Y(y);$$

$$\frac{dG_{Y,y}}{dt}(0) = 1, \qquad \frac{dG_{Y,y}}{dt}(t_0) = 1;$$

$$\frac{d^k G_{Y,y}}{dt^k}(0) = 0, \qquad \frac{d^k G_{Y,y}}{dt^k}(t_0) = 0, \qquad k = 2, \ldots, r+1.$$

We have $G_{Y,y}(t) = t + a_1 t^{r+2} + \cdots + a_{r+2} t^{2r+3}$ where

$$a_j = \frac{\beta_Y(y) A_{1j} - t_0 A_{1j}}{\det A}, \qquad j = 1, \ldots, r+2,$$

$$A = \begin{bmatrix} t_0^{r+2} & t_0^{r+3} & \cdots & t_0^{2r+3} \\ (r+2)t_0^{r+1} & (r+3)t_0^{r+2} & \cdots & (2r+3)t_0^{2r+2} \\ \vdots & \vdots & & \vdots \\ (r+2)! t_0 & \frac{(r+3)!}{2!} t_0^2 & \cdots & \frac{(2r+3)!}{(r+2)!} t_0^{r+2} \end{bmatrix}$$

and A_{1j} is the cofactor of the entry $a_{1j} = t_0^{r+j+1}$ in A.

It now follows that $H_Y(y, t) = dG_{Y,y}/dt(t)$ satisfies the following conditions:

(a) $H_Y(y, 0) = H_Y(y, t_0) = 1$ for $y \in \Sigma$;
(b) $H_Y(y, t) = 1$ for $y \notin \tilde{U}$;
(c) $D^k H_Y(y, 0) = D^k H_Y(y, t_0) = 0$ for all $y \in \Sigma$ and $k = 1, \ldots, r$.

Condition (c) follows easily from the previous equalities.

Thus $H | \Sigma \times [0, t_0]$ extends to a C^r map, $H: \Sigma \times \mathbb{R} \to \mathbb{R}$, with $H = 1$ outside $\Sigma \times [0, t_0]$.

For small $\mathscr{V}$ and $\tilde{U}$, the map $G_{Y,y}: [0, t_0] \to [0, \beta_Y(y)]$ is a diffeomorphism since $G_{Y,y}$ is the identity for $y \notin \tilde{U}$. Also, the map $\varphi_Y: \Sigma \times [0, t_0] \to M$, defined by $\varphi_Y(y, t) = Y_{t^*}(y)$ where $t^* = G_{Y,y}(t)$, is a C^r diffeomorphism. Let W be $\varphi_Y(\Sigma \times [0, t_0]) \subset M$ and let us define $\rho_Y: M \to \mathbb{R}$ to be equal to 1 outside W and to be, on W, the composition of the maps

$$q \xmapsto{\varphi_Y^{-1}} (y, t) \xmapsto{dG_{Y,y}} \frac{dG_{Y,y}}{dt}(t) = H_Y(y, t).$$

By the construction above ρ_Y is of class C^r. Let us consider the vector field $Y^* = \rho_Y \cdot Y$. We claim that $Y^*_{t_0}(y) \in \Sigma'_Y$ for any $y \in \Sigma$. In fact, let $y \in U$ and let $\psi: [0, \beta_Y(y)] \to M$ be the integral curve of Y through the point y. Thus, $\psi(t) = Y_t(y)$ and $\psi(\beta_Y(y)) \in \Sigma'_Y$. Now let $\psi^*: [0, t_0] \to M$ be the map $\psi^* = \psi \circ G_{Y,y}$. We check that ψ^* is the integral curve of Y^* through y. We have

$$\frac{d\psi^*}{dt}(t) = \frac{d\psi}{dt}(G_{Y,y}(t)) \cdot \frac{dG_{Y,y}}{dt}(t)$$

$$= \frac{dG_{Y,y}}{dt}(t) \cdot Y(\psi \circ G_{Y,y}(t)) = \rho_Y(\psi^*(t)) \cdot Y(\psi^*(t)) = Y^*(\psi^*(t)).$$

Thus, $Y^*_{t_0}(y) = \psi^*(t_0) = \psi(\beta_Y(y)) \in \Sigma'_Y$.

We complete the proof by defining $\mu(Y) = Y^*$. □

Remark. Let $X \in \mathfrak{X}^r(M)$, $p_0 \in M$ and let Σ be a section transversal to X containing p_0. Let $\Sigma' = X_{t_1}(\Sigma)$ and $p_1 = X_{t_1}(p_0)$. From the proof of the lemma it follows that we can reparametrize all the vector fields near X so that they take the section Σ to the section Σ' in time t_1. Such a reparametrization can be concentrated in a neighbourhood of a point $p = X_t(p_0)$ for $0 < t < t_1$.

1.4 Proposition. *If γ is a hyperbolic closed orbit of a vector field $X \in \mathfrak{X}^r(M)$ then X is locally stable at γ.*

PROOF. Let ω be the period of γ and let Σ be a section transversal to X through a point $p \in \gamma$. Then p is a hyperbolic fixed point of the Poincaré map P_X. For Y close to X the Poincaré map P_Y is close to P_X. Thus, by the local stability of a hyperbolic fixed point, there exist neighbourhoods $\mathscr{V}$ of X and $U \subset \Sigma$ of p with the following property: for each $Y \in \mathscr{V}$ we can find a homeomorphism h_Y from U to a neighbourhood of p in Σ conjugating P_X and P_Y, that is $h_Y P_X = P_Y h_Y$. Let us extend h_Y to a neighbourhood of γ. Let $\mu: \mathscr{V} \to \mathfrak{X}^r(M)$ be the map obtained in Lemma 1.3. Let V be a neighbourhood of γ such that, if $y \in V$, there exists $0 \leq t \leq \omega$ with $X^*_t(y) \in U$ where $X^* = \mu(X)$. Then we define $h_Y(y) = Y^*_{-t} h_Y X^*_t(y)$. It is easy to see that, if V is small enough, h_Y is well defined, is a homeomorphism and conjugates the flows X^*_t and Y^*_t. As X^* and Y^* have the same orbits as X and Y, respectively, we conclude that h_Y takes orbits of X to orbits of Y. □

If γ is a hyperbolic closed orbit of a vector field $X \in \mathfrak{X}^r(M)$ we define the *stable* and *unstable manifolds* of γ by

$$W^s(\gamma) = \{y \in M; \omega(y) = \gamma\},$$
$$W^u(\gamma) = \{y \in M; \alpha(y) = \gamma\}.$$

There exists a neighbourhood V of γ such that if $X_t(q) \in V$, for all $t \geq 0$, then $q \in W^s(\gamma)$. This follows from an analogous property of the Poincaré map of γ. Let us consider the sets

$$W^s_V(\gamma) = \{y \in V; X_t(y) \in V \text{ for all } t \geq 0\}$$
$$W^u_V(\gamma) = \{y \in V; X_t(y) \in V \text{ for all } t \leq 0\}.$$

We have $W^s(\gamma) = \bigcup_{n \in \mathbb{N}} X_{-n} W^s_V(\gamma)$ and $W^u(\gamma) = \bigcup_{n \in \mathbb{N}} X_n W^u_V(\gamma)$. □

1.5 Proposition. *Let γ be a hyperbolic closed orbit of a vector field $X \in \mathfrak{X}^r(M)$. If V is a small neighbourhood of γ then $W^s_V(\gamma)$ and $W^u_V(\gamma)$ are C^r submanifolds of M, $W^s_V(\gamma)$ is transversal to $W^u_V(\gamma)$ and $W^s_V(\gamma) \cap W^u_V(\gamma) = \gamma$.*

PROOF. Let Σ be a section transversal to X through a point $p \in \gamma$. If U is a small neighbourhood of p in Σ we denote by $W^s_U(p)$ and $W^u_U(p)$ the stable and unstable manifolds, respectively, of the Poincaré map P_X. As p is a hyperbolic fixed point of P_X we see that $W^s_U(p)$ and $W^u_U(p)$ are C^r-submanifolds which are transversal to each other in Σ and $W^s_U(p) \cap W^u_U(p) = \{p\}$. Therefore, if ω is the period of γ then $\bigcup_{t \in (0, 2\omega)} X_t(W^s_U(p))$ and $\bigcup_{t \in (0, 2\omega)} X_{-t}(W^u_U(p))$ are C^r submanifolds which intersect each other transversally along γ. If V is a small neighbourhood of γ then $W^s_V(\gamma)$ and $W^u_V(\gamma)$ are open neighbourhoods of γ in $\bigcup_{t \in (0, 2\omega)} (W^s_U(p))$ and in $\bigcup_{t \in (0, 2\omega)} X_{-t}(W^u_U(p))$ respectively. This proves the proposition. □

Corollary. *$W^s(\gamma)$ and $W^u(\gamma)$ are immersed submanifolds of M of class C^r.* □

We leave it to the reader to prove the next statement. Let γ be a hyperbolic closed orbit of a vector field $X \in \mathfrak{X}^r(M)$. Show that there exists a neighbourhood $\mathscr{V}$ of X and, for each $Y \in \mathscr{V}$, a neighbourhood $W_Y \subset W^s(\gamma_Y)$ of γ_Y such that the map $Y \mapsto W_Y$ is continuous. That is given $\varepsilon > 0$ and $Y_0 \in \mathscr{V}$, there exists $\delta > 0$ such that, if $\|Y - Y_0\| < \delta$, then W_Y is ε C^r-close to W_{Y_0}.

§2 Genericity of Vector Fields Whose Closed Orbits Are Hyperbolic

In the last chapter we showed that the set $\mathscr{G}_1 \subset \mathfrak{X}^r(M)$, consisting of vector fields whose singularities are hyperbolic, is open and dense in $\mathfrak{X}^r(M)$. In this section we shall show that the set $\mathscr{G}_{12} \subset \mathscr{G}_1$, of vector fields in $\mathscr{G}_1$ whose closed orbits are hyperbolic, is residual. For this it is enough to show that, if $T > 0$ is any integer, then the set $\mathfrak{X}(T) \subset \mathscr{G}_1$ of those vector fields whose closed orbits of period $\leq T$ are hyperbolic is open and dense. As $\mathscr{G}_{12} = \bigcap_{T \geq 1} \mathfrak{X}(T)$ it will then follow that $\mathscr{G}_{12}$ is residual.

2.1 Lemma. *Let $p \in M$ be a hyperbolic singularity of $X \in \mathfrak{X}^r(M)$. Given $T > 0$ there exist neighbourhoods $U \subset M$ of p and $\mathcal{U} \subset \mathfrak{X}^r(M)$ of X and a continuous map $\rho \colon \mathcal{U} \to U$ such that*

(i) *if $Y \in \mathcal{U}$ then $\rho(Y)$ is the unique singularity of Y in U and it is hyperbolic;*
(ii) *every closed orbit of a field $Y \in \mathcal{U}$ that passes through U has period $> T$.*

PROOF. Part (i) was proved in the last chapter.

By the Grobman–Hartman Theorem there exists a neighbourhood V of p such that there is no closed orbit of $Y \in \mathcal{U}$ entirely contained in V. As $X(p) = 0$ we can find a neighbourhood U of p contained in V such that, if $q \in U$, then $X_t(q) \in V$ for $t \le 2T$. Shrinking $\mathcal{U}$ if necessary we shall have $Y_t(q) \in V$ for $t \le T$ if $q \in U$ and $Y \in \mathcal{U}$. As Y does not have a closed orbit entirely contained in V, the lemma is proved. $\square$

2.2 Lemma. *Let $T > 0$ and let γ be a hyperbolic closed orbit of a vector field $X \in \mathfrak{X}^r(M)$. Then there exist neighbourhoods $U \subset M$ of γ and $\mathcal{U} \subset \mathfrak{X}^r(M)$ of X such that:*

(i) *each $Y \in \mathcal{U}$ has a hyperbolic closed orbit $\gamma_Y \subset U$ and each closed orbit of Y distinct from γ_Y that passes through U has period greater than T;*
(ii) *the orbit γ_Y depends continuously on Y.*

PROOF. Take a transversal section Σ through a point p of γ and let $P_X \colon V \to \Sigma$ be the Poincaré map associated to X. Let τ be the period of γ and n a positive integer such that $n\tau > 2T$. For sufficiently small $V \subset \Sigma$ we have P_X^n defined on V. As the Poincaré map depends continuously on the vector field there exists a neighbourhood $\mathcal{U}$ of X such that, if $Y \in \mathcal{U}$, P_Y^n is defined on V. As p is a hyperbolic fixed point of P_X, there exists, for possibly smaller $\mathcal{U}$ and V, a continuous map $\rho \colon \mathcal{U} \to V$ that associates to each $Y \in \mathcal{U}$ the unique fixed point $\rho(Y)$ of P_Y in V and this fixed point is hyperbolic. If γ_Y is the orbit of Y through $\rho(Y)$ it follows that γ_Y is a hyperbolic closed orbit and γ_Y clearly depends continuously on Y. By the Grobman–Hartman Theorem for diffeomorphisms and by the continuous dependence of the Poincaré map on the vector field there exists a neighbourhood $\tilde{V} \subset V$ of p and a neighbourhood $\mathcal{U}$ of X such that, for all $Y \in \mathcal{U}$ and all $q \in \tilde{V}$, $P_Y^k(q) \in V$ for $k = 1, \ldots, n$. Therefore, for possibly smaller $\mathcal{U}$, every closed orbit of $Y \in \mathcal{U}$, other than γ_Y, through a point $q \in \tilde{V}$ has period $> T$, because q is not a periodic point of P_Y of period less than or equal to n. It is now sufficient to take $U = \bigcup_{t \in [0, \tau + \varepsilon]} X_t \tilde{V}$, where $\varepsilon > 0$ is small enough. $\square$

Corollary. *Let $X \in \mathcal{G}_{12}$, that is, all the singularities and closed orbits of X are hyperbolic. Given $T > 0$ there are only a finite number of closed orbits of period $\le T$. In particular, X has at most a countable number of closed orbits.*

PROOF. Suppose, if possible, that X has an infinite number of closed orbits of period $\le T$ and let γ_n be a sequence of them with $\gamma_n \ne \gamma_{n'}$ if $n \ne n'$. Take $p_n \in \gamma_n$. As M is compact we can suppose, by passing to a subsequence if

§2 Genericity of Vector Fields Whose Closed Orbits Are Hyperbolic   101

necessary, that p_n converges to some point p. Thus the orbit of p is in the closure of the set which consists of an infinite number of closed orbits of period $\leq T$. By Lemma 2.1, p cannot be a singularity and, by Lemma 2.2, the orbit of p cannot be closed. Thus, the orbit of p is regular. Set $p' = X_{-T}(p)$ and $p'' = X_T(p)$. Let F be a flow box containing the arc $p'p''$ of the orbit of p. If p_n is close enough to p then $X_t(p_n) \in F$ for $t \in [-\frac{1}{2}T, \frac{1}{2}T]$ so that the orbit of p_n has period $> T$ which is absurd. □

2.3 Lemma. *Let $X \in \mathfrak{X}^r(M)$ and let $K \subset M$ be a compact set such that X has no singularities in K and the closed orbits of X through points of K have period $> T$. Then there exists a neighbourhood $\mathcal{U} \subset \mathfrak{X}^r(M)$ of X such that each $Y \in \mathcal{U}$ has no singularities in K and the closed orbits of Y through points of K have period $> T$.*

PROOF. As K is compact and $X \neq 0$ in K there exists a neighbourhood $\mathcal{N} \subset \mathfrak{X}^r(M)$ of X such that each vector field $Y \in \mathcal{N}$ has no singularities in K. Take $p \in K$. As the orbit of X through p is either regular or has period greater than T, it follows that there exists $\varepsilon > 0$ and a neighbourhood U_p of p such that, for $q \in U_p$, we have $X_t(q) \notin U_p$ for all $t \in [\varepsilon, T + \varepsilon]$. As the flow depends continuously on the vector field, there exists a neighbourhood $\mathcal{U}_p \subset \mathcal{N}$ of X such that the same property holds for all $Y \in \mathcal{U}_p$. In particular, each closed orbit of $Y \in \mathcal{U}_p$ through a point of U_p has period greater than T. Let $U_{p_1}, \ldots, U_{p_k}$ be a finite cover of the compact set K and let $\mathcal{U} = \bigcap_{i=1}^{k} \mathcal{U}_{p_i}$. It is clear that each closed orbit of $Y \in \mathcal{U}$ through a point of K has period greater than T. □

Let X be a C^∞ vector field on M, γ a closed orbit of X and Σ a transversal section through a point $p \in \gamma$. Let $\mathcal{U} \subset \mathfrak{X}^r(M)$ be a neighbourhood of X and let $V \subset \Sigma$ be a neighbourhood of p such that, for all $Y \in \mathcal{U}$, the Poincaré map of Y is defined on V.

2.4 Lemma. *In the above conditions there exists a neighbourhood $U \subset M$ of γ with the following property: given $\varepsilon > 0$ there is a C^∞ vector field $Y \in \mathcal{U}$, with $\|Y - X\|_r < \varepsilon$, such that P_Y has only a finite number of fixed points in $\Sigma \subset U$ and they are all elementary.*

PROOF. Let (F, f) be a tubular flow with centre p such that $f^{-1}(\{0\} \times I^{m-1}) = \Sigma \cap F$ and also $f_* X$ is the unit vector field C on $[-b, b] \times I^{m-1}$. Let $\tilde{C}$ be a C^∞ vector field defined on $f(F) \subset \mathbb{R}^m$ such that $\tilde{C}$ is transversal to $\{-b\} \times I^{m-1}$ and $\{b\} \times I^{m-1}$ and each orbit of $\tilde{C}$ through a point of $\{-b\} \times I^{m-1}$ meets $\{b\} \times I^{m-1}$. Then we can define a map $L_{\tilde{C}} \colon \{-b\} \times I^{m-1} \to \{b\} \times I^{m-1}$ which associates to each point of $\{-b\} \times I^{m-1}$ the intersection of its orbit with $\{b\} \times I^{m-1}$. By the Tubular Flow Theorem $L_{\tilde{C}}$ is a diffeomorphism. Let

$$A = [-b, -\tfrac{1}{2}b] \times I^{m-1} \cup [\tfrac{1}{2}b, b] \times I^{m-1} \cup [-\tfrac{1}{2}b, \tfrac{1}{2}b] \times (I^{m-1} - I^{m-1}_{3/4}).$$

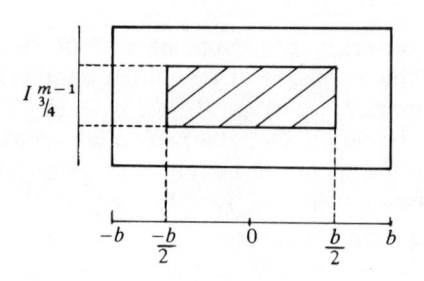

Figure 8

We claim that, given $\varepsilon > 0$, there exists $\varepsilon_1 > 0$ such that, for all $v \in \mathbb{R}^m$ with $\|v\| < \varepsilon_1$ we can choose the C^∞ vector field $\tilde{C}$ such that $\|\tilde{C} - C\|_r < \varepsilon$ on $[-b, b] \times I^{m-1}$, $\tilde{C} = C$ on A and $L_{\tilde{C}}(-b, y) = (b, y + v)$ if $y \in I^{m-1}_{1/4}$. In fact, let $\psi: [-b, b] \to \mathbb{R}^+$ be a C^∞ function such that $\psi(t) = 0$ if $t \in [-b, -\frac{1}{2}b] \cup [\frac{1}{2}b, b]$ and $\psi(t) > 0$ if $t \in (-\frac{1}{2}b, \frac{1}{2}b)$. Take $\varphi: I^{m-1} \to \mathbb{R}^+$ such that $\varphi(y) = 0$ if $\|y\| > \frac{3}{4}$ and $\varphi(y) = 1$ if $\|y\| \le \frac{1}{2}$. We define $\tilde{C}(x, y) = (1, \rho\varphi(y)\psi(x)v)$ and look for a real number ρ for which $\tilde{C}$ has the required properties. The differential equation associated with $\tilde{C}$ can be written as

$$\begin{cases} \dfrac{dx}{dt} = 1, \\ \dfrac{dy}{dt} = \rho\varphi(y)\psi(x)v. \end{cases}$$

The solution of this equation with initial conditions $x(0) = -b$, $y(0) = y_0 \in I_{1/4}$ can be written as $x(t) = t - b$, $y(t) = y_0 + (\int_0^t \rho\varphi(y(s))\psi(s-b)\,ds)v$. Let us take $1/\rho = \int_0^{2b} \psi(s - b)\,ds$. It is easy to see that, if $\|v\|$ is small enough, we shall have $\|y(t)\| < \frac{1}{2}$ for all $t \in [0, 2b]$. Thus, $\varphi(y(s)) = 1$ and $y(t) = y_0 + (\rho\int_0^t \psi(s - b)\,ds)v$. Thus $L_{\tilde{C}}(-b, y_0) = (b, y_0 + v)$, which proves our claim.

Let Y be the vector field on M which is equal to X outside $f^{-1}([-b, b] \times I^{m-1})$ and equal to $(f^{-1})_*\tilde{C}$ on $f^{-1}([-b, b] \times I^{m-1})$. It is clear that Y is C^∞ and $\|Y - X\|_r < \varepsilon$. We write Σ_1 for the transversal section defined by $f^{-1}(\{b\} \times I^{m-1})$ and let $V \subset \Sigma_1$ be a small neighbourhood of $p_1 = f^{-1}(b, 0)$ on which the Poincaré maps P_X and P_Y are defined. The expression for P_Y in the local chart $f|\Sigma_1$ is

$$P_Y(b, y) = P_X(b, y) + v$$

if $V \subset f^{-1}(\{b\} \times I_{1/4})$ and $y \in I_{1/4}$.

By Proposition 3.3 of Chapter 1 we can choose $v \in \mathbb{R}^{m-1}$, $\|v\| < \varepsilon_1$, such that the map $y \mapsto (y, P_Y(y))$ is transversal to the diagonal in $\mathbb{R}^{m-1} \times \mathbb{R}^{m-1}$. With this choice the fixed points of P_Y in V are all elementary which proves the lemma. □

2.5 Lemma.
Let γ be a closed orbit of a C^∞ vector field X. Given $\varepsilon > 0$, there exists a C^∞ vector field Y on M such that $\|Y - X\|_r < \varepsilon$ and γ is a hyperbolic closed orbit of Y.

PROOF. Let (F, f) be a tubular flow with centre a point of γ such that $f_* X$ is the unit field C on $[-b, b] \times I^{m-1}$. We shall show that, given $\varepsilon > 0$, there exists $\varepsilon_1 > 0$ such that, for any $0 < \delta < \varepsilon_1$, we can find a C^∞ vector field $\tilde{C}$ such that $\|\tilde{C} - C\|_r < \varepsilon$ on $[-b, b] \times I^{m-1}$, $\tilde{C} = C$ on A and $L_{\tilde{C}}(-b, y) = (b, (1 + \delta)y)$ if $y \in I^{m-1}_{1/4}$. Here we are using the same notation as in the proof of Lemma 2.4.

In fact, let $\psi: [-b, b] \to \mathbb{R}^+$ be a C^∞ function such that $\psi(t) = 0$ if $t \in [-b, -\frac{1}{2}b] \cup [\frac{1}{2}b, b]$ and $\psi(t) > 0$ if $t \in (-\frac{1}{2}b, \frac{1}{2}b)$. Take $\varphi: I^{m-1} \to \mathbb{R}^+$ such that $\varphi(y) = 0$ if $\|y\| \geq \frac{3}{4}$, $\varphi(y) = 1$ if $\|y\| \leq \frac{1}{2}$ and $\varphi(y) > 0$ if $\frac{1}{2} < \|y\| < \frac{3}{4}$. We define $\tilde{C}(t, y) = (1, \rho\varphi(y)\psi(t)y)$ and look for ρ for which $\tilde{C}$ satisfies the required conditions. It is immediate from the definition that $\tilde{C} = C$ on A. The differential equation associated with $\tilde{C}$ can be written as

$$\begin{cases} \dfrac{dx}{dt} = 1, \\ \dfrac{dy}{dt} = \rho\varphi(y)\psi(x)y. \end{cases}$$

Let $y_0 \in I^{m-1}$ satisfy $\|y_0\| \leq \frac{1}{4}$. We have $\varphi(y_0) = 1$ and $\varphi(y) = 1$ in a neighbourhood of y_0. The solution of the above equation with initial conditions $x(0) = -b$, $y(0) = y_0$ can be written as $x(t) = t - b$, $y(t) = y_0 + \int_0^t \rho\varphi(y(s))\psi(s - b)y(s)\, ds$. By the continuity of $y(s)$, there exists $l > 0$ such that $\varphi(y(s)) = 1$ for all $s \in [0, -\frac{1}{2}b + l]$. Hence $y(t) = y_0 \times \exp(\rho\int_0^t \psi(s - b)\, ds)$ in $[0, -\frac{1}{2}b + l]$ as we can check by differentiating. Let $\mu(t) = \exp(\rho\int_0^t \psi(s - b)\, ds)$. Then $\mu(0) = 1$. Now, for $0 < \rho < \varepsilon_1 = \log 2 / \int_0^{2b} \psi(s - b)\, ds$ we have μ an increasing function, $0 < \mu(2b) \leq 2$ and $\|y(s)\| \leq 2\|y_0\| \leq \frac{1}{2}$. Therefore, $\varphi(y(s)) = 1$ for all $s \in [0, 2b]$ so that $y(s) = \mu(s)y_0$ and $L_{\tilde{C}}(-b, y_0) = (b, \mu(2b)y_0)$. It is easy to see that we can choose ρ to make $\mu(2b) = 1 + \delta$ where $0 < \delta < \varepsilon_1$, so that $L_{\tilde{C}}(-b, y_0) = (b, (1 + \delta)y_0)$. Moreover, by taking ρ small enough we clearly get $\|C - \tilde{C}\|_r < \varepsilon$.

Let Y be the vector field on M which is equal to X outside $f^{-1}([-b, b] \times I^{m-1})$ and to $(f^{-1})_* \tilde{C}$ on $f^{-1}([-b, b] \times I^{m-1})$. It is clear that Y is C^∞, $\|Y - X\|_r < \varepsilon$ and γ is still a closed orbit of Y. We show that, for δ small enough, γ is a hyperbolic closed orbit of Y. In fact, put $\Sigma = f^{-1}(\{b\} \times I^{m-1})$. The expression for the Poincaré map in the local chart $f|\Sigma$ is

$$P_Y(b, y) = (1 + \delta)P_X(b, y) \quad \text{if } \|y\| < \tfrac{1}{4}.$$

Thus, $D(P_Y)_{(b, 0)} = (1 + \delta)D(P_X)_{(b, 0)}$. For small $\delta > 0$ the eigenvalues of $D(P_Y)_{(b, 0)}$ will have absolute value different from 1. □

Remark. In Lemmas 2.4 and 2.5 we started from a vector field X of class C^∞ and made the perturbation in the C^r topology. The proofs of these lemmas cannot be made as above if X is of class C^r. This is because the tubular flow (F, f) is then of class C^r and the vector field $(f^{-1})_* \tilde{C}$ is only C^{r-1}. However, the reader can prove that given $\varepsilon > 0$ there exists a C^∞ vector field $\tilde{X}$ which is εC^r near X and has a closed orbit $\tilde{\gamma}$ near γ. We shall return to this question at the end of this chapter.

2.6 Theorem. *The set $\mathscr{G}_{12}$, which consists of the vector fields whose critical elements (that is, singularities and closed orbits) are hyperbolic, is residual (and therefore dense) in $\mathfrak{X}^r(M)$.*

PROOF. Take $T > 0$ and consider the set $\mathfrak{X}(T) = \{X \in \mathscr{G}_1;$ the closed orbits of X with period $\leq T$ are hyperbolic$\}$. We shall show that $\mathfrak{X}(T)$ is open and dense in $\mathfrak{X}^r(M)$ and then $\mathscr{G}_{12}$ will be residual since it is equal to $\bigcap_{n \in \mathbb{N}} \mathfrak{X}(n)$.

Part 1. $\mathfrak{X}(T)$ is open in $\mathfrak{X}^r(M)$.

Let $X \in \mathfrak{X}(T)$. By the corollary of Lemma 2.2, X has only a finite number of closed orbits of period $\leq T$. Choose $p \in M$. We have three cases to consider:

(a) p is a singularity of X;
(b) $\mathcal{O}(p)$ is regular or closed with period $> T$;
(c) $\mathcal{O}(p)$ is closed with period $\leq T$ (and is therefore hyperbolic).

In case (a) there exist, by Lemma 2.1, neighbourhoods U_p of p in M and $\mathcal{N}_p$ of X in $\mathfrak{X}^r(M)$ such that every vector field $Y \in \mathcal{N}_p$ has only one singularity $p(Y) \in U_p$, which is hyperbolic, and any closed orbit of Y that intersects U_p has period $> T$.

In case (b) there exists, by the Tubular Flow Theorem, a neighbourhood U_p of p in M such that any closed orbit of X that intersects $\bar{U}_p$ has period $> T$ and X has no singularities in $\bar{U}_p$. By Lemma 2.3, there exists a neighbourhood $\mathcal{N}_p$ of X in $\mathfrak{X}^r(M)$ such that every vector field $Y \in \mathcal{N}_p$ has no singularities in $\bar{U}_p$ and the closed orbits of Y through points of $\bar{U}_p$ have period $> T$.

In case (c) there exist, by Lemma 2.2, neighbourhoods U_p of $\mathcal{O}(p)$ in M and $\mathcal{N}_p$ of X in $\mathfrak{X}^r(M)$ such that any $Y \in \mathcal{N}_p$ has only one closed orbit in U_p, γ_Y, it is hyperbolic and all other closed orbits of Y that intersect U_p have period $> T$. Moreover Y has no singularities in U_p.

Now $\{U_p; p \in M\}$ is an open cover of M. Choose a finite subcover $U_1, \ldots, U_l$, let $\mathcal{N}_1, \ldots, \mathcal{N}_l$ be the corresponding neighbourhoods of X in $\mathfrak{X}^r(M)$ obtained in cases (a), (b) and (c) and put $\mathcal{U} = \mathcal{N}_1 \cap \cdots \cap \mathcal{N}_l$. It is now easy to see that any vector field $Y \in \mathcal{U}$ has its closed orbits of period $\leq T$ near the corresponding closed orbits of X and they are still hyperbolic. Also Y has the same number of singularities as X (again still hyperbolic), which proves the first part of the theorem.

Part 2. $\mathfrak{X}(T)$ is dense in $\mathfrak{X}^r(M)$.

It is sufficient to prove that $\mathfrak{X}(T)$ is dense in $\mathscr{G}_1$ so choose $X \in \mathscr{G}_1$.

§2 Genericity of Vector Fields Whose Closed Orbits Are Hyperbolic   105

Claim 1. There exists $\tau > 0$ such that every closed orbit of X has period $\geq \tau$.

Suppose, if possible, that there exists a sequence $\{\gamma_n\}$ of distinct closed orbits and that the sequence $\{t_n\}$ of their periods decreases and tends to 0. Take a sequence of points $p_n \in \gamma_n$. We can suppose, by passing to a subsequence if necessary, that p_n converges to some $p \in M$. The point p must be a singularity of X for, if not, there would be a flow box containing p and the closed orbits intersecting this box could not have arbitrarily small period. Since $X \in \mathscr{G}_1$, p is a hyperbolic singularity. By Lemma 2.1, there exists a neighbourhood U of p in M such that every closed orbit of X that intersects U has period greater than one and this contradiction proves Claim 1.

Now consider the set $\Gamma = \Gamma(\tau, 3\tau/2) = \{p \in M; \mathcal{O}(p) \text{ is closed with period } t \text{ and } \tau \leq t \leq 3\tau/2\}$.

Claim 2. Γ is compact.

It is enough to prove that Γ is closed. Let p_n be a sequence in Γ with $p_n \to p$. As noted above p cannot be a singularity of X. If the orbit of p is regular or closed with period greater than $3\tau/2$ then any closed orbit of X through a point near p has period greater than $3\tau/2$, which is a contradiction. Thus, $\mathcal{O}(p)$ is closed with period between τ and $3\tau/2$ which proves the claim.

Given $\varepsilon > 0$, we want to find $Y \in \mathfrak{X}(T)$ with $\|X - Y\|_r < \varepsilon$ to conclude the proof. First, we outline the construction of Y. Express T as $\frac{1}{2}n\tau + q$ where $0 \leq q < \frac{1}{2}\tau$. Initially, we approximate X by a C^∞ vector field $\tilde{X}$ with $\|X - \tilde{X}\|_r < \varepsilon/2n$. Next, we approximate $\tilde{X}$ by a C^∞ field $Y_1 \in \mathfrak{X}(3\tau/2)$ such that $\|\tilde{X} - Y_1\|_r < \varepsilon/2n$. The next step is to approximate Y_1 by a C^∞ field $Y_2 \in \mathfrak{X}(2\tau)$ with $\|Y_1 - Y_2\|_r < \varepsilon/n$. We carry out the process used in approximating Y_1 by Y_2 $n - 1$ times and obtain C^∞ fields $Y_1, Y_2, \ldots, Y_n$ with $Y_j \in \mathfrak{X}(\frac{1}{2}j\tau + \tau)$ and $\|Y_{j+1} - Y_j\|_r < \varepsilon/n$. Then putting $Y = Y_n$ we have $Y \in \mathfrak{X}(T)$ and $\|Y - X\|_r < \varepsilon$.

Approximating $\tilde{X}$ by $Y_1 \in \mathfrak{X}(3\tau/2)$.

Let $p_1, \ldots, p_s$ be the singularities of $\tilde{X}$. By Lemma 2.1 there exist neighbourhoods $U_1, \ldots, U_s$ of $p_1, \ldots, p_s$ and $\mathcal{N}_1 \subset \mathscr{G}_1$ of $\tilde{X}$ in $\mathfrak{X}^r(M)$ such that, for all $Y \in \mathcal{N}_1$, the closed orbits of Y that intersect $U = U_1 \cup \cdots \cup U_s$ have period greater than T. Moreover, we can suppose that any $Y \in \mathcal{N}_1$ has no singularities in $M - U$. From now on we restrict ourselves to fields in $\mathcal{N}_1$.

Let γ be a closed orbit of $\tilde{X}$ in Γ and Σ_γ a transversal section through $p \in \gamma$. Let us consider neighbourhoods $V_\gamma \subset \Sigma_\gamma$ of p and $\mathcal{N}_\gamma$ of $\tilde{X}$ such that P_Y^2 is defined on V_γ for all $Y \in \mathcal{N}_\gamma$ and the positive orbit of Y through a point of V_γ first intersects Σ_γ at a time $t > 3\tau/4$. We also consider a neighbourhood W_γ of γ such that the positive orbit of Y through any point of W_γ intersects Σ_γ at least twice.

The open sets W_γ for closed orbits $\gamma \subset \Gamma$ cover the compact set Γ. Let $W_1, \ldots, W_k$ be a finite subcover and let $\mathcal{N}_{\gamma_1}, \ldots, \mathcal{N}_{\gamma_k}$ be the corresponding neighbourhoods of $\tilde{X}$. Put $\mathcal{N}_2 = \mathcal{N}_{\gamma_1} \cap \cdots \cap \mathcal{N}_{\gamma_k}$ and $W = W_1 \cup \cdots \cup W_k$.

Now consider the compact set $K = M - (U \cup W)$. Since $\tilde{X}$ has no singularities in K and every closed orbit of $\tilde{X}$ through points of K has period greater than $3\tau/2$, there exists, by Lemma 2.3, a neighbourhood $\mathcal{N}_3 \subset \mathcal{N}_2$

of $\tilde{X}$ such that the closed orbits of any $Y \in \mathcal{N}_3$ through points of K have period greater than $3\tau/2$. From now on we restrict attention to fields in $\mathcal{N}_3$. For each $j = 1, \ldots, k$ and for each $Y \in \mathcal{N}_3$ consider the Poincaré map $P_{jY}: V_j \to \Sigma_j$. By Lemma 2.4, we can approximate $\tilde{X}$ by a C^∞ field $\tilde{Y}_1$ such that $P_{1\tilde{Y}_1}$ only has elementary fixed points in $W_1 \cap \Sigma_1$. As we saw in the last chapter, every field close enough to $\tilde{Y}_1$ has the same property. Thus we can approximate $\tilde{Y}_1$ by a field $\tilde{Y}_2$ such that $P_{1\tilde{Y}_2}$ and $P_{2\tilde{Y}_2}$ only have elementary fixed points in $W_1 \cap \Sigma_1$ and $W_2 \cap \Sigma_2$. By repeating this argument we obtain a field $\tilde{Y}_k$ arbitrarily close to $\tilde{X}$ for which the Poincaré maps $P_{j\tilde{Y}_k}$ only have elementary fixed points in $W_j \cap \Sigma_j$ for $j = 1, \ldots, k$. Let $\mu_1, \ldots, \mu_l$ be the closed orbits of $\tilde{Y}_k$ corresponding to the fixed points of these Poincaré maps. The other closed orbits of $\tilde{Y}_k$ have period $> 3\tau/2$. Moreover, there exists a neighbourhood $\mathcal{N}(\tilde{Y}_k)$ such that each field $Y \in \mathcal{N}(\tilde{Y}_k)$ has its closed orbits of period $\leq 3\tau/2$ elementary and they are near $\mu_1, \ldots, \mu_l$.

Using Lemma 2.5 repeatedly we approximate $\tilde{Y}_k$ by a C^∞ field Y_1 whose closed orbits of length less than or equal to $3\tau/2$ are the same as those of $\tilde{Y}_k$ and they are hyperbolic for Y_1. Thus, we have $Y_1 \in \mathfrak{X}(3\tau/2)$.

As $\mathfrak{X}(3\tau/2)$ is open, there exists a neighbourhood $\mathcal{N}_4$ of Y_1 contained in $\mathfrak{X}(3\tau/2)$. Consider neighbourhoods $U_{s+1}, \ldots, U_{s+l}$ of the closed orbits of Y_1 of period less than or equal to $3\tau/2$ as in Lemma 2.2. Put $U = \bigcup_{i=1}^{s+l} U_i$. Thus every closed orbit of a field near Y_1 through a point of U is hyperbolic or has period greater than $3\tau/2$.

Approximating Y_1 by $Y_2 \in \mathfrak{X}(2\tau)$.

From the compactness of $\Gamma = \Gamma(3\tau/2, 2\tau)$ and Lemmas 2.4 and 2.5 as before we obtain a neighbourhood W of Γ such that Y_1 can be approximated by a field Y_2 whose closed orbits through points of W are either hyperbolic or have period greater than 2τ. As the closed orbits of Y_1 in the compact set $K = M - (U \cup W)$ have period greater than 2τ it follows that we can choose Y_2 to be a C^∞ field in $\mathfrak{X}(2\tau)$. Similarly we can obtain $Y_3, \ldots, Y_n$ and this completes the proof of density. $\square$

§3 Transversality of the Invariant Manifolds

In this section we shall complete the proof of the Kupka–Smale Theorem.

We say that a vector field $X \in \mathfrak{X}^r(M)$ is *Kupka–Smale* if it satisfies the following properties:

(a) the critical elements of X (the singularities and closed orbits) are hyperbolic, that is, $X \in \mathcal{G}_{12}$;
(b) if σ_1 and σ_2 are critical elements of X then the invariant manifolds $W^s(\sigma_1)$ and $W^u(\sigma_2)$ are transversal.

We write $\mathcal{G}_{123}$ or K–S for the set of Kupka–Smale vector fields.

§3 Transversality of the Invariant Manifolds

3.1 Theorem (Kupka–Smale). K–S *is residual in* $\mathfrak{X}^r(M)$.

We have already shown that $\mathscr{G}_{12}$ is residual in $\mathfrak{X}^r(M)$ so it only remains to show that K–S is residual in $\mathscr{G}_{12}$. We shall divide the proof of this fact into several lemmas.

To simplify the notation we make the convention that a singularity of $X \in \mathfrak{X}^r(M)$ is a critical element of period zero. Thus, if $T > 0$ and $X \in \mathfrak{X}(T)$ then X has only a finite number of critical elements of period $\leq T$ and they are hyperbolic. Let $\bar{\mathfrak{X}}(T)$ be the set of vector fields $X \in \mathfrak{X}(T)$ for which $W^s(\sigma_1)$ is transversal to $W^u(\sigma_2)$ whenever σ_1 and σ_2 are critical elements of period $\leq T$.

3.2 Lemma. *If* $\bar{\mathfrak{X}}(T)$ *is residual in* $\mathfrak{X}^r(M)$ *for all* $T \geq 0$ *then* K–S *is residual.*

PROOF. As K–S $= \bigcap_{n \in \mathbb{N}} \bar{\mathfrak{X}}(n)$ and each $\bar{\mathfrak{X}}(n)$ is residual it follows that K–S is residual. □

3.3 Lemma. *Let* E *be a separable Baire space and* $F \subset E$ *a dense subset. A subset* $U \subset E$ *is residual if and only if each* $x \in F$ *has a neighbourhood* V_x *such that* $U \cap V_x$ *is residual in* V_x.

PROOF. Let $V_{x_1}, \ldots, V_{x_n}, \ldots$ be a countable cover of F such that $U \cap V_{x_i}$ is residual in V_{x_i} for all i. Then $U \cap V_{x_i} \supset \bigcap_{j=1}^{\infty} U_{ij}$ where U_{ij} is open and dense in V_{x_i}. Let $V = \bigcup_{i=1}^{\infty} V_{x_i}$ and $W_{ij} = U_{ij} \cup (V - \bar{V}_{x_i})$. Then V and W_{ij} are open and dense. It is easy to see that U contains $\bigcap_{i=1}^{\infty} \bigcap_{j=1}^{\infty} W_{ij}$. Hence U is residual. The reciprocal implication is trivial. □

Corollary. *If, for all* $T \geq 0$ *and for all* $X \in \mathfrak{X}(T)$, *there exists a neighbourhood* $\mathcal{N}$ *of* X *such that* $\bar{\mathfrak{X}}(T)$ *is residual in* $\mathcal{N}$ *then* K–S *is residual.*

PROOF. This follows from Lemmas 3.2 and 3.3 and the density of $\mathfrak{X}(T)$ in $\mathfrak{X}^r(M)$. □

Let $X \in \mathfrak{X}(T)$ and let $\sigma_1, \ldots, \sigma_s$ be the critical elements of X with period $\leq T$. For each i let us take compact neighbourhoods $W_0^s(\sigma_i)$ and $W_0^u(\sigma_i)$ of σ_i in $W^s(\sigma_i)$ and $W^u(\sigma_i)$, respectively, such that the boundaries of $W_0^s(\sigma_i)$ and $W_0^u(\sigma_i)$ are fundamental domains for $W^s(\sigma_i)$ and $W^u(\sigma_i)$. Let Σ_i^s be a codimension 1 submanifold of M that is transversal to the vector field X and to the local stable manifold of σ_i which it meets in the fundamental domain $\partial W_0^s(\sigma_i)$, see Figure 9.

For each Y in a small enough neighbourhood $\mathcal{N}$ of X, Y is transversal to each Σ_i^s and the critical elements of Y of period less than or equal to T are hyperbolic and are near the corresponding critical elements of X. Thus, if $\sigma_1(Y), \ldots, \sigma_s(Y)$ are the critical elements of Y of period less than or equal to T then there exists a compact neighbourhood $W_0^s(\sigma_i, Y)$ of $\sigma_i(Y)$ in $W^s(\sigma_i, Y)$ whose boundary is the intersection of Σ_i^s with $W_0^s(\sigma_i, Y)$. By the Stable

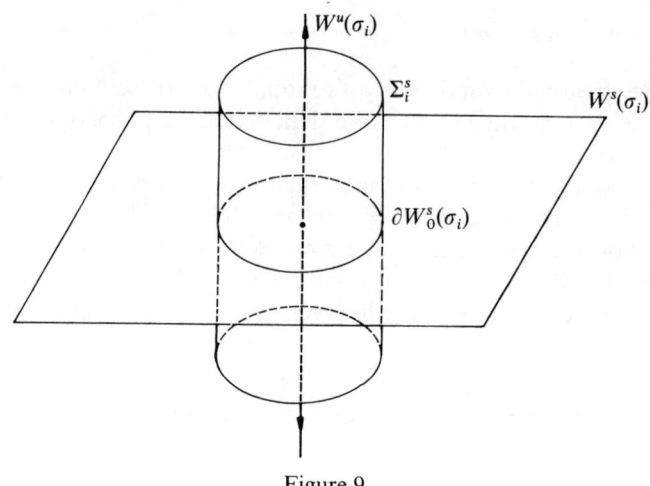

Figure 9

Manifold Theorem the map $Y \mapsto W_0^s(\sigma_i, Y)$ is continuous; that is, given $Y_0 \in \mathcal{N}$ and $\varepsilon > 0$ there exists $\delta > 0$ such that if $\|Y - Y_0\|_r < \delta$ then $W_0^s(\sigma_i, Y)$ is ε C^r-close to $W_0^s(\sigma_i, Y_0)$. Similarly, for each $Y \in \mathcal{N}$ and each $i = 1, \ldots, s$, we construct a compact neighbourhood $W_0^u(\sigma_i, Y)$ of $\sigma_i(Y)$ in $W^u(\sigma_i, Y)$ so that the map $Y \mapsto W_0^u(\sigma_i, Y)$ is continuous.

For each positive integer n we define $W_n^s(\sigma_i, Y) = Y_{-n}(W_0^s(\sigma_i, Y))$ and $W_n^u(\sigma_i, Y) = Y_n(W_0^u(\sigma_i, Y))$. It is clear that the maps $Y \mapsto W_n^s(\sigma_i, Y)$ and $Y \mapsto W_n^u(\sigma_i, Y)$ are continuous since Y_n and Y_{-n} are diffeomorphisms that depend continuously on Y. Moreover, $W_n^s(\sigma_i, Y)$ and $W_n^u(\sigma_i, Y)$ are compact submanifolds with boundary, $W^s(\sigma_i, Y) = \bigcup_{n \geq 0} W_n^s(\sigma_i, Y)$ and $W^u(\sigma_i, Y) = \bigcup_{n \geq 0} W_n^u(\sigma_i, Y)$. Let $\bar{\mathfrak{X}}_n(T)$ be the set of vector fields $Y \in \mathcal{N}$ such that $W_n^s(\sigma_i, Y)$ is transversal to $W_n^u(\sigma_j, Y)$ for all i and j.

3.4 Lemma. *Let $X \in \mathfrak{X}(T)$ and let $\mathcal{N}$ be a neighbourhood of X as above. If, for all $n \in \mathbb{N}$, $\bar{\mathfrak{X}}_n(T)$ is open and dense in $\mathcal{N}$ then $\bar{\mathfrak{X}}(T) \cap \mathcal{N}$ is residual in $\mathcal{N}$.*

PROOF. If suffices to observe that $\bar{\mathfrak{X}}(T) \cap \mathcal{N} = \bigcap_{n=1}^{\infty} \bar{\mathfrak{X}}_n(T)$. □

3.5 Lemma. *Let $X \in \bar{\mathfrak{X}}(T)$ and let $\mathcal{N}$ be a neighbourhood of X as above. Then, for all $n \in \mathbb{N}$, $\bar{\mathfrak{X}}_n(T)$ is open and dense in $\mathcal{N}$.*

PROOF. Let $\sigma_1, \ldots, \sigma_s$ be the critical elements of X of period less than or equal to T. We write $\bar{\mathfrak{X}}_{n, i, j}(T)$ for the set of vector fields $Y \in \mathcal{N}$ such that $W_n^s(\sigma_i, Y)$ is transversal to $W_n^u(\sigma_j, Y)$. It is clear that $\bar{\mathfrak{X}}_n(T) = \bigcap_{i, j=1}^{s} \bar{\mathfrak{X}}_{n, i, j}(T)$. Thus it is enough to show that each $\bar{\mathfrak{X}}_{n, i, j}(T)$ is open and dense in $\mathcal{N}$.

Part 1. Openness of $\bar{\mathfrak{X}}_{n, i, j}(T)$.

Let $\tilde{X} \in \bar{\mathfrak{X}}_{n, i, j}(T)$. As $W_n^s(\sigma_i, \tilde{X})$ is transversal to $W_n^u(\sigma_j, \tilde{X})$ and the maps $Y \mapsto W_n^s(\sigma_i, Y)$ and $Y \mapsto W_n^u(\sigma_j, Y)$ are continuous, it follows that there

§3 Transversality of the Invariant Manifolds

exists a neighbourhood $\mathcal{N}_{ij}$ of $\tilde{X}$ in $\mathcal{N}$ such that, for all $Y \in \mathcal{N}_{ij}$, $W_n^s(\sigma_i, Y)$ is transversal to $W_n^u(\sigma_j, Y)$. Therefore $\mathcal{N}_{ij} \subset \bar{\mathfrak{X}}_{n,i,j}(T)$, which proves this set is open.

Part 2. Density of $\bar{\mathfrak{X}}_{n,i,j}(T)$.

Let $\tilde{X} \in \mathfrak{X}(T) \cap \mathcal{N}$. We shall show that there exists a neighbourhood $\tilde{\mathcal{N}}$ of $\tilde{X}$ in $\mathcal{N}$ such that $\bar{\mathfrak{X}}_{n,i,j}(T) \cap \tilde{\mathcal{N}}$ is open and dense in $\tilde{\mathcal{N}}$. In particular, $\tilde{X}$ can be approximated arbitrarily well by elements of $\bar{\mathfrak{X}}_{n,i,j}(T)$.

Let us consider the compact set $K = W_n^s(\sigma_i, \tilde{X}) \cap W_n^u(\sigma_j, \tilde{X})$. If $x \in K$ there exist a tubular flow $(F_x, f_{\tilde{X},x})$ containing x and a positive real number b_x such that $f_{\tilde{X},x}(F_x) \supset [-b_x, b_x] \times I^{m-1}$ and the vector field $(f_{\tilde{X},x})_* \tilde{X}$ coincides with the unit field on $[-b_x, b_x] \times I^{m-1}$. Let $A_x \subset F_x$ be an open neighbourhood of x such that the closure of A_x is contained in the interior of $(f_{\tilde{X},x})^{-1}([-b_x, b_x] \times I_{1/4}^{m-1})$. By shrinking F_x if necessary we can suppose that $W_{2n}^s(\sigma_i, \tilde{X}) \cap F_x$ and $W_{2n}^u(\sigma_j, \tilde{X}) \cap F_x$ have only one connected component each. Let $A_1, \ldots, A_l$ be a finite cover of K by such open sets and let us write $(F_k, f_{\tilde{X},k})$ for the corresponding tubular flows. Then $(f_{\tilde{X},k})^{-1}([-b_k, b_k] \times I_{1/4}^{m-1})$ contains A_k. As the maps $Y \mapsto W_n^s(\sigma_i, Y)$, $Y \mapsto W_n^u(\sigma_j, Y)$ are continuous, there exists a neighbourhood $\tilde{\mathcal{N}}$ of $\tilde{X}$ in $\mathcal{N}$ such that $W_n^s(\sigma_i, Y) \cap W_n^u(\sigma_j, Y) \subset \bigcup_{k=1}^l A_k$ for all $Y \in \tilde{\mathcal{N}}$. By shrinking $\tilde{\mathcal{N}}$ if necessary we can even suppose that, for each $Y \in \tilde{\mathcal{N}}$ and $k = 1, \ldots, l$, there exists a tubular flow $(F_{Y,k}, f_{Y,k})$ for Y with $F_{Y,k} \supset A_k$, $f_{Y,k}(F_{Y,k}) \supset [-b_k, b_k] \times I^{m-1}$ and such that the interior of $(f_{Y,k})^{-1}([-b_k, b_k] \times I_{1/4}^{m-1})$ contains the closure of A_k. This is possible because the flow depends continuously on the vector field. Now let $\tilde{\mathfrak{X}}_k$ be the set of those vector fields $Y \in \tilde{\mathcal{N}}$ such that $W_n^s(\sigma_i, Y)$ is transversal to $W_n^u(\sigma_j, Y)$ at all points of $\bar{A}_k$. Clearly $\tilde{\mathfrak{X}}_k$ is an open subset of $\tilde{\mathcal{N}}$ and it will be sufficient for our purposes to show that $\tilde{\mathfrak{X}}_k$ is dense in $\tilde{\mathcal{N}}$ since $\bar{\mathfrak{X}}_{n,i,j}(T) \cap \tilde{\mathcal{N}} = \bigcap_{k=1}^l \tilde{\mathfrak{X}}_k$.

We shall now show that $\tilde{\mathfrak{X}}_k$ is dense in $\tilde{\mathcal{N}}$. Let us take a C^∞ vector field $Y \in \tilde{\mathcal{N}}$. We denote by $S_+(Y)$ the intersection of $f_{Y,k}(W_n^s(\sigma_i, Y) \cap F_{Y,k})$ with $\{b_k\} \times I^{m-1}$ and by $U_+(Y)$ the intersection of $f_{Y,k}(W_n^u(\sigma_j, Y) \cap F_{Y,k})$ with $\{b_k\} \times I^{m-1}$. It is easy to see that if $S_+(Y)$ is transversal to $U_+(Y)$ in $\{b_k\} \times I_{1/4}^{m-1}$ then $W_n^s(\sigma_i, Y)$ is transversal to $W_n^u(\sigma_j, Y)$ in A_k. See Figure 10.

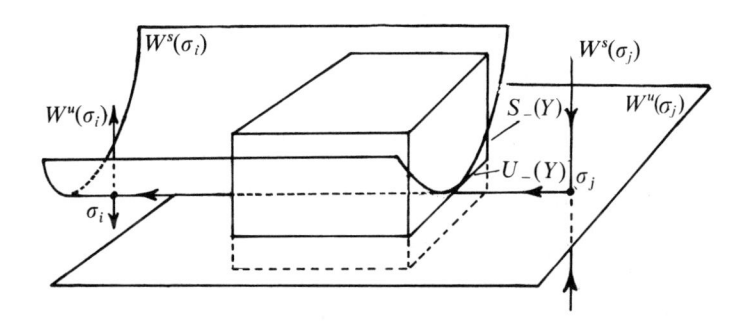

Figure 10

By the same argument as in the proof of Lemma 2.4, given $\varepsilon > 0$ and $v \in \mathbb{R}^{m-1}$ with $\|v\|$ small enough, we find a C^∞ vector field $\tilde{Y}$ on M with $\|Y - \tilde{Y}\|_r < \varepsilon$ such that

(a) $\tilde{Y} = Y$ outside $f_{Y,k}^{-1}([-b_k, b_k] \times I^{m-1})$,
(b) $L_{\tilde{Y}}(-b, y) = (b, y + v)$ for all $y \in I_{1/4}^{m-1}$.

Here $b = b_k$ and $L_{\tilde{Y}}$ is the map from $\{-b\} \times I^{m-1}$ to $\{b\} \times I^{m-1}$ which associates to each $(-b, y)$ the point of intersection with $\{b\} \times I^{m-1}$ of the orbit of $(f_{Y,k})_* \tilde{Y}$ through the point $(-b, y)$. On the other hand, by Proposition 3.3 of Chapter 1, we can choose v small so that $S_+(Y)$ will be transversal to $U_+(Y) + v$. Then $f_{Y,k}^{-1}(S_+(Y))$ is the intersection of $W_{2n}^s(\sigma_i, \tilde{Y})$ with the transversal section $f_{Y,k}^{-1}(\{b_k\} \times I^{m-1})$ and $f_{Y,k}^{-1}(U_+(Y) + v)$ is the intersection of $W_{2n}^u(\sigma_j, \tilde{Y})$ with the section $f_{Y,k}^{-1}(\{b_k\} \times I_{1/4}^{m-1})$. Thus, we conclude that these two submanifolds are transversal in $f^{-1}([-b_k, b_k] \times I_{1/4}^{m-1})$. Consequently $W_n^s(\sigma_i, \tilde{Y})$ is transversal to $W_n^u(\sigma_j, \tilde{Y})$ on $\bar{A}_k$. This shows that $\tilde{Y} \in \tilde{\mathfrak{X}}_k$.

Thus any C^∞ field in $\tilde{\mathcal{N}}$ can be approximated by a field in $\tilde{\mathfrak{X}}_k$. As any field in $\tilde{\mathcal{N}}$ can be approximated by a C^∞ field it follows that $\tilde{\mathfrak{X}}_k$ is dense in $\tilde{\mathcal{N}}$ which completes the proof of the lemma. □

Theorem 3.1 is now an immediate consequence of Lemmas 3.2, 3.3, 3.4 and 3.5. □

Remark. It is important to note that K–S is not open in $\mathfrak{X}^r(M)$. In fact, consider an irrational flow X_t on the torus T^2. The vector field $X \in$ K–S since it has no critical elements. However, X can be approximated by vector fields Y for which Y_t is a rational flow. All the orbits of Y are closed and non-hyperbolic. Thus $Y \notin$ K–S.

Now we shall state the Kupka–Smale theorem for diffeomorphisms. The proof is similar to the case of vector fields and will be left as an exercise for the reader. We shall, however, present two separate sketches of proofs.

A diffeomorphism $f \in \text{Diff}^r(M)$ is said to be *Kupka–Smale* if

(a) the periodic points of f are hyperbolic, and
(b) if p and q are periodic points of f then $W^s(p)$ is transversal to $W^u(q)$.

We shall denote the set of Kupka–Smale diffeomorphisms by K–S too.

3.6 Theorem *K–S is residual in $\text{Diff}^r(M)$.*

SKETCH OF PROOF. (1) If $f \in \text{Diff}^r(M)$ and $k \in \mathbb{N}$ we write $\tilde{f}^k: M \to M \times M$ for the map given by $\tilde{f}^k(p) = (p, f^k(p))$. If p is a periodic point of f of period k then $\tilde{f}^k(p) = (p, p)$ which belongs to the diagonal $\Delta \subset M \times M$. Such a point p is elementary if and only if $\tilde{f}^k$ is transversal to Δ at p. Let $n \in \mathbb{N}$ and let $\mathcal{D}^n$ be the set of diffeomorphisms $f \in \text{Diff}^r(M)$ such that $\tilde{f}^k$ is transversal to Δ for $k = 1, \ldots, n$. Then $\mathcal{D}^n$ is open and dense.

§3 Transversality of the Invariant Manifolds   111

(2) Let $f \in \mathscr{D}^n$ and let p be a periodic point of f of period less than or equal to n. It is easy to approximate f by $g \in \mathscr{D}^n$ so that $g = f$ outside a neighbourhood of p and p is a hyperbolic periodic point of g. Thus the set $\tilde{\mathscr{D}}^n \subset \mathscr{D}^n$ of those diffeomorphisms whose periodic points of period $k = 1, \ldots, n$ are all hyperbolic is open and dense.

(3) Let $\mathscr{G}^n$ be the set of those diffeomorphisms in $\tilde{\mathscr{D}}^n$ whose stable and unstable manifolds of periodic points of period up to n are pairwise transversal. $\mathscr{G}^n$ is residual. As K–S $= \bigcap_{n \in \mathbb{N}} \mathscr{G}^n$ it follows that K–S is residual. □

Another proof of the Kupka–Smale Theorem for diffeomorphisms can be obtained from the corresponding theorem for vector fields. For this we shall need a construction that enables us to relate diffeomorphisms on a manifold M with vector fields on a manifold $\tilde{M}$ of dimension one higher. This construction is called the suspension of a diffeomorphism.

Let $X \in \mathfrak{X}^r(\tilde{M})$ and let $\tilde{\Sigma} \subset \tilde{M}$ be a compact submanifold of codimension 1. We say that $\tilde{\Sigma}$ is a *global transversal section* for X if (a) X is transversal to $\tilde{\Sigma}$, and (b) the positive orbit of X through each point of $\tilde{\Sigma}$ returns to intersect $\tilde{\Sigma}$ again.

If $\tilde{\Sigma}$ is a global transversal section for $X \in \mathfrak{X}^r(\tilde{M})$ then the flow of X induces a diffeomorphism $\tilde{f}: \tilde{\Sigma} \to \tilde{\Sigma}$ which associates to each point $\tilde{p} \in \tilde{\Sigma}$ the point $\tilde{f}(\tilde{p})$ where the positive orbit of $\tilde{p}$ first intersects $\tilde{\Sigma}$. The diffeomorphism $\tilde{f}$ is called the Poincaré map associated with $\tilde{\Sigma}$.

It is easy to see that if $X \in \mathfrak{X}^r(M)$ admits a global transversal section $\tilde{\Sigma}$ then the saturation of $\tilde{\Sigma}$ by the flow of X coincides with $\tilde{M}$, that is, $\bigcup_{t \in \mathbb{R}} X_t(\tilde{\Sigma}) = \tilde{M}$. In particular, X has no singularities.

Lastly, let us remark that the orbit structure of X is determined by the orbit structure of the Poincaré map $\tilde{f}$ and vice-versa. In effect, the following facts are immediate:

(i) $\tilde{p} \in \tilde{\Sigma}$ is a periodic point of $\tilde{f}$ if and only if $\mathcal{O}_X(\tilde{p})$ is closed;
(ii) $\tilde{p} \in \tilde{\Sigma}$ is a hyperbolic periodic point of $\tilde{f}$ if and only if $\mathcal{O}_X(\tilde{p})$ is a hyperbolic closed orbit;
(iii) if $\tilde{p}_1$ and $\tilde{p}_2$ are hyperbolic periodic points of $\tilde{f}$ then $W^s(\tilde{p}_1)$ is transversal to $W^u(\tilde{p}_2)$ if and only if $W^s(\mathcal{O}_X(\tilde{p}_1))$ is transversal to $W^u(\mathcal{O}_X(\tilde{p}_2))$;
(iv) $\tilde{q} \in \omega(\tilde{p})$ if and only if $\mathcal{O}_X(\tilde{q}) \subset \omega(\mathcal{O}_X(\tilde{p}))$.

The next proposition shows that every diffeomorphism is the Poincaré map associated with a global transversal section of some vector field.

3.7 Proposition. *Let $f \in \text{Diff}^r(M)$ where M is a compact manifold. Then there exist a manifold $\tilde{M}$, a vector field $X \in \mathfrak{X}^{r-1}(\tilde{M})$ admitting a global transversal section $\tilde{\Sigma}$ and a C^r diffeomorphism $h: M \to \tilde{\Sigma}$ that conjugates f and the Poincaré map $\tilde{f}: \tilde{\Sigma} \to \tilde{\Sigma}$.*

PROOF. Consider the following equivalence relation on $M \times \mathbb{R}$:

$$(p, s) \sim (q, t) \Leftrightarrow s - t = n \in \mathbb{Z} \quad \text{and} \quad q = f^n(p).$$

Let $\tilde{M}$ be the quotient space $M \times \mathbb{R}/\sim$ and $\pi\colon M \times \mathbb{R} \to \tilde{M}$ the natural projection. Let $\tilde{\Sigma} \subset \tilde{M}$ denote the image of $M \times \{0\}$ by π. For each $t_0 \in \mathbb{R}$ the restriction of π to $M \times (t_0, t_0 + 1)$ is a one–one correspondence between $M \times (t_0, t_0 + 1)$ and $\tilde{M} - \pi(M \times \{t_0\})$. Moreover, $\pi(p, 1) = \pi(f(p), 0)$. On $\tilde{M}$ we use the topology induced by π, that is, $A \subset \tilde{M}$ is open if and only if $\pi^{-1}(A)$ is open.

We shall show that $\tilde{M}$ has a natural differentiable manifold structure and that π is a C^r local diffeomorphism. Let $x_i\colon U_i \to U_0 \subset \mathbb{R}^m$, $i = 1, \ldots, s$ be local charts on M such that $\bigcup_{i=1}^{s} U_i = M$. Then $\tilde{U}_i = \pi(U_i \times (-\frac{1}{2}, \frac{1}{2}))$ and $\tilde{V}_i = \pi(U_i \times (\frac{1}{4}, \frac{5}{4}))$ are open in $\tilde{M}$. We define $\tilde{x}_i\colon \tilde{U}_i \to U_0 \times (-\frac{1}{2}, \frac{1}{2})$ and $\tilde{y}_i\colon \tilde{V}_i \to U_0 \times (\frac{1}{4}, \frac{5}{4})$ by $\tilde{x}_i(\pi(p, t)) = (x_i(p), t)$ and $\tilde{y}_i(\pi(p, t)) = (x_i(p), t)$. Clearly $\tilde{x}_i$ and $\tilde{y}_i$ are homeomorphisms. We claim that $\{(\tilde{x}_i, \tilde{U}_i), (\tilde{y}_i, \tilde{V}_i); i = 1, \ldots, s\}$ is a C^r atlas on M. In fact, $\tilde{x}_i \tilde{x}_j^{-1}(u, t) = (x_i x_j^{-1}(u), t)$, $\tilde{y}_i \tilde{y}_j^{-1}(u, t) = (x_i x_j^{-1}(u), t)$ and $\tilde{x}_i \tilde{y}_j^{-1}(u, t) = (x_i f x_j^{-1}(u), t - 1)$ are C^r diffeomorphisms, which proves the claim.

In fact we can consider $\tilde{M}$ as a C^∞ manifold since, by Theorem 0.19 of Chapter 1, there is a C^∞ manifold structure on $\tilde{M}$ such that $\tilde{x}_i$ and $\tilde{y}_i$ are C^r diffeomorphisms.

As $\tilde{x}_i \circ \pi \circ (x_i^{-1} \times \text{id})$ is the identity on $U_0 \times (-\frac{1}{2}, \frac{1}{2})$ and $\tilde{y}_i \circ \pi \circ (x_i^{-1} \times \text{id})$ is the identity on $U_0 \times (\frac{1}{4}, \frac{5}{4})$ it follows that π is a C^r local diffeomorphism. Let $\partial/\partial t$ be the unit vector field on $M \times \mathbb{R}$ whose orbits are the lines $\{p\} \times \mathbb{R}$, $p \in M$. Let $X(\pi(p, t)) = d\pi(p, t) \cdot (\partial/\partial t(p, t))$. It is easy to see that $X(\pi(p, t)) = X(\pi(f(p), t - 1))$. Thus X is a C^{r-1} vector field on $\tilde{M}$. The field X is transversal to $\tilde{\Sigma}$ and its orbit through the point $\tilde{p} = \pi(p, t)$ is $\pi(\{p\} \times \mathbb{R})$. Thus the positive orbit of X through a point $\tilde{p} = \pi(p, 0) \in \tilde{\Sigma}$ returns to intersect $\tilde{\Sigma}$ for the first time again at the point $\tilde{q} = \pi(p, 1) = \pi(f(p), 0)$. The Poincaré map associated to $\tilde{\Sigma}$, $\tilde{f}\colon \tilde{\Sigma} \to \tilde{\Sigma}$ is, therefore, defined by $\tilde{f}(\pi(p, 0)) = \pi(f(p), 0)$. The map $h\colon M \to \tilde{\Sigma}$ given by $h(p) = \pi(p, 0)$ is a C^r diffeomorphism and $\tilde{f} \circ h = h \circ f$ which completes the proof. $\square$

Remark. By Proposition 3.7 the suspension of a C^r diffeomorphism is a C^{r-1} vector field. The above construction is modified in [80] to give a C^r vector field as the suspension of a C^r diffeomorphism but we shall not make use of this fact.

We shall now show the density of the Kupka–Smale diffeomorphisms using the method of suspension. Take $f_0 \in \text{Diff}^r(M)$. First we approximate f_0 by a C^∞ diffeomorphism f. In order to approximate f by a Kupka–Smale diffeomorphism we consider the C^∞ vector field X on $\tilde{M}$ obtained by suspending f and let $h\colon M \to \tilde{\Sigma}$ be the diffeomorphism which conjugates f with the Poincaré map $\tilde{f}\colon \tilde{\Sigma} \to \tilde{\Sigma}$ of X. We approximate X in the C^r topology by a C^∞ Kupka–Smale vector field Y and write $\tilde{g}\colon \tilde{\Sigma} \to \tilde{\Sigma}$ for the Poincaré map of Y. As $\tilde{g}$ is a Kupka–Smale diffeomorphism C^r close to $\tilde{f}$ it follows that $g = h^{-1} \circ \tilde{g} \circ h$ is a Kupka–Smale diffeomorphism C^r close to f. $\square$

Exercises

EXERCISES

1. Let X_t be the flow generated by a vector field $X \in \mathfrak{X}^r(M)$. Let γ be a hyperbolic closed orbit of X of period λ. Consider the diffeomorphism $f = X_\lambda$ and a point $p \in \gamma$.
   (a) Show that $T_p M$ is the direct sum of two subspaces H and H_0, each invariant for df_p, such that H_0 has dimension 1 and contains the vector $X(p)$.
   (b) Show that if $S \subset M$ is a submanifold whose tangent space at p is H and $\pi: U \subset S \to S$ is the Poincaré map associated to γ then $d\pi(p) = df(p)|H$.

2. Show that, if X is a Kupka–Smale vector field on S^2 then the ω-limit of any orbit is a critical element.

3. Show that the set of Kupka–Smale vector fields is open in $\mathfrak{X}^r(S^1)$.

4. Show that the set of Kupka–Smale vector fields is open in $\mathfrak{X}^r(S^2)$.

5. Let γ and $\tilde\gamma$ be closed orbits of vector fields X and $\tilde X$, respectively. Show that if there exists a homeomorphism h from a neighbourhood of γ to a neighbourhood of $\tilde\gamma$ taking orbits of X to orbits of $\tilde X$ and preserving the orientation of the orbits then the Poincaré maps associated to these closed orbits are conjugate. If h is a diffeomorphism then the Poincaré maps are conjugate by a diffeomorphism.

6. Show that any Kupka–Smale vector field on a compact manifold of dimension two has a finite number of closed orbits.

   Hint. Any compact manifold of dimension two is diffeomorphic, for some n, to the sphere with n handles, if it is orientable, and to the projective plane or the Klein bottle with n handles, if it is nonorientable. The Klein bottle is diffeomorphic to the projective plane with one cross-cap.

7. Show that a structurally stable vector field on a compact manifold of dimension two is Kupka–Smale.

8. Sketch a vector field on S^2 with infinitely many hyperbolic critical elements.

9. Let $C \subset S^2$ be a circle. Consider the set $\mathfrak{X}_C \subset \mathfrak{X}^r(S^2)$ of vector fields that are tangent to C.
   (a) Show that $\mathfrak{X}_C \cap$ K–S is not dense in $\mathfrak{X}_C$.
   (b) Consider the set $KS_C \subset \mathfrak{X}_C$ of vector fields with the following properties: (i) the singularities and closed orbits are hyperbolic; (ii) if γ is an orbit such that $\alpha(\gamma)$ and $\omega(\gamma)$ are saddles then $\gamma \subset C$. Show that KS_C is open and dense in $\mathfrak{X}_C$.

10. Let $\mathrm{Grad}^r(M) \subset \mathfrak{X}^r(M)$ be the set of gradient vector fields on M, that is, $X \in \mathrm{Grad}^r(M)$ if and only if there exist $f \in C^{r+1}(M)$ and a Riemannian metric g such that $X = \mathrm{grad}\, f$ in the metric g. Show that the set of Kupka–Smale vector fields is residual in $\mathrm{Grad}^r(M)$.

    Hint. Show that, if $X = \mathrm{grad}\, f$ in the metric g, F is a flow box for X and Y is a vector field near X that coincides with X on $M - F$, then there exists a metric $\tilde g$ such that Y is the gradient of f in the metric $\tilde g$. Also use Exercise 8 from Chapter 2.

11. (a) Show that, if $X \in \mathfrak{X}^r(M)$ is a Kupka–Smale vector field which has a global transversal section, then the associated Poincaré map is a Kupka–Smale diffeomorphism.

(b) Show that the suspension of a Kupka–Smale diffeomorphism is a Kupka–Smale vector field.

12. (a) Show that if $f \in \text{Diff}^r(M), r \geq 1$, is structurally stable then its periodic orbits are hyperbolic.
    (b) Show that if $f \in \text{Diff}^1(M)$ is structurally stable then f is a Kupka–Smale diffeomorphism. Notice that in [98], C. Robinson proved this result for $f \in \text{Diff}^r(M)$, $r \geq 2$.
    (c) Show that a structurally stable vector field $X \in \mathfrak{X}^1(M)$ is Kupka–Smale.

    Remark. It is an open problem to show that a structurally stable vector field $X \in \mathfrak{X}^r(M), r \geq 3$, has its closed orbits hyperbolic.

13. Let $X \in \mathfrak{X}^r(M)$ be a vector field with a closed orbit. Let X_t be the flow generated by X. Show that the diffeomorphism X_1 is not structurally stable.

14. Show that on any compact manifold M^n, $n \geq 1$, the set of Kupka–Smale diffeomorphisms is not open in $\text{Diff}^r(M), r \geq 1$.

    Hint. Show that any manifold has a Kupka–Smale vector field with closed orbits.

15. Show that on any compact manifold M^n, $n \geq 3$, the set of Kupka–Smale vector fields is not open in $\mathfrak{X}^r(M), r \geq 1$.

Chapter 4

Genericity and Stability of Morse–Smale Vector Fields

As we have emphasized before, the central objective of the Theory of Dynamical Systems is the description of the orbit structures of the vector fields on a differentiable manifold. There exist, however, fields with extremely complicated orbit structures as the example in Section 3 of Chapter 2 shows. Thus the strategy this programme must adopt is to restrict the study to a subset of the space of vector fields. It is desirable that this subset should be open and dense (or as large as possible) and that its elements should be structurally stable with simple enough orbit structures for us to be able to classify them. As far as the local aspect is concerned this problem is completely solved as we saw in Chapter 2.

In this chapter, in Sections 1 and 2, we show that the global aspect of the above programme can be achieved on compact manifolds of dimension two. This result, due to Peixoto [81], [85], is one of the early landmarks in the recent development of the theory. Besides the previous fundamental work of Andronov–Pontryagin, followed by that of De Baggis, on the disc D^2 or the sphere S^2 (see [5], [47]), we also mention that Pliss [87] obtained the same result as Peixoto's for vector fields without singularities on the torus T^2.

In higher dimensions the structurally stable fields are still plentiful, but they are not dense. There exist richer and more complicated phenomena that persist under small perturbations of the original field. Even for the stable fields, the orbit structures of the limit sets are not always completely understood and their description is still an active area of research. These facts are discussed in Sections 3 and 4.

In this context we should emphasize again the importance of studying generic properties, that is, properties satisfied by almost all (a Baire subset of) vector fields. This was the case with the Kupka–Smale Theorem in the last chapter.

Finally we remark that the above programme can be posed for certain subsets of the space of vector fields of special interest. One of the most relevant examples is the gradient vector fields on a compact manifold. In this case the structurally stable vector fields form an open dense subset [79], [106]. We shall indicate in Section 3 some of the basic properties of their orbit structures.

In Section 4 we present a collection of general results on structural stability. In particular, we exhibit structurally stable systems with infinitely many periodic orbits.

§1 Morse–Smale Vector Fields; Structural Stability

Here we define a class of vector fields which play an important role in the Theory of Dynamical Systems. This class, called Morse–Smale systems, forms a nonempty open subset and its elements are structurally stable. Although these results hold on compact manifolds of any dimension [75], [79], we shall only study in this chapter the case of dimension two, where the class is also dense.

We begin the section by defining Morse–Smale vector fields. Then we prove that a Morse–Smale field on M^2 is structurally stable. The proof we give was introduced in [75], is different from the original one [85] and can be generalized to higher dimensions.

Before presenting the definition of Morse–Smale vector fields formally we shall motivate it by means of some examples. As our aim is to find a class of structurally stable fields we must certainly require the singularities and closed orbits to be hyperbolic. Without this requirement the fields would not even be locally structurally stable. We also emphasize that the intersections of the stable and unstable manifolds of critical elements (singularities and closed orbits) have to be preserved by any topological equivalence. Therefore, it is natural to require these intersections to be transversal since this will guarantee that they persist under small perturbations of the vector field.

EXAMPLE 1. Consider the torus $T^2 \subset \mathbb{R}^3$ and let $X = \operatorname{grad} h$ where h is the height function of points of T^2 above the horizontal plane in Figure 1. This vector field has four singularities p_1, p_2, p_3, p_4 where p_1 is a sink, p_2 and p_3 are saddles and p_4 is a source. The stable manifold of p_2 intersects the unstable manifold of p_3 nontransversally. As in Chapter 1 we can destroy this intersection with a small perturbation of the field X. The resulting field Y will therefore not be equivalent to X.

The discussion so far leads us to define Morse–Smale fields as a subset of the Kupka–Smale fields. At this stage it is fundamental to notice that saying a vector field is Kupka–Smale gives no information about the α- and ω-limit sets of a general orbit. As a topological equivalence between vector fields preserves the α- and ω-limit sets of corresponding orbits we ought to impose some specific conditions on these limit sets.

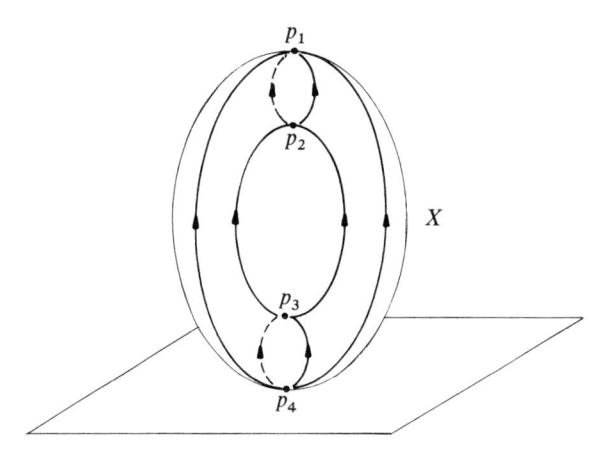

§1 Morse–Smale Vector Fields; Structural Stability 117

Figure 1

EXAMPLE 2. Consider the vector field X that induces an irrational flow on the torus T^2. The α- and ω-limit of any orbit of X is the whole torus T^2. In particular X does not have singularities or closed orbits. Consequently X is a Kupka–Smale field. However X is not structurally stable because it can be approximated by a field Y that induces a rational flow (see Section 4 of Chapter 1). All the orbits of Y are closed so this is a radical change in the α- and ω-limit sets.

Before defining the set of Morse–Smale fields we still need some new concepts and notation.

Let $X \in \mathfrak{X}^r(M)$. Consider the sets $L_\alpha(X) = \{p \in M; \, p \in \alpha(q) \text{ for some } q \in M\}$ and $L_\omega(X) = \{p \in M; \, p \in \omega(q) \text{ for some } q \in M\}$. These sets are invariant by the flow generated by X and the orbit of any point is "born" in L_α and "dies" in L_ω.

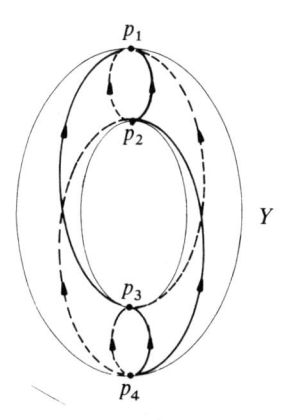

Figure 2

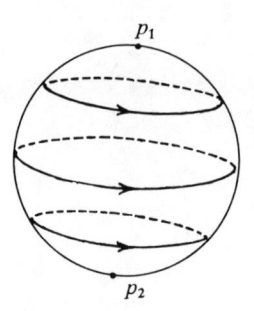

Figure 3

Definition. Let $X \in \mathfrak{X}^r(M)$. We say that $p \in M$ is a *wandering point* for X if there exists a neighbourhood V of p and a number $t_0 > 0$ such that $X_t(V) \cap V = \emptyset$ for $|t| > t_0$. Otherwise we say that p is *nonwandering*.

We write $\Omega(X)$ for the set of nonwandering points of X. The following properties follow immediately from this definition:

(a) $\Omega(X)$ is compact and invariant by the flow X_t;
(b) $\Omega(X) \supset L_\alpha(X) \cup L_\omega(X)$. In particular, $\Omega(X)$ contains the critical elements of X;
(c) if $X, Y \in \mathfrak{X}^r(M)$ and $h: M \to M$ is a topological equivalence between X and Y then $h(\Omega(X)) = \Omega(Y)$.

The next example shows that, in general, Ω contains $L_\alpha \cup L_\omega$ strictly.

EXAMPLE 3. Consider a C^∞ vector field X on S^2 with two singularities p_1 and p_2 such that all the other orbits are closed, see Figure 3. We now multiply the field X by a nonnegative C^∞ function $\varphi: S^2 \to \mathbb{R}$ that only vanishes at one point p where p is distinct from p_1 and p_2. Let $Y = \varphi X$. The field Y is C^∞, has three singularities p_1, p_2 and p and all other orbits are closed except for one orbit γ with $\alpha(\gamma) = \omega(\gamma) = p$. No point x of γ belongs to $L_\alpha \cup L_\omega$ but $x \in \Omega$ since it is accumulated by closed orbits of Y. See Figure 4.

Definition. Let M be a compact manifold of dimension n and let $X \in \mathfrak{X}^r(M)$. We say that X is a *Morse–Smale vector field* if:

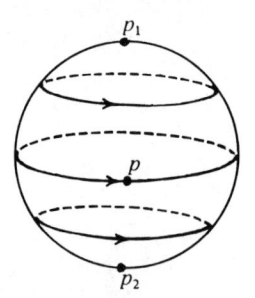

Figure 4

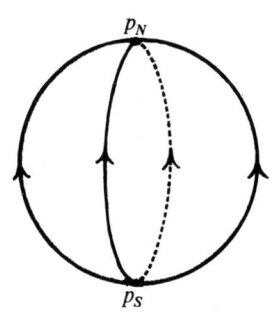

§1 Morse–Smale Vector Fields; Structural Stability   119

Figure 5

(1) X has a finite number of critical elements (singularities and closed orbits) all of which are hyperbolic;
(2) if σ_1 and σ_2 are critical elements of X then $W^s(\sigma_1)$ is transversal to $W^u(\sigma_2)$;
(3) $\Omega(X)$ is equal to the union of the critical elements of X.

Next we give some examples of Morse–Smale fields. Examples 4, 5, 6 and 7 are on S^2, and Examples 8 and 9 are on T^2. A critical element is called an *attractor* or *repellor* if its index is the maximum possible or zero respectively. Otherwise it is called a *saddle*.

EXAMPLE 4. Any vector field on S^2 with the following characteristics is called a north pole–south pole field:

p_N, p_S are hyperbolic singularities;

p_N is an attractor;

p_S is a repellor;

if $x \in S^2 - \{p_N, p_S\}$ then $\omega(x) = p_N$ and $\alpha(x) = p_S$.

EXAMPLE 5. p_N, p_S are hyperbolic repelling singularities, γ is a hyperbolic attracting closed orbit and if $x \in S^2 - \{p_N, p_S\} - \gamma$ then $\omega(x) = \gamma$ and $\alpha(x) = p_N$ or p_S.

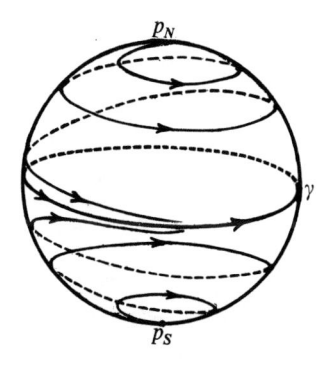

Figure 6

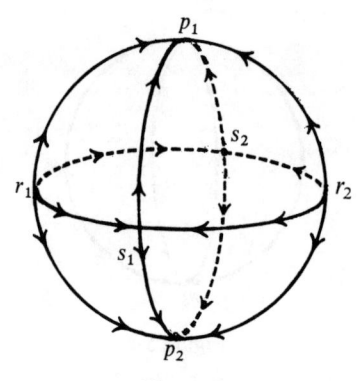

Figure 7

EXAMPLE 6. p_1, p_2 are hyperbolic attractors, r_1, r_2 are hyperbolic repellors, s_1, s_2 are hyperbolic saddles and $\Omega(X) = \{p_1, p_2, r_1, r_2, s_1, s_2\}$.

EXAMPLE 7. r_1, r_2, r_3 are hyperbolic repellors, s is a hyperbolic saddle, γ_1, γ_2 are hyperbolic attracting closed orbits and $\Omega(X) = \{r_1, r_2, r_3, s\} \cup \gamma_1 \cup \gamma_2$.

The orbits of this field on the cylinder bounded by γ_1 and γ_2 are as in Figure 9.

EXAMPLE 8. p is a hyperbolic attractor, r is a hyperbolic repellor, s_1, s_2 are hyperbolic saddles and $\Omega(X) = \{p, r, s_1, s_2\}$. See Figure 10.

EXAMPLE 9. γ_1 is an attracting hyperbolic closed orbit, γ_2 is a repelling hyperbolic closed orbit and $\Omega(X) = \gamma_1 \cup \gamma_2$. See Figure 11.

We shall write M–S for the set of Morse–Smale fields. The following proposition gives a simpler characterization of Morse–Smale fields on two-dimensional manifolds. A *saddle-connection* is an orbit whose α- and ω-limits are saddles.

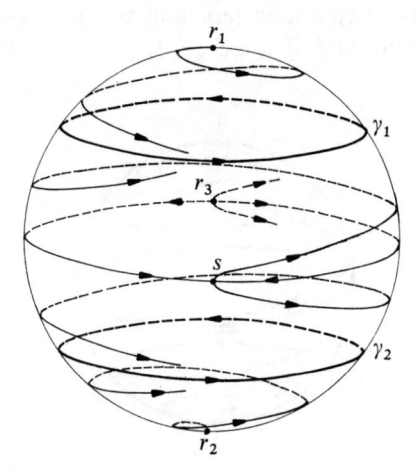

Figure 8

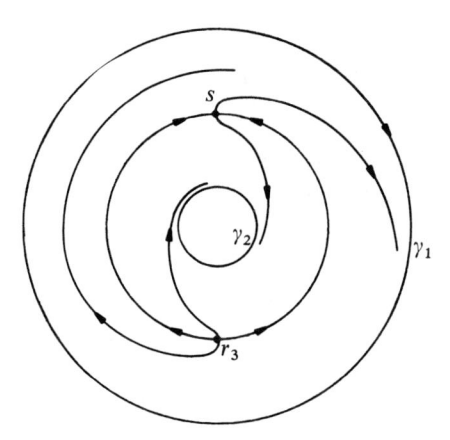

§1 Morse–Smale Vector Fields; Structural Stability   121

Figure 9

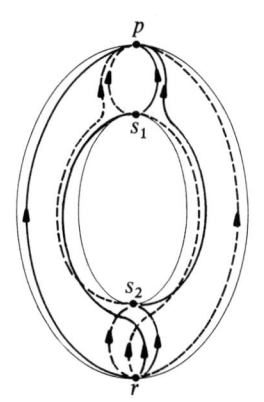

Figure 10

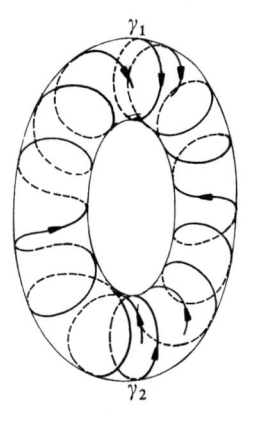

Figure 11

1.1 Proposition. *Let M be a compact manifold of dimension two. A vector field $X \in \mathfrak{X}^r(M)$ is Morse–Smale if and only if:*

(a) *X has a finite number of critical elements, all hyperbolic;*
(b) *there are no saddle-connections; and*
(c) *each orbit has a unique critical element as its α-limit and has a unique critical element as its ω-limit.*

PROOF. Clearly, if $X \in$ M–S then X satisfies conditions (a), (b) and (c) above. Let us show the converse. Take $X \in \mathfrak{X}^r(M)$ satisfying (a), (b) and (c). As the stable manifold of a sink and the unstable manifold of a source are two-dimensional, the transversality condition can only be broken by the stable and unstable manifolds of saddles. This does not happen because there are no saddle-connections. Thus, it is sufficient to prove that $\Omega(X)$ consists of the critical elements.

First let us show that the stable manifold of a sink consists of wandering points except for the sink itself. Suppose that the sink is a singularity p. As we have already seen there exists a disc $D \subset W^s(p)$ containing p, whose boundary is a circle C transversal to X. As $W^s(p) - \{p\} = \bigcup_{t \in \mathbb{R}} X_t(C)$ and the set of wandering points is invariant, it is enough to show that the points of C are wandering. Consider the discs $D_1 = X_1(D)$ contained in the interior of D and $D_{-1} = X_{-1}(D)$ whose interior contains D. Take $x \in C$. Let V be a neighbourhood of x disjoint from D_1 and $M - D_{-1}$. Then $X_t(V) \cap V = \emptyset$ for $|t| > 2$. This proves that x is wandering. Now suppose that the sink is a closed orbit γ. In this case there also exists a neighbourhood U of γ whose boundary S is transversal to X. If γ is an orientable curve then U is homeomorphic to an annulus and S is the disjoint union of two circles. If γ is not orientable then U is homeomorphic to a Möbius band and S is a circle. Moreover, $W^s(\gamma) - \gamma = \bigcup_{t \in \mathbb{R}} X_t(S)$. By the same argument as before S, and therefore $W^s(\gamma) - \gamma$ too, consists of wandering points. In an entirely analogous manner we can show that $W^u(\sigma) - \sigma$ consists of wandering points when σ is a repelling critical element. Finally, if $x \in M$ is not a singularity and does not belong to a closed orbit then either $\omega(x)$ or $\alpha(x)$ is a repellor, since there are no saddle-connections. Thus, x is wandering, which completes the proof. □

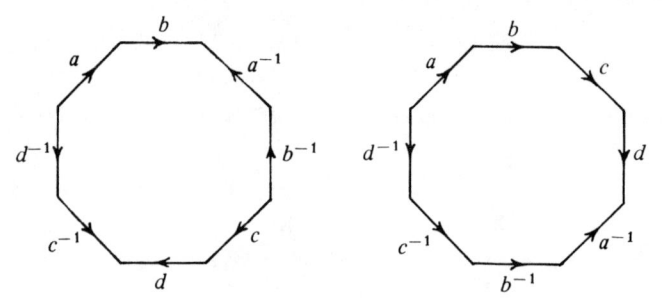

Figure 12

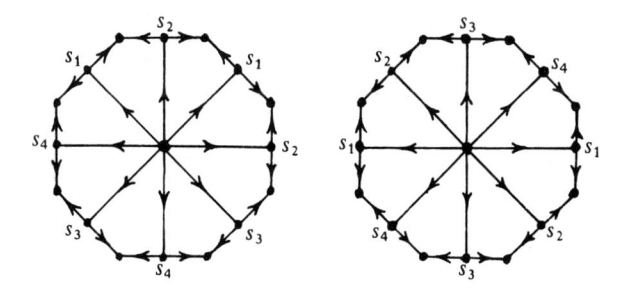

§1 Morse–Smale Vector Fields; Structural Stability 123

Figure 13

EXAMPLE 10. Polar fields on the pretzel.

We present some examples of polar Morse–Smale vector fields on the pretzel, that is, vector fields without closed orbits and with just one source and one sink. The pretzel (or the "torus with two holes" or sphere with two handles) can be represented [59] by an octagon with its edges identified in pairs in the following way:

(1) two edges to be identified do not have a vertex in common;
(2) the diffeomorphism that identifies the two edges reverses orientation;
(3) the vertices are all identified to one point.

In Figure 12 we represent the pretzel in two ways.

Given a representation of the pretzel as an octagon we can construct a polar Morse–Smale field by putting a source at the centre of the octagon, a saddle at the mid-point of each edge and one sink at the vertices. In Figure 13 we sketch the Morse–Smale fields on the pretzel corresponding to its representations above. Conversely, let X be a polar Morse–Smale field on the pretzel. By cutting the pretzel along the unstable manifolds of the saddles we obtain a field like the one described above.

Similar constructions enable us to exhibit polar Morse–Smale fields on any compact manifold of dimension two.

Definition. Given a Morse–Smale field X we define the *phase diagram* Γ of X to be the set of critical elements of X with the following partial order: $\sigma_1, \sigma_2 \in \Gamma$, $\sigma_1 \leq \sigma_2$ if $W^u(\sigma_1) \cap W^s(\sigma_2) \neq \emptyset$. That is, there exists an orbit which is born in σ_1 and dies in σ_2.

The relation $\leq$ is a partial order since there are no saddle connections. We may also remark that, as dim $M = 2$, the phase diagram of any Morse–Smale field has, at most, three levels.

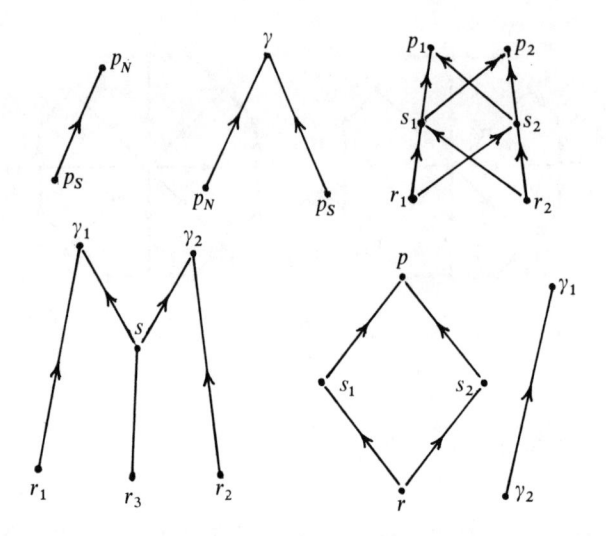

Figure 14

EXAMPLES. The phase diagrams for Examples 4 to 9 of this section are shown in Figure 14.

Definition. Let X, $\tilde{X} \in$ M–S and let Γ, $\tilde{\Gamma}$ be their phase diagrams. We say that Γ and $\tilde{\Gamma}$ are *isomorphic* if there exists a one to one correspondence $h: \Gamma \to \tilde{\Gamma}$ such that

(a) $x \in \Gamma$ is a singularity if and only if $h(x) \in \tilde{\Gamma}$ is a singularity;
(b) for $x_1, x_2 \in \Gamma$, $x_1 \leq x_2$ if and only if $h(x_1) \leq h(x_2)$.

We are going to show that small perturbations of a Morse–Smale field give rise to Morse–Smale fields with isomorphic phase diagrams. For this we shall use the concept of a filtration associated to a Morse–Smale field. We observe that such a concept can be usefully applied to a more general kind of vector field [14], [109].

Definition. Let $X \in$ M–S. A *filtration* for X is a sequence $M_0 = \emptyset$, $M_1 \subset M_2 \subset \cdots \subset M_k = M$ of compact submanifolds M_i (with boundary for $0 < i < k$) such that:

(a) X is transversal to the boundary of M_i and $X_t(M_i) \subset$ Interior M_i for $t > 0$;
(b) in $M_{i+1} - M_i$, the maximal invariant set of the flow X_t is just one critical element σ_{i+1}, that is, $\bigcap_{t \in \mathbb{R}} X_t(M_{i+1} - \text{int } M_i) = \sigma_{i+1}$.

1.2 Lemma. *Let $X \in \mathfrak{X}^r(M^2)$ be a Morse–Smale field. Then there exists a filtration for X.*

PROOF. Let $\sigma_1, \sigma_2, \ldots, \sigma_j$ be the attractors of X. Let us take disjoint neighbourhoods $V_1, V_2, \ldots, V_j$ with boundaries transversal to X as in the proof of Proposition 1.1. We define $M_1 = V_1$, $M_2 = M_1 \cup V_2, \ldots, M_j = M_{j-1} \cup V_j$.

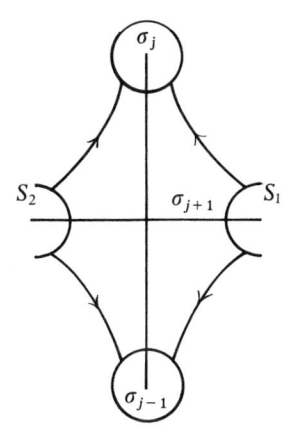

§1 Morse–Smale Vector Fields; Structural Stability

Figure 15

Let $\sigma_{j+1}, \ldots, \sigma_s$ be the saddles of X. Let us consider σ_{j+1} and the components of $W^u(\sigma_{j+1}) - \sigma_{j+1}$ (which intersect ∂M_j transversally). In a neighbourhood of σ_{j+1} we construct two sections, S_1 and S_2, transversal to $W^s(\sigma_{j+1}) - \sigma_{j+1}$. If this neighbourhood is small enough then the trajectories of X through the end points of S_1 and S_2 cut ∂M_j transversally. Near these arcs of trajectories we construct curves c_1, c_2, c_3 and c_4 transversal to X joining the end-points of S_1 and S_2 to ∂M_j. We can construct these curves to be tangent to the submanifolds S_1, S_2 and ∂M_j, by using tubular flows containing these arcs of trajectories as indicated in Figure 16.

Let V_{j+1} be the region containing σ_{j+1} and bounded by $S_1, S_2, c_1, c_2, c_3, c_4$ and part of ∂M_j. Put $M_{j+1} = M_j \cup V_{j+1}$. It is easy to check that $\bigcap_{t \in \mathbb{R}} X_t(V_{j+1}) = \sigma_{j+1}$ and that M_{j+1} satisfies the required conditions. We repeat the construction for each saddle and thus obtain the sequence of submanifolds $\varnothing = M_0 \subset M_1 \subset \cdots \subset M_{j+1} \subset \cdots \subset M_s$. Finally, let $\sigma_{s+1}, \ldots, \sigma_k$ be the sources of X. We consider neighbourhoods $V_{s+1}, V_{s+2}, \ldots, V_k$ of these sources with their boundaries transversal to X, as in the case of the sinks. Then we define

$$M_{s+1} = M - (\text{int } V_{s+2} \cup \cdots \cup \text{int } V_k),$$
$$M_{s+2} = M - (\text{int } V_{s+3} \cup \cdots \cup \text{int } V_k),$$

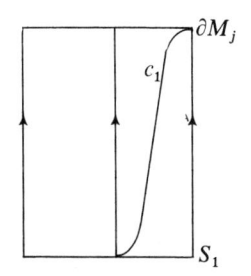

Figure 16

[Figure 17]

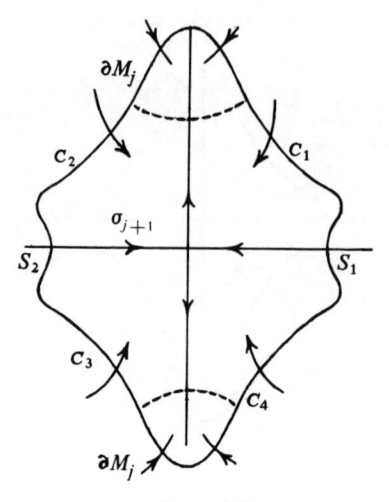

Figure 17

and so on up to $M_k = M$. It is easy to check that $\emptyset = M_0 \subset M_1 \subset \cdots \subset M_k$ is a filtration for X. □

The next two theorems are true in higher dimensions [75], [79]. We have adapted their proofs to the much simpler two-dimensional case.

1.3 Theorem. *Let $X \in$ M–S. Then there exists a neighbourhood $\mathscr{U}$ of X in $\mathfrak{X}^r(M^2)$ such that if $Y \in \mathscr{U}$ then $Y \in$ M–S and its phase diagram is isomorphic to that of X.*

PROOF. As the critical elements of X are hyperbolic, for each $\sigma_i \in \Omega(X)$ there exist neighbourhoods $\mathscr{U}_i \subset \mathfrak{X}^r(M^2)$ of X and U_i of σ_i such that any $Y \in \mathscr{U}_i$ has a unique critical element $\sigma_i(Y) \subset U_i$. Moreover, by the Grobman–Hartman Theorem we can, by shrinking the neighbourhoods $\mathscr{U}_i$, U_i if necessary, suppose that $\sigma_i(Y)$ is the unique set invariant under the flow Y_t that is entirely contained in U_i. Put $\mathscr{U} = \bigcap_i \mathscr{U}_i$. Now let us consider a filtration $\emptyset = M_0 \subset M_1 \subset M_2 \subset \cdots \subset M_k = M$ for X. We shall show that, for $\mathscr{U}$ small enough, $\emptyset = M_0 \subset M_1 \subset \cdots \subset M_k = M$ is also a filtration for any $Y \in \mathscr{U}$. This will imply that $\Omega(Y)$ consists of the critical elements $\sigma_i(Y)$ defined above. First, we remark that, as X is transversal to the compact set ∂M_i, so is any Y near X. Now shrink the neighbourhoods U_i of σ_i until $U_i \subset M_i - M_{i-1}$. As $\bigcap_{t \in \mathbb{R}} X_t(M_i - \text{int } M_{i-1}) = \sigma_i$, there exists $T > 0$ such that $\bigcap_{t=-T}^{T} X_t(M_i - \text{int } M_{i-1}) \subset U_i$. The same fact holds for $Y \in \mathscr{U}$ if $\mathscr{U}$ is small enough. As $\bigcap_{t \in \mathbb{R}} Y_t(U_i) = \sigma_i(Y)$, it follows that $\bigcap_{t \in \mathbb{R}} Y_t(M_i - \text{int } M_{i-1}) = \sigma_i(Y)$. Hence $\emptyset = M_0 \subset M_1 \subset \cdots \subset M_k = M$ is a filtration for all $Y \in \mathscr{U}$. We claim that $\Omega(Y) \cap (M_i - \text{int } M_{i-1}) = \sigma_i(Y)$. In fact, any orbit γ distinct from $\sigma_i(Y)$ through a point of $M_i - \text{int } M_{i-1}$ must intersect ∂M_i or ∂M_{i-1} (or both). This is because the only orbit entirely

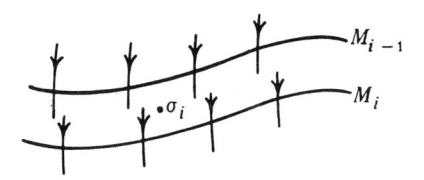

§1 Morse–Smale Vector Fields; Structural Stability

Figure 18

contained in $M_i - \text{int } M_{i-1}$ is $\sigma_i(Y)$. As $Y_t(M_{i-1}) \subset M_{i-1}$ for $t > 0$ and $Y_t(M - M_i) \subset M - M_i$ for $t < 0$, γ is a wandering orbit. Thus, $\sigma_i(Y)$ is the only orbit of $\Omega(Y)$ in $M_i - \text{int } M_{i-1}$. Therefore, $\Omega(Y) = \bigcup_i \sigma_i(Y)$ and $\sigma_i(Y)$ is hyperbolic for each i.

In order to conclude that $Y \in \mathcal{U}$ is a Morse–Smale field it is enough to show that, for small $\mathcal{U}$, there are no saddle-connections. Thus, let $\sigma_i = \sigma_i(X)$ be a saddle and suppose that one component γ of $W^u(\sigma_i) - \sigma_i$ has the sink $\sigma = \sigma(X)$ as its ω-limit. Let V be a neighbourhood of σ as in the construction of the filtration. As compact parts of $W^u(\sigma_i(Y))$ are near $W^u(\sigma_i)$, one component of $W^u(\sigma_i(Y)) - \sigma_i(Y)$ also intersects ∂V transversally. Thus, its ω-limit is $\sigma(Y)$. The same reasoning applies to all the components of stable and unstable manifolds of saddles. Thus, for small enough $\mathcal{U}$, if $Y \in \mathcal{U}$ then $Y \in$ M–S and the above correspondence $\sigma_i(X) \mapsto \sigma_i(Y)$ is an isomorphism of phase diagrams. This proves the theorem. □

1.4 Theorem. *If $X \in \mathfrak{X}^r(M^2)$ is a Morse–Smale field then X is structurally stable.*

PROOF. By the previous theorem we know there exists a neighbourhood $\mathcal{U} \subset \mathfrak{X}^r(M^2)$ of X such that if $Y \in \mathcal{U}$ then $Y \in$ M–S and there exists an isomorphism $\sigma_i(X) \mapsto \sigma_i(Y)$ of phase diagrams.

Part 1. Let us suppose initially that X has no closed orbits. Consider a sink σ of X and the corresponding sink $\sigma(Y)$ with $Y \in \mathcal{U}$. Let V be a disc in $W^s(\sigma)$ containing σ as before. That is, ∂V is transversal to X and to all $Y \in \mathcal{U}$. Also $\sigma(Y) \in V \subset W^s(\sigma(Y))$. Let $\sigma_1, \sigma_2, \ldots$ be the saddles of X such

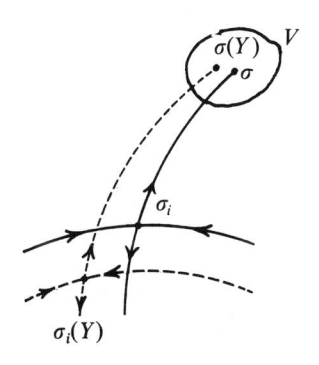

Figure 19

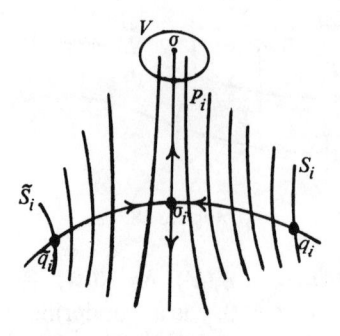

Figure 20

that $\sigma_i \leq \sigma$. Let $p_1, p_2, \ldots$ be the points at which the unstable separatrices of the saddles σ_i (that is, the components of $W^u(\sigma_i) - \sigma_i$) intersect ∂V. Let $p_1(Y), p_2(Y), \ldots$ be the corresponding points for Y. For each σ_i let us consider sections S_i, $\tilde{S}_i$ transversal to the stable separatrices of σ_i through points q_i, $\tilde{q}_i$ as in Figure 20.

Saturating S_i, $\tilde{S}_i$ by the flow X_t we obtain a tubular family for $W^u(\sigma_i)$ as in Section 7 of Chapter 2. The fibres of this family are $X_t(S_i)$ and $X_t(\tilde{S}_i)$ for each $t \in \mathbb{R}$ and also $W^u(\sigma_i)$. The projection π_i, which associates the point $f \cap W^s(\sigma_i)$ to each fibre f, is continuous. Moreover π_i is a homeomorphism from a neighbourhood I_i of p_i in ∂V to a neighbourhood of σ_i in $W^s(\sigma_i)$. We make the same construction for the field Y. Now we begin to define the topological equivalence h between X and Y. We put $h(\sigma) = \sigma(Y)$, $h(\sigma_i) = \sigma_i(Y)$, $h(p_i) = p_i(Y)$, $h(q_i) = q_i(Y)$ and $h(\tilde{q}_i) = \tilde{q}_i(Y)$. We extend h to $W^s(\sigma_i)$ by the equation $hX_t(q_i) = Y_t h(q_i) = Y_t q_i(Y)$, $hX_t(\tilde{q}_i) = Y_t \tilde{q}_i(Y)$. Now we define h on I_i in the following way: for $x \in I_i$, $hx = [\pi_i(Y)]^{-1} h\pi_i x$. In this way h has been defined on a finite number of disjoint intervals I_i in ∂V. Notice that if the neighbourhood $\mathcal{U}$ of X is small then $h|I_i$ is near the identity. Thus, we can extend h to the whole circle ∂V. We repeat the same construction for all the sinks. Finally, we define h on the whole of M^2 by the equation $hX_t z =$

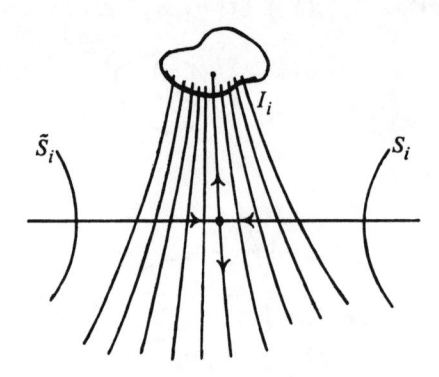

Figure 21

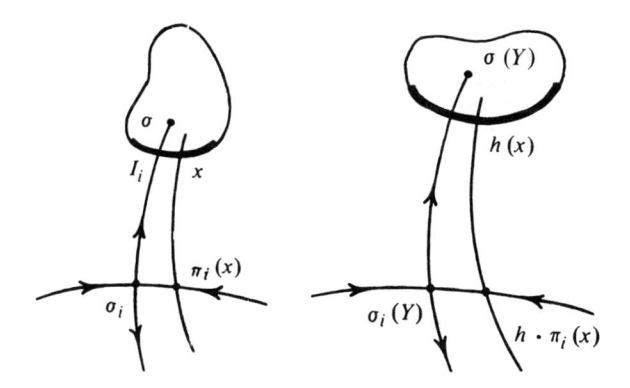

§1 Morse–Smale Vector Fields; Structural Stability

Figure 22

$Y_t hz$. It is easy to see from the construction that h has an inverse h^{-1}, which can be defined exactly as h was by interchanging the roles of X and Y.

Thus it remains to prove the continuity of h. This is obvious at the sinks and sources and on the stable manifolds of the sinks. We shall analyse the case of the stable manifolds of the saddles. Take $x \in W^s(\sigma_i)$ where σ_i is a saddle. Recall that h takes fibres of the tubular family for σ_i to fibres of the tubular family for $\sigma_i(Y)$, that is, $\pi_i(Y)hz = h\pi_i(X)z$. Consider any sequence $x_n \to x$. We want to show that $hx_n \to hx$. By the above remark, the fibre through hx_n converges to the fibre through hx. That is, $\pi_i(Y)hx_n \to hx$. It remains to prove that hx_n converges to $W^s(\sigma_i(Y))$. For this we construct tubular families for $W^s(\sigma_i)$ and $W^s(\sigma_i(Y))$. This is done by starting from segments I_i, $I_i(Y)$ in ∂V and saturating them by the flows X_t and Y_t. As $h(I_i) = I_i(Y)$, we see that h takes fibres of the tubular family of $W^s(\sigma_i)$ to fibres of the tubular family of $W^s(\sigma_i(Y))$. Therefore, if $\tilde{\pi}_i$ and $\tilde{\pi}_i(Y)$ are the respective projections onto $W^u(\sigma_i)$ and $W^u(\sigma_i(Y))$ then $h\tilde{\pi}_i(z) = \tilde{\pi}_i(Y)h(z)$. As $\tilde{\pi}_i$ is continuous and $x_n \to x \in W^s(\sigma_i)$ we see that $\tilde{\pi}_i x_n \to \tilde{\pi}_i x = \sigma_i$. Also, the restriction of h to $W^u(\sigma_i)$ is continuous, so that $h(\tilde{\pi}_i x_n) \to h(\sigma_i) = \sigma_i(Y)$.

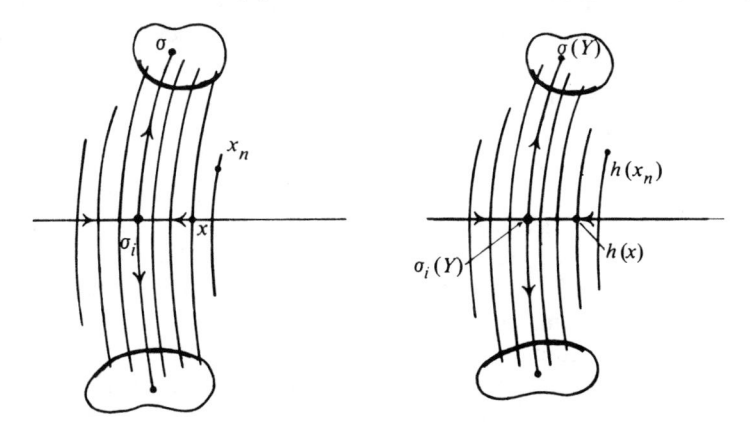

Figure 23

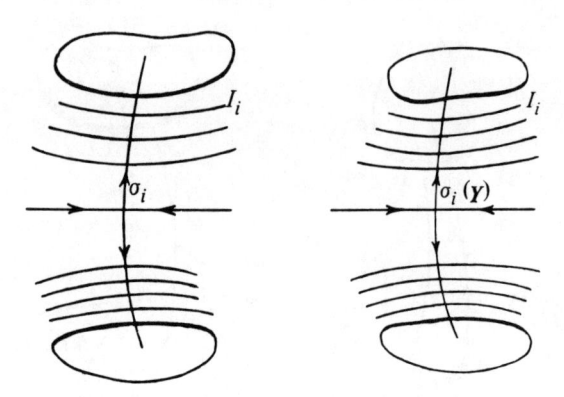

Figure 24

On the other hand, $h(\tilde{\pi}_i(x_n)) = \tilde{\pi}_i(Y)h(x_n)$, and therefore $\tilde{\pi}_i(Y)h(x_n) \to \sigma_i(Y)$. This shows that $h(x_n)$ converges to the stable manifold of $\sigma_i(Y)$, hence $h(x_n) \to h(x)$. This completes the proof in the case in which X has no closed orbit.

Part 2. Now suppose that X does have closed orbits. These closed orbits must be attractors or repellors because they are hyperbolic and dim $M = 2$. As we have already remarked, there exist fields Y arbitrarily close to X such that the flows X_t and Y_t are not conjugate. For this it is enough to alter the period of one of the closed orbits of X by a small perturbation. We shall avoid this difficulty by defining a conjugacy h between flows $\tilde{X}_t$ and $\tilde{Y}_t$ that are reparametrizations of X_t and Y_t. As the orbits of X_t and $\tilde{X}_t$ are the same and so are the orbits of Y_t and $\tilde{Y}_t$, it follows that h will be an equivalence between the fields X and Y.

Using Lemma 1.3 of Chapter 3 we can suppose right from the beginning that the closed orbits of X and Y all have the same period τ and admit invariant transversal sections. To simplify the exposition we shall consider two subcases.

(2.a) Consider, first, the case in which all the closed orbits are attractors. We shall try to imitate the construction of the conjugacy made in the case where there were no closed orbits. Around each attracting singularity σ_i, $\sigma_i(Y)$ we consider a circle C_i transversal to X and Y. For each closed orbit σ_j, $\sigma_j(Y)$ we take an invariant transversal section Σ_j and fundamental domains I_j, $I_j(Y)$ in Σ_j for the associated Poincaré maps. As before we construct unstable tubular families associated to the saddles σ_k, $\sigma_k(Y)$ of X and Y: we take sections S_k and $\tilde{S}_k$ transversal to $W^s(\sigma_k)$ and $W^s(\sigma_k(Y))$ and use the families $X_t(S_k)$, $X_t(\tilde{S}_k)$ and $Y_t(S_k)$, $Y_t(\tilde{S}_k)$. The homeomorphism we want to construct will have to take each fibre of the tubular family of σ_k to a fibre of the tubular family of $\sigma_k(Y)$. Moreover, it will preserve the transversal circles C_i and the transversal sections Σ_j. Thus, by defining a conjugacy between $X|W^s(\sigma_k)$ and $Y|W^s(\sigma_k(Y))$ for each saddle σ_k, we shall induce a homeomorphism h on a finite number of subintervals of C_i and I_j. These subintervals contain the intersections of the unstable manifolds of the saddles

§1 Morse–Smale Vector Fields; Structural Stability 131

with C_i and I_j. This homeomorphism is near the identity, if Y is near X, and can, therefore, be extended to all of C_i and I_j. For each singularity σ_l, we define $h(\sigma_l) = \sigma_l(Y)$ and, for each closed orbit σ_j, we define $h(\Sigma_j \cap \sigma_j) = \Sigma_j \cap \sigma_j(Y)$. Finally we extend h to the whole of M using the conjugacy equation $h = Y_t h X_{-t}$ as in the first part of the proof. It follows, then, that h is one to one and onto. The continuity of h at the singularities and the stable manifolds of the saddles can be checked as in the first part. The continuity of h at the closed orbits follows from the invariance of the sections Σ_j as we saw in the local stability of hyperbolic closed orbits (Section 1 of Chapter 3).

(2.b) Finally let us suppose that X has an attracting closed orbit and a repelling closed orbit. This case becomes entirely analogous to the previous one after a further reparametrization of the fields X and Y. This reparametrization is necessary to enable us to extend the homeomorphism constructed in (2.a) to the repelling closed orbits. For this let us take transversal invariant sections $\tilde{\Sigma}_l$ associated to the repelling closed orbits $\tilde{\sigma}_l$ and let $\tilde{I}_l$ be the corresponding fundamental domains. Each $\tilde{I}_l$ decomposes as a union of closed subintervals whose end-points interior to $\tilde{I}_l$ are the intersections of the stable manifolds of the saddles with int $\tilde{I}_l$. Notice that all the points in each of these open subintervals have the same attractor as ω-limit, as in Figure 25. Let $p \in W^s(\sigma_k) \cap \tilde{I}_l$ be one of the end-points of one of these subintervals. Let us consider a small interval $[a, b] \subset \tilde{I}_l$ around p such that the orbit through every point of $[a, b]$ intersects the transversal section S_k. Using Lemma 1.3 of Chapter 3, we make a reparametrization of X in such a way that all the points of $[a, b]$ reach S_k at the same time 1. Let X also denote this reparametrized field. Let $[a', b'] \subset (a, b)$ be an interval containing p. By a new reparametrization we can ensure that all points of $[X_1(a), X_1(a')] \subset S_k$ reach Σ_j at time 1. Similarly we can do the same for $[X_1(b), X_1(b')]$ reaching C_i. Thus $X_2[a, a'] \subset \Sigma_j$ and $X_2[b, b'] \subset C_i$. We repeat this construction for

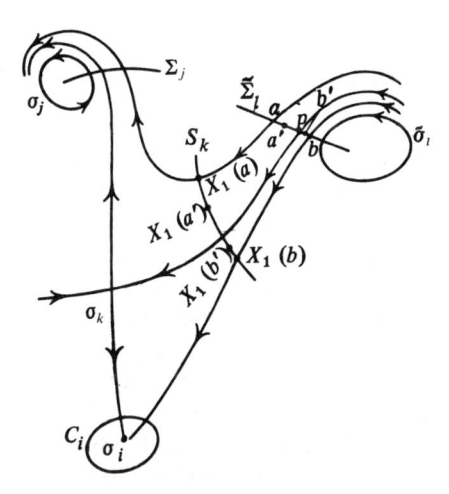

Figure 25

the various saddles whose stable manifolds intersect $\tilde{I}_l$. Finally we reparametrize X so that all the points in the complement in $\tilde{I}_l$ of the union of the intervals $[a, b]$ above reach the various sections C_i and Σ_j at the same time $t = 2$.

We also make the same reparametrizations for fields Y near X. The conjugacy h between X_t and Y_t is now constructed exactly as in (2.4) with the extra requirement $h(\tilde{\Sigma}_l \cap \tilde{\sigma}_l) = \tilde{\Sigma}_l \cap \tilde{\sigma}_l(Y)$. □

§2 Density of Morse–Smale Vector Fields on Orientable Surfaces

In this section we show that M–S is dense in $\mathfrak{X}^r(M^2)$ for an orientable surface M^2. We use the Kupka–Smale Theorem, which permits some simplification of Peixoto's original proof [81], although we must remark that Peixoto's work came earlier and served as motivation for that theorem. At the end of the section we analyse the case where M^2 is nonorientable and discuss the corresponding results for diffeomorphisms.

We begin the section by proving the density theorem for the sphere S^2, which is much simpler and yet illustrates the general case.

Definition. Let γ be an orbit of $X \in \mathfrak{X}^r(M)$. We say that γ is *recurrent* if $\omega(\gamma) \supset \gamma$ or $\alpha(\gamma) \supset \gamma$.

A critical element of X is always recurrent. In this case we say that the recurrent orbit is *trivial*. Any orbit of the irrational flow on the torus is recurrent and nontrivial.

By the Poincaré–Bendixson Theorem every recurrent orbit of a vector field $X \in \mathfrak{X}^r(S^2)$ is trivial. This fact simplifies considerably the proof that M–S is dense in $\mathfrak{X}^r(S^2)$.

2.1 Theorem. *If $X \in \mathfrak{X}^r(S^2)$ is a Kupka–Smale vector field then X is a Morse–Smale field.*

PROOF. As X is Kupka–Smale it has a finite number of singularities, all hyperbolic. By the Poincaré–Bendixson theorem the ω- and α-limit of any orbit is a singularity or a closed orbit. This is because if the ω-limit of an orbit γ contains more than one singularity then these singularities must be saddles and $\omega(\gamma)$ must also contain a regular orbit joining these saddles. As X is Kupka–Smale it has no saddle-connections, which proves the above statement. The closed orbits are hyperbolic attractors or repellors, so it remains to prove that there is only a finite number of them. To get a contradiction suppose X has infinitely many closed orbits. Let $x_1, x_2, \ldots, x_n, \ldots$ be a sequence of points in distinct closed orbits. By taking a subsequence we can suppose that x_n converges to some $x \in S^2$. Clearly, $\omega(x)$ is a saddle since there cannot be a closed orbit in the stable manifold of an attracting singu-

§2 Density of Morse–Smale Vector Fields on Orientable Surfaces

larity or closed orbit. Similarly, $\alpha(x)$ is a saddle. Thus x is itself a saddle for otherwise its orbit would be a saddle-connection. On the other hand, the ω-limits of the unstable separatrices of x are sinks because, again, there are no saddle-connections. This gives a contradiction because an orbit through a point close to x but not in its stable separatrices has one of these sinks as ω-limit. Therefore, x cannot be accumulated by closed orbits, which proves the theorem. ☐

As the set of Kupka–Smale fields is dense in $\mathfrak{X}^r(M)$ we have the next corollary.

Corollary. M–S *is dense in* $\mathfrak{X}^r(S^2)$. ☐

We shall follow the same line of argument in proving that M–S is dense in $\mathfrak{X}^r(M^2)$. However, the proof has to be more delicate because of the presence of nontrivial recurrent orbits. The irrational flow on the torus provides the simplest example of nontrivial recurrence. We suggest that the reader tries to show that the vector field generating an irrational flow can be approximated by a Morse–Smale field. Let us consider some examples of vector fields exhibiting nontrivial recurrence on other two-dimensional manifolds.

EXAMPLE 11. We shall give a vector field on the pretzel (or sphere with two handles) that has nontrivial recurrence. The following construction can be generalized to give examples of vector fields with nontrivial recurrence on the sphere with r handles for $r > 2$. Let $\lambda: \mathbb{R}^3 \to \mathbb{R}^3$ be the reflection in the plane $x_3 = 0$, that is, $\lambda(x_1, x_2, x_3) = (x_1, x_2, -x_3)$. Let us consider the torus T^2 embedded in $\mathbb{R}^3$ in such a way that $\lambda(T^2) = T^2$ and $T^2 \cap \{x; x_3 = 0\}$ is the union of two circles.

Let X be the gradient field of the height function measured above the plane $x_3 = 0$. Clearly, X is a symmetric field, that is, $\lambda_* X = -X$. The singularities of X are the source r, the sink a and the saddles s_1 and s_2 as in Figure 26. Consider a circle $C_1 \subset T^2$ orthogonal to X and bounding a disc

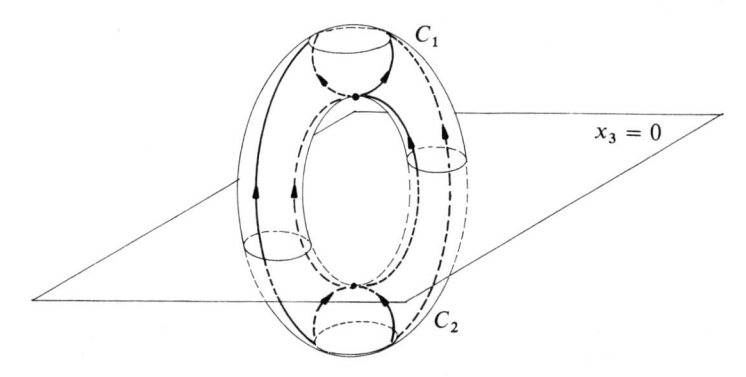

Figure 26

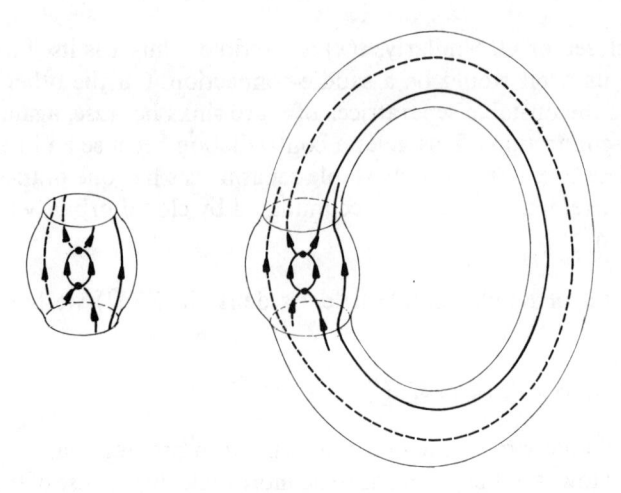

Figure 27

D_1 that contains the sink. Let $C_2 = \lambda C_1$ and then $D_2 = \lambda(D_1)$ is the disc bounded by C_2. By the symmetry of the field, the orbit through any point $p \in C_2$ (not in a separatrix of a saddle) intersects C_1 in the point $q = \lambda(p)$.

Let $h: C_1 \to C_2$ be the diffeomorphism defined by $h = R_\alpha \circ \lambda$, where R_α is an irrational rotation of C_2. On $T^2 - (D_1 \cup D_2)$ consider the equivalence relation that identifies C_1 and C_2 according to h. Let M be the quotient manifold and $P: T^2 - (D_1 \cup D_2) \to M$ the canonical projection. Let $Y = P_* X$. Clearly, M is diffeomorphic to the pretzel. Moreover, all the orbits of Y are dense except for the two saddles s_1 and s_2 and the unstable separatrices of s_1. In fact all the other orbits intersect $P(C_2)$ and so it suffices to prove that the intersection of each orbit with $P(C_2)$ is dense in $P(C_2)$. Let $p \in P(C_2)$ and take $q \in C_2$ such that $P(q) = p$. The orbit of X through the point q intersects C_1 in the point $\lambda(q)$ but this point is identified with $h\lambda(q) = R_\alpha(q)$. Thus, the positive orbit through p intersects $P(C_2)$ for the first time at the point $PR_\alpha(q)$. By induction we see that the positive orbit through p intersects $P(C_2)$ for the nth time at the point $PR_\alpha^n(q)$. As α is irrational, $\{R_\alpha^n(q); n \in \mathbb{N}\}$ is dense in C_2. This shows that the positive orbit of p is dense in $P(C_2)$ and, therefore, it is dense in the pretzel.

EXAMPLE 12. We shall now describe a C^∞ vector field X on the pretzel with the following properties:

(a) X is a Kupka–Smale field;
(b) X has only two singularities s_1 and s_2 which are saddles;
(c) every regular orbit is dense in the pretzel.

We shall see in Lemma 2.5 of this section that X can be approximated by a vector field that has a saddle connection. (This fact can be verified directly.) Consequently, the set of Kupka–Smale fields is not open on the pretzel.

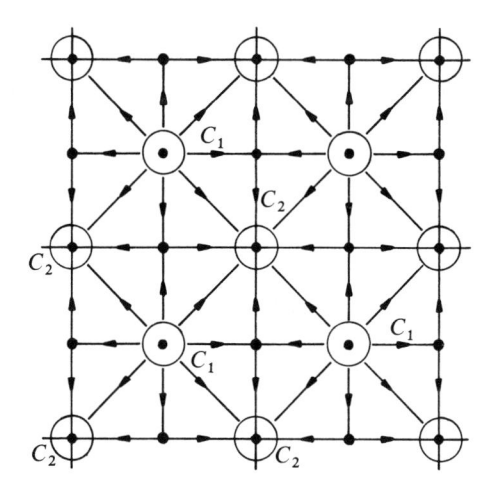

§2 Density of Morse–Smale Vector Fields on Orientable Surfaces 135

Figure 28

Similar examples can be obtained on the sphere with k handles for any $k \geq 2$. We construct X starting from a Morse–Smale field Y on the torus which has one source, one sink and two saddles. We cut out a disc around the sink and a disc around the source and then identify the boundaries of these discs by a convenient diffeomorphism, which is equivalent to glueing a handle onto T^2. Let $\pi\colon \mathbb{R}^2 \to T^2$ denote the canonical projection. First we describe a field $\tilde{Y}$ on $\mathbb{R}^2$ and then consider the field $Y = \pi_* \tilde{Y}$. In order that $\tilde{Y}$ projects to a field on T^2 we shall make $\tilde{Y}(x) = \tilde{Y}(y)$ if the coordinates of $x, y \in \mathbb{R}^2$ differ by integers. Therefore, it suffices to describe $\tilde{Y}$ on the square in $\mathbb{R}^2$ with vertices $(0, 0)$, $(1, 0)$, $(1, 1)$ and $(0, 1)$. The field $\tilde{Y}$ has a source at the point $(\frac{1}{2}, \frac{1}{2})$, a sink at the origin and saddles at the points $(0, \frac{1}{2})$ and $(\frac{1}{2}, 0)$. Let C_1 and C_2 be circles of radius $\delta < \frac{1}{4}$ around the sources and sinks, respectively, as in Figure 28. If $\alpha \in (0, \frac{1}{2}\pi) \subset C_1$ then the positive orbit through α meets C_2 at a point $\varphi_1(\alpha)$. Thus, we have a diffeomorphism $\varphi_1\colon (0, \frac{1}{2}\pi) \subset C_1 \to (-\pi, -\frac{1}{2}\pi)$. Similarly, we define $\varphi_2\colon (\frac{1}{2}\pi, \pi) \to (-\frac{1}{2}\pi, 0)$, $\varphi_3\colon (\pi, 3\pi/2) \to (-2\pi, -3\pi/2)$ and $\varphi_4\colon (3\pi/2, 2\pi) \to (-3\pi/2, -\pi)$. By constructing the field $\tilde{Y}$ symmetrically we have $\varphi_i(\alpha) = -\alpha - \frac{1}{2}\pi$ if $i = 1, 3$ and $\varphi_i(\alpha) = -\alpha + \frac{1}{2}\pi$ if $i = 2, 4$.

Let $\varphi\colon C_2 \to C_1$ be the diffeomorphism $\varphi(\alpha) = -\alpha + \varepsilon$ where ε/π is irrational. Let D_1 and D_2 be the open discs whose boundaries are the circles C_1 and C_2, respectively. We obtain the pretzel T_2 by using the diffeomorphism φ to identify the circles C_1 and C_2 which form the boundary of $T^2 - (D_1 \cup D_2)$. Let X be the field induced on T_2 by the field Y via this identification. That is, $X = P_* Y$ where $P\colon T^2 - (D_1 \cup D_2) \to T_2 = T^2 - (D_1 \cup D_2)/\sim$ is the projection. See Figure 29.

We shall show that every regular orbit of X is dense in the pretzel. In the following argument C_1 denotes the circle C_1 in the torus as well as the circle

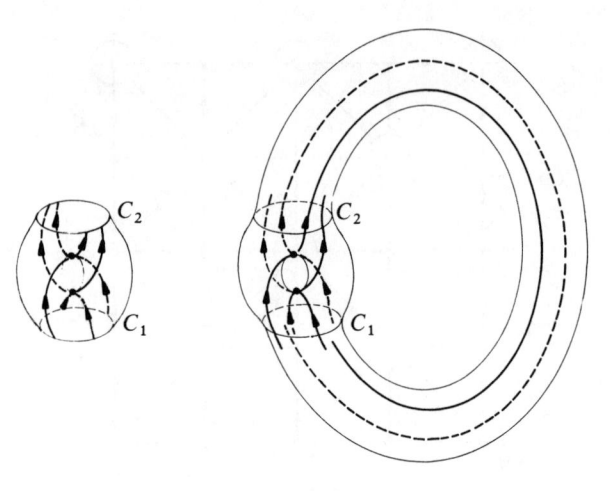

Figure 29

$P(C_1)$ in the pretzel. As every regular orbit meets the circle C_1 it is enough to show that the intersection of a regular orbit with C_1 is dense in C_1. Let $D = \{i(\pi/2); i = 0, 1, 2, 3\}$. The positive orbit of X through a point $\alpha \in C_1 - D$ meets C_1 again, for the first time, at the point $\psi(\alpha)$, where $\psi: C_1 \to C_1$ is defined by

$$\psi(\alpha) = \begin{cases} \alpha + \varepsilon + \dfrac{\pi}{2}, & \text{if } \alpha \in \left[0, \dfrac{\pi}{2}\right) \cup \left[\pi, \dfrac{3\pi}{2}\right) \\ \alpha + \varepsilon - \dfrac{\pi}{2}, & \text{if } \alpha \in \left[\dfrac{\pi}{2}, \pi\right) \cup \left[\dfrac{3\pi}{2}, 2\pi\right) \end{cases}$$

Let $\theta_+(\alpha) = \{\psi^m(\alpha); m \geq 0\}$ be the positive ψ-orbit of α and $\theta_-(\alpha) = \{\psi^m(\alpha); m \leq 0\}$ its negative ψ-orbit. If $\theta_+(\alpha) \cap D \neq \emptyset$ then α belongs to the stable manifold of a saddle point of X and if $(\theta_-(\alpha) - \alpha) \cap D \neq \emptyset$ to one of the unstable manifolds. If $\theta_+(\alpha) \cap D = \emptyset$ and $\theta_+(\alpha)$ is dense in C_1 then the positive X-orbit of α is dense in T_2. Similarly for the negative ψ- and X-orbits. We will show that all positive (and negative) ψ-orbits are dense and so the same is true for the X-orbits. In particular, X exhibits nontrivial recurrence.

Let us show that the positive ψ-orbits are dense in C_1. First, notice that ψ is 1-1, onto, discontinuous only at a finite set $D \subset C_1$ and it preserves lengths of segments (Lebesgue measure), i.e. the image of an interval of length L is a finite union of intervals whose lengths add up to L. We also claim that ψ has no periodic points and each orbit of ψ intersects D in at most one point. In fact, let $x \in C_1$ (resp. $x \in D$) and let $m \neq 0$ be an integer such that $\psi^m(x) = x$ (resp. $\psi^m(x) \in D$). But $\psi^m(x) = x + m\varepsilon + n_m\pi/2$ for some integer $0 \leq n_m \leq 3$. Thus $x + m\varepsilon + n_m\pi/2 = x + 2k\pi$, $k \in \mathbb{Z}$ (or $x + m\varepsilon + n_m\pi/2 = j\pi/2$, $j \in \mathbb{Z}$) which is a contradiction because ε/π is irrational. Next we show, following the ideas in [45], that if ψ is a mapping of the circle

§2 Density of Morse–Smale Vector Fields on Orientable Surfaces

with the above properties then each positive (negative) ψ-orbit is dense. This is an immediate consequence of the statements (1) and (2) proved below.

(1) If $F \subset C_1$ is a finite union of closed intervals such that $\psi(F) = F$ then $F = C_1$. Suppose, if possible, that $F \neq C_1$ and let $x \in C_1$ be a boundary point of F. Since $\psi(F) = F$ and the restriction of ψ to $C_1 - D$ is a homeomorphism it follows that $\psi^{-1}(x)$ is either a boundary point of F or an element of D. Hence there is a positive integer k such that $\psi^{-k}(x) \in D$ because F has finitely many boundary points and ψ has no periodic points. Similarly either $\psi(x)$ is a boundary point of F or $x \in D$. Thus there is a nonnegative integer j such that $\psi^j(x) \in D$. As a consequence we get $\psi^{k+j}(y) \in D$, where $y = \psi^{-k}(x)$. This contradicts one of the properties of ψ listed above. We conclude that $F = C_1$.

(2) If $I \subset C_1$ is a closed interval then there exists a positive integer n such that $\bigcup_{i=0}^n \psi^{-i}(I) = C_1$. In particular, $\theta_+(x)$ is dense for every $x \in C_1$. The proof that $\theta_-(x)$ is dense is entirely similar. Let $B = (\partial I) \cup D$, where ∂I is the boundary of I. For each $x \in B$ we set

$$\beta(x) = \begin{cases} +\infty, & \text{if } \psi^n(x) \notin I - \partial I \text{ for all } n \geq 0, \\ \inf\{n \geq 0; \psi^n(x) \in I - \partial I\}, & \text{otherwise.} \end{cases}$$

We can write $I = \bigcup_{j=1}^n I_j$, where the I_js are closed intervals with pairwise disjoint interiors and

$$\bigcup_{j=1}^n \partial I_j = (\partial I) \cup \{\psi^{\beta(x)}(x); x \in B, \beta(x) < \infty\}.$$

For each j, let $n_j = \inf\{n > 0; \psi^{-n}I_j \cap I \neq \varnothing\}$. We claim that n_j is finite. Suppose not. Then $\psi^{-m}I_j \cap \psi^{-n}I_j = \varnothing$ for all $0 \leq n < m$ because, if not, $\psi^{-(m-n)}I_j \cap I_j \neq \varnothing$ and since $I_j \subset I$ we would have n_j finite. But $\psi^{-n}I_j$, $n \geq 0$, cannot be all disjoint since the length of each of these sets (finite union of intervals) is the same as that of I_j. We conclude that n_j must be finite. From the first part of this argument we also conclude that $I_j, \psi^{-1}I_j, \ldots, \psi^{-(n_j-1)}I_j$ are pairwise disjoint. Finally, we claim that $\psi^{-n_j}I_j \subset I$. Otherwise, since $\psi^{-n_j}I_j \cap I \neq \varnothing$, there is a point $x \in \partial I$ in the interior of $\psi^{-n_j}I_j$. From the definition of n_j, we have that $\psi^i(x) \notin I$, for $0 \leq i < n_j$, and $\psi^{n_j}(x)$ is in the interior of $I_j \subset I$. Since $x \in B$, this implies that $n_j = \beta(x)$ which is a contradiction because $\psi^{\beta(x)}(x)$ would be a boundary point of some I_k, $1 \leq k \leq n$, and thus it could not belong to the interior of I_j. Now we set $F = \bigcup_{j=1}^n \bigcup_{k=0}^{n_j-1} \psi^{-k}I_j$. Clearly $\psi^{-1}F \subset F$. Since ψ^{-1} preserves length of intervals it follows that $\psi^{-1}F = F$. By (1) we have that $F = C_1$ proving (2). We suggest that the reader show this vector field X can be approximated by Morse–Smale fields.

EXAMPLE 13. We are now going to describe very briefly an example due to Cherry of a C^∞ (or even analytic) vector field on the torus T^2 that has nontrivial recurrence and also a source. The construction of the vector field is quite complicated and will be made in the Appendix at the end of this

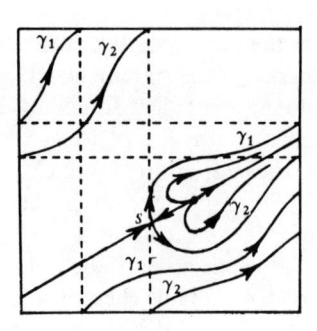

Figure 30

chapter. Let us represent the torus by a square in the plane with its opposite sides identified. The Cherry field X has one source f and one saddle s. The unstable separatrices γ_1 and γ_2 of the saddle are ω-recurrent. In fact the ω-limit of γ_1 contains γ_1 and γ_2. Moreover, X has no periodic orbit. Therefore X is a Kupka–Smale field. As we shall see at the end of this section, X can be approximated by a field that has a saddle-connection. Figure 30 shows the Cherry field on T^2.

Let us consider a circle C transversal to X and bounding a disc D that contains the source f. Let M be a compact two-dimensional manifold and Y a vector field on M that has a hyperbolic attractor. Let $\tilde{D}$ be a disc containing a sink of Y with its boundary $\tilde{C}$ transversal to Y. By glueing $T^2 - D$ into $M - \tilde{D}$ by means of a diffeomorphism $h: C \to \tilde{C}$ we obtain a manifold $\tilde{M}$. The vector field $\tilde{X}$ induced on $\tilde{M}$ by X, Y and this identification has a nontrivial recurrent orbit. In this way we can construct vector fields with non-trivial recurrence on any two-dimensional manifold except the sphere, the projective plane and the Klein bottle where all recurrence is trivial. This is true in the sphere and projective plane by the Poincaré–Bendixson Theorem and in the Klein bottle by [56].

Definition. Let $X \in \mathfrak{X}^r(M)$. We say that $K \subset M$ is a *minimal set* for X if K is closed, nonempty and invariant by X_t and there does not exist a proper subset of K with these properties. If K is a critical element of X we say that K is a *trivial minimal set*.

We remark that if K is minimal and γ is an orbit contained in K then γ is recurrent. This is because $\omega(\gamma)$ is closed, nonempty and invariant by X_t and $\omega(\gamma) \subset K$. Thus $\omega(\gamma) = K \supset \gamma$. Similarly, $\alpha(\gamma) \supset \gamma$.

2.2 Lemma. *Let $F \subset M$ be closed, nonempty and invariant by X_t where $X \in \mathfrak{X}^r(M)$. Then there exists a minimal set $K \subset F$.*

PROOF. Let $\mathscr{F}$ be the set of closed subsets of F that are invariant by X_t and let us consider in $\mathscr{F}$ the following partial order: if $A, B \in \mathscr{F}$ then $A \leq B$ if

§2 Density of Morse–Smale Vector Fields on Orientable Surfaces

$A \subset B$. Now let $\{A_i\}$ be a totally ordered family in $\mathscr{F}$. By the Bolzano–Weierstrass Theorem $\bigcap_i A_i$ is nonempty. As $\bigcap_i A_i$ is closed and invariant by X_t, it belongs to $\mathscr{F}$ and is thus a lower bound for $\{A_i\}$. By Zorn's Lemma [46] there exists a minimal element in $\mathscr{F}$. □

We must mention the following important facts about minimal sets even though they will not be used in the text, except in the Appendix where a complete description of Cherry's flow is presented. In [16] Denjoy exhibited a C^1 vector field on the torus T^2 with a nontrivial minimal set distinct from T^2. On the other hand, Denjoy [16] and Schwartz [100] showed that a minimal set of a C^2 field on M^2 is either trivial or is the whole of M^2 and, in this case, M^2 is the torus. Consequently the ω-limit of an orbit will either contain singularities or be a closed orbit or be T^2. We refer the reader to [31] for a converse of the Denjoy–Schwartz Theorem.

Definition. A *graph* for $X \in \mathfrak{X}^r(M^2)$ is a connected closed subset of M consisting of saddles and separatrices such that:

(1) the ω-limit and α-limit of each separatrix of the graph are saddles;
(2) each saddle in the graph has at least one stable and one unstable separatrix in the graph.

EXAMPLES. Figures 31 and 32 give four examples of graphs for vector fields on a chart of M^2.

2.3 Proposition. *Let $X \in \mathfrak{X}^r(M^2)$ be a vector field whose singularities are all hyperbolic. If X only possesses trivial recurrent orbits then the ω-limit of any orbit is a critical element or a graph. Similarly for the α-limit.*

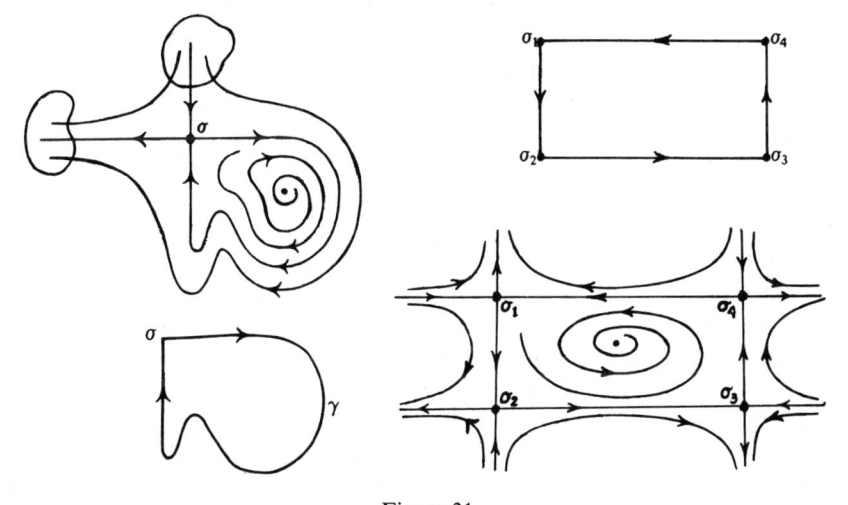

Figure 31

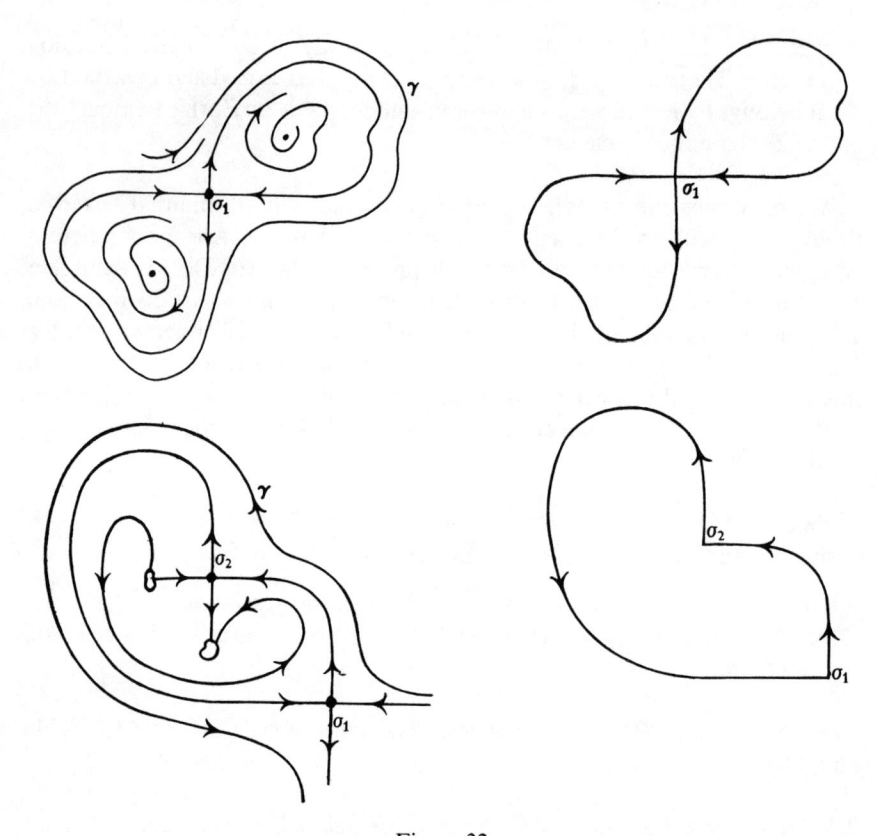

Figure 32

PROOF. Pick any trajectory γ of X and suppose that $\omega(\gamma)$ is not a critical element. It is clear that $\omega(\gamma)$ cannot contain an attracting singularity or a closed orbit for then it would reduce to one of these elements. On the other hand, $\omega(\gamma)$ must contain a saddle. This is because a minimal set in $\omega(\gamma)$ must be a critical element and we have already dealt with the possibility of an attracting singularity or a closed orbit. Thus, $\omega(\gamma)$ contains saddles and, as it is not just one singularity, it contains separatrices of saddles. Clearly, the number of separatrices in $\omega(\gamma)$ is finite.

Let us first suppose that each separatrix in $\omega(\gamma)$ has a unique saddle as its ω-limit. We shall show that $\omega(\gamma)$ is a graph. We claim that there exists a graph contained in $\omega(\gamma)$. In fact, let $\sigma_1 \in \omega(\gamma)$ be a saddle and γ_1 an unstable separatrix of σ_1 contained in $\omega(\gamma)$. Let $\sigma_2 = \omega(\gamma_1)$ and let γ_2 be an unstable separatrix of σ_2 contained in $\omega(\gamma)$ and so on. As we only have a finite number of separatrices this process will give a sequence $\gamma_i, \gamma_{i+1}, \ldots, \gamma_l = \gamma_i$ of separatrices in $\omega(\gamma)$ such that $\omega(\gamma_j) = \alpha(\gamma_{j+1})$ and this defines a graph.

Let G be a maximal graph in $\omega(\gamma)$, that is, there does not exist any graph in $\omega(\gamma)$ containing G. We claim that $\omega(\gamma) = G$. If not, there exists a saddle $\tilde{\sigma}_1 \in G$ and an unstable separatrix $\tilde{\gamma}_1$ of $\tilde{\sigma}_1$ not belonging to G. Consider the saddle $\tilde{\sigma}_2 = \omega(\tilde{\gamma}_1)$ and a separatrix of $\tilde{\sigma}_2$, $\tilde{\gamma}_2 \subset \omega(\gamma)$. By continuing this argument we obtain a sequence $\tilde{\sigma}_1, \ldots, \tilde{\sigma}_k$ such that $\tilde{\sigma}_k = \tilde{\sigma}_j$ for some $j < k$ or $\tilde{\sigma}_k \in G$. Either way we obtain a graph $\tilde{G}$ in $\omega(\gamma)$ properly containing G, namely, $\tilde{G}$ is the union of G with the saddles $\tilde{\sigma}_1, \ldots, \tilde{\sigma}_k$ and the separatrices $\tilde{\gamma}_1, \ldots, \tilde{\gamma}_{k-1}$. This is a contradiction since G is maximal. Thus $\omega(\gamma) = G$ as claimed.

Now let us suppose that there exists a separatrix $\tilde{\gamma}_1 \subset \omega(\gamma)$ whose ω-limit is not just one saddle. Then $\omega(\tilde{\gamma}_1) \subset \omega(\gamma)$ and $\omega(\tilde{\gamma}_1)$ does not contain $\tilde{\gamma}_1$ because all recurrence is trivial. If $\omega(\tilde{\gamma}_1)$ only contains separatrices that have a unique saddle as ω-limit then, by the previous argument, $\omega(\tilde{\gamma}_1)$ is a graph. This graph will also be the ω-limit of γ, which is absurd since $\omega(\gamma) \supset \tilde{\gamma}_1$. Thus, there must exist $\tilde{\gamma}_2 \subset \omega(\tilde{\gamma}_1)$ whose ω-limit is not a unique singularity. Moreover, $\omega(\tilde{\gamma}_2) \subset \omega(\tilde{\gamma}_1) \subset \omega(\gamma)$. By continuing this argument we shall necessarily find a separatrix $\bar{\gamma} \subset \omega(\gamma)$ such that $\omega(\bar{\gamma}) \supset \bar{\gamma}$ since the number of separatrices is finite. This is absurd since there is no nontrivial recurrence. □

Corollary. *If $X \in \mathfrak{X}^r(M^2)$ is a Kupka–Smale field with all recurrent orbits trivial then X is a Morse–Smale field.*

PROOF. By the last proposition, the ω-limit of any orbit is a critical element (and so is its α-limit) because there cannot be any graphs as there are no saddle-connections. It remains to show that there only exist a finite number of closed orbits. This can be proved by the argument for the case $M = S^2$ in Theorem 2.1. □

We now move on to the proof that any $X \in \mathfrak{X}^r(M^2)$ can be approximated by a Morse–Smale field, provided M is orientable. For this we shall exhibit a field Y near X with the following property: there exists a neighbourhood $\mathscr{U} \subset \mathfrak{X}^r(M^2)$ of Y such that any $Z \in \mathscr{U}$ only has trivial recurrence. Then we approximate Y by a Kupka–Smale field $Z \in \mathscr{U}$. By the corollary above Z is Morse–Smale.

We shall use the next two lemmas. Their proofs will be given at the end of this section.

2.4 Lemma. *If $X \in \mathfrak{X}^r(M^2)$ is a vector field without singularities then X can be approximated by a field Y that contains a closed orbit.*

2.5 Lemma. *Suppose M^2 is orientable. If $X \in \mathfrak{X}^r(M^2)$ has singularities all of which are hyperbolic and there is a nontrivial recurrent orbit then X can be approximated by a field Y that has one more saddle-connection than X has.*

2.6 Theorem. *The set of Morse–Smale fields is dense in $\mathfrak{X}^r(M^2)$ for M^2 orientable.*

PROOF. Let $X \in \mathfrak{X}^r(M^2)$. By perturbing X, if necessary, we can suppose that it has only hyperbolic singularities.

Case 1. X has no singularities. As M^2 is orientable it must be the torus T^2.

By Lemma 2.4, X can be approximated by a field X_1 that has a closed orbit. By Lemma 2.5 of Chapter 3 we can approximate X_1 by Y that has a hyperbolic closed orbit γ. As Y has no singularities, γ does not bound a disc in T^2 and, therefore, $T^2 - \gamma$ is a cylinder. Thus Y has only trivial recurrence. Each field near Y enjoys the same property since it also has a closed orbit. We now take a Kupka–Smale field Z near Y. By the corollary of Proposition 2.3, Z is a Morse–Smale field.

Case 2. X has singularities, all of which are hyperbolic.

Let γ be an unstable separatrix of a saddle. We say that γ is *stabilized* if $\omega(\gamma)$ is a hyperbolic attractor (singularity or closed orbit). Similarly, a stable separatrix of a saddle is stabilized if its α-limit is a hyperbolic repellor.

(1) If X has all the separatrices of its saddles stabilized then X can be approximated by a Morse–Smale field.

In fact, there exists a neighbourhood $\mathcal{U}$ of X such that any $Y \in \mathcal{U}$ also has all its separatrices stabilized. Thus, by Lemma 2.5, these fields have only trivial recurrence. Now approximate X by a Kupka–Smale field $Y \in \mathcal{U}$. By the corollary of Proposition 2.3, Y is Morse–Smale.

(2) If X has separatrices that are not stabilized then X can be approximated by a field Y that has one more stabilized separatrix than X.

Proving this claim will complete the theorem since there are only a finite number of separatrices and by stabilizing them one by one we arrive at (1).

Let $\mathcal{U}$ be a neighbourhood of X such that each $Y \in \mathcal{U}$ has at least as many stabilized separatrices as X. By Lemma 2.5, we can approximate X by $Y \in \mathcal{U}$ which has only trivial recurrence because there are only a finite number of saddles that can be connected.

There are four possibilities to consider:

(a) Y has no saddle-connections. Let γ be a separatrix of a saddle for Y that is not stabilized. (Y already satisfies (1) if there is no such γ.) Then $\omega(\gamma)$ (or $\alpha(\gamma)$) is a nonhyperbolic closed orbit. We can approximate Y by $Z \in \mathcal{U}$ to make this orbit hyperbolic and thus stabilize the separatrix of Z corresponding to γ.

(b) Y has a graph that is the ω-limit (or α-limit) of some orbit. Consider a transversal section S through a regular point p of the graph. As there is a trajectory γ whose ω-limit (or α-limit) contains p, γ must intersect S in a sequence $a_n \to p$. For n large enough, the arc of the trajectory γ between a_n and a_{n+1}, the segment (a_n, a_{n+1}) of S and the graph bound an open region $A \subset M^2$ that is homeomorphic to an annulus. Let F be a small flow box containing p and let ΔY be a C^r-small vector field on F that is transversal to Y at all points of the interior of F and vanishes outside F as in Figure 33. If

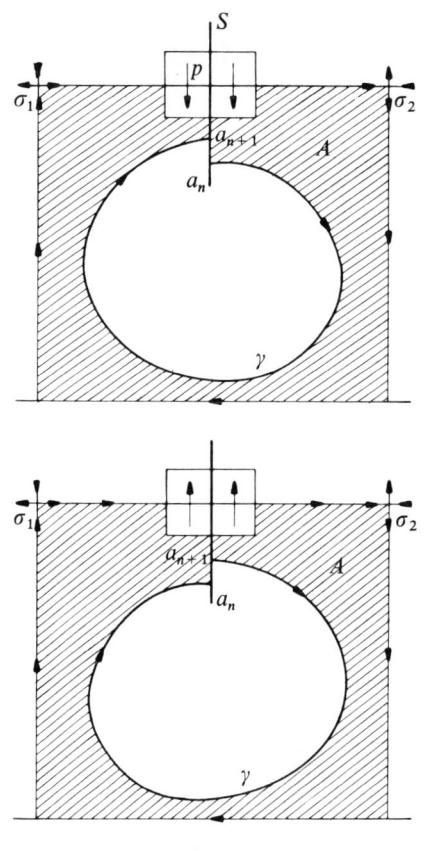

§2 Density of Morse–Smale Vector Fields on Orientable Surfaces 143

Figure 33

$Z = Y + \Delta Y$ then $Z_t(A) \subset A$ for $t \geq 0$ in the case $p \in \omega(\gamma)$. The separatrix of σ_1 (or σ_2 in the second case), which is initially contained in the graph, penetrates the annulus after the perturbation. Thus the ω-limit (or α-limit) of this separatrix becomes a certain closed orbit of the field Z that forms in the annulus. This comes from the Poincaré–Bendixson Theorem since Z has no singularities in A. Now by a further perturbation of the field Z we make this closed orbit hyperbolic and, in this way, stabilize another separatrix. (If $p \in \alpha(\gamma)$ we consider Z_t, for $t \leq 0$.)

(c) Y has a graph that is accumulated by closed orbits, as in Figure 34.

The annulus A bounded by the graph and a closed orbit close enough to it makes (c) entirely analogous to case (b).

(d) The last possibility will now be analysed. Let γ be a saddle-connection and S a transversal section through a point $p \in \gamma$, as in Figure 35.

Let us consider a small open interval $(a, p) \subset S$. All the points of (a, p) have the same ω-limit which is an attracting singularity or a closed orbit. We remark that if this does not happen for sufficiently small (a, p) then we must have the situation described in (b) or (c). In fact, if Y does not come

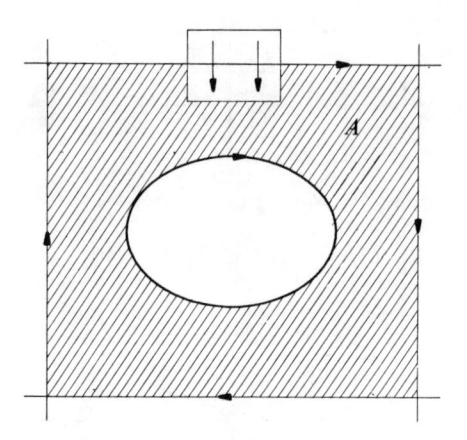

Figure 34

under case (b) then the ω-limit of each point of (a, p) is a singularity or a closed orbit. Moreover, there is no stable separatrix of a saddle, except γ, that accumulates on p since the α-limit of such a separatrix would contain γ and we should be in case (b). Thus the ω-limit of each point of (a, p) is an attracting singularity or a closed orbit. As p is not accumulated by closed orbits, which corresponds to (c), it follows that all points in (a, p) have the same ω-limit, which we shall denote by σ.

Proceeding as before, we perturb the field so that the ω-limit of the unstable separatrix of σ_1 becomes σ. If σ is an attracting singularity we have stabilized one more separatrix. If σ is a closed orbit we make it hyperbolic by a further perturbation, if necessary, and again obtain one more stabilized separatrix. This concludes the proof of the theorem. □

Before proving Lemmas 2.4 and 2.5 we shall make some comments on nontrivial recurrent orbits.

Let $X \in \mathfrak{X}^r(M^2)$ and let γ be a nontrivial ω-recurrent orbit. We claim that there exists a circle transversal to X through any point $p \in \gamma$.

In fact let us consider a flow box F_1 containing p. Let ab and cd denote the

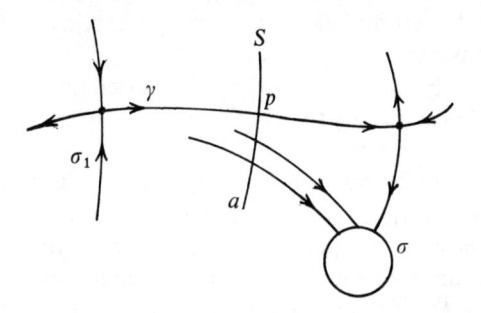

Figure 35

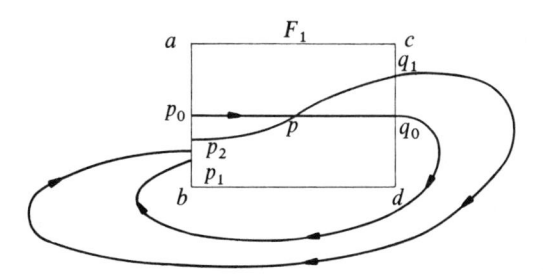

§2 Density of Morse–Smale Vector Fields on Orientable Surfaces

Figure 36

sides of F_1 transversal to X. As $p \in \omega(\gamma)$, γ intersects ab infinitely many times. Let p_1 be the first occasion that γ returns to intersect ab. Let us take a flow box F_2 containing the arc of the trajectory $q_0 p_1$. We shall suppose that p_1 is below p_0 in ab, see Figure 36. The construction is similar in the other case. If M^2 is orientable we can find in F_2 an arc of a trajectory of X starting from a point q_1 of cd above q_0 and intersecting ab at p_2 above p_1. In F_2 we can take an arc $q_1 p_3$, with p_3 above p_2 but below p_0, that is transversal to X and has positive slope at the ends as in Figure 37. Now we complete the required circle by joining p_3 and q_1 by an arc in F_1 that contains p and is transversal to X. This arc must have the same slope as the previous one at p_3 and q_1.

We leave it to the reader to construct a transversal circle as above in the case where M^2 is not orientable. In this case it is necessary to consider consecutive intersections of γ with ab.

We now denote by C a circle transversal to X through $p \in \gamma$. Let $D \subset C$ be the subset of those points whose positive trajectories return to intersect C again. We define the Poincaré map $P: D \to C$ to be the map that associates to each $x \in D$ the first point in which its positive trajectory intersects C. By the Tubular Flow Theorem, D is open. Thus, $D = C$ or else D is a union of open intervals. Let us suppose that $D \neq C$ and let (a_1, a_2) be a maximal interval in D. We show that $\omega(a_1)$ is a saddle and so is $\omega(a_2)$. If $\omega(a_1)$ is a singularity it must be a saddle since a sink would also attract orbits near that of a_1 whereas every orbit beginning in (a_1, a_2) returns to intersect C. Therefore it will suffice to show that $\omega(a_1)$ does not contain regular points. Let us suppose the contrary and let x be a regular point in $\omega(a_1)$ and S a section transversal to X through x. The positive orbit of a_1 intersects S infinitely many times since $x \in \omega(a_1)$. On the other hand, if $q \in (a_1, a_2)$ the

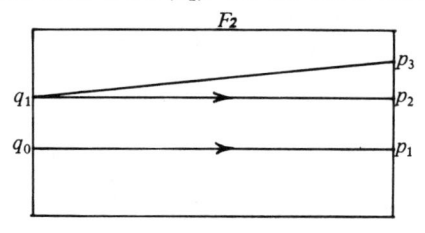

Figure 37

number of times N that the piece of trajectory $qP(q)$ meets S is finite, since S is transversal to X and $qP(q)$ is compact. By the Tubular Flow Theorem, this number is constant on a neighbourhood of q and as (a_1, a_2) is connected it is constant on the whole of that interval. By applying the Tubular Flow Theorem to an arc of the positive orbit of a_1 that cuts S some number of times $n > N$ we find points in (a_1, a_2) whose positive orbit intersects S at least n times before returning to C. This is a contradiction, which proves that $\omega(a_1)$ is a saddle. Similarly $\omega(a_2)$ is a saddle.

Thus D is a finite union of open intervals in C whose end-points belong to the stable separatrices of saddles. If we consider the inverse P^{-1} of P (the Poincaré map for $-X$) its domain is a finite union of open intervals whose end-points belong to the unstable separatrices of saddles.

We should remark that if the Poincaré map is defined on the whole circle C then M^2 is the torus T^2 or the Klein bottle K^2. In fact, it is easy to see that the set saturated by C, $\bigcup_{t \in \mathbb{R}} X_t(C)$, is open and closed and thus coincides with M^2. Therefore, the vector field X has no singularities, which proves our statement, and we have $M^2 = T^2$ or $M^2 = K^2$ depending on whether P preserves or reverses the orientation of C. If $P: C \to C$ reverses orientation then P has a fixed point. This fixed point corresponds to a closed orbit γ of X. This closed orbit does not bound a disc in K^2 and so $K^2 - \gamma$ is a Möbius band, which shows that X has only trivial recurrent orbits.

PROOF OF LEMMA 2.4. If X has a closed orbit there is nothing to prove. If X has no closed orbit then it has a recurrent orbit γ and $M^2 = T^2$. Take $p \in \gamma$ and let C be a transversal circle through p. Put $C_1 = X_{-\delta}(C)$ and $C_2 = X_\delta(C)$, where $\delta > 0$ is small. Clearly, C_1 and C_2 are transversal to X. Consider a flow box F containing p whose sides transversal to X lie in C_1 and C_2.

We define $P: C_2 \to C_1$ by associating to each point x of C_2 the first point where its positive trajectory intersects C_1. As we have seen P is well defined and preserves orientation. By this we mean that, given an orientation of C_2, the maps P and $X_{-2\delta}: C_2 \to C_1$ induce the same orientation on C_1. Now let

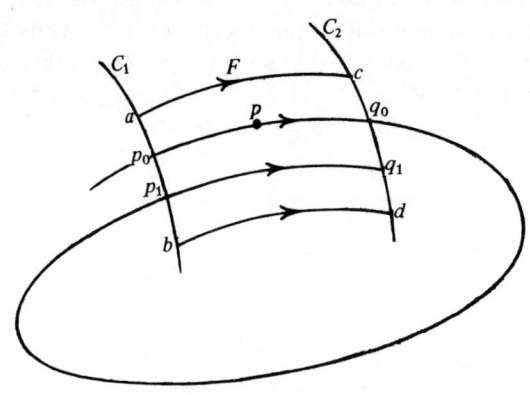

Figure 38

§2 Density of Morse–Smale Vector Fields on Orientable Surfaces

$p_i = P^i(q_0) = P(q_{i-1})$ where $q_0 = X_\delta(p)$. There exists a sequence n_i such that $p_{n_i} \to p_0$ and we can suppose that each p_{n_i} lies below p_0 on ab, see Figure 38.

Let us consider the family of fields $Z(u) = X + \varepsilon u Y$, where $\varepsilon > 0$, $0 \le u \le 1$ and Y is a field that is transversal to X on the interior of F, points upwards and vanishes outside F. If ε is small then $Z(u)$ is near X for all $0 \le u \le 1$.

Now fix a closed interval I in ab with p_0 in its interior. For each point $x \in I$ we consider the vertical distance in F between $x \in I$ and the point where the positive orbit of $Z(1)$ through x intersects C_2. As I is compact this distance has a minimum $\rho > 0$. We claim that, for some $0 < u \le 1$, the orbit of $Z(u)$ through q_0 is closed. In fact, as $Z(u) = X$ outside F for any u, we can define $p_k(u) = P(q_{k-1}(u))$ for $k \ge 1$, where $q_0(u) = q_0$ and $q_{k-1}(u)$ is the point where the positive orbit of $p_{k-1}(u) \in C_1$ meets C_2 for the first time.

Let us fix a number i such that $p_i = P^i(q_0)$ is below p_0 and its distance from p_0 is less then ρ.

We should remark that $p_i(u)$ and $q_i(u)$ depend continuously on u. For small u, $p_i(u)$ is in I and below p_0 and its height increases with u. In the same way $q_i(u)$ is below q_0 and its height increases with u. Thus, either $p_i(u_0) = p_0$ for some $u_0 \in (0, 1]$ or $p_i(u) \in I$ for all $u \in (0, 1]$. In the first case $q_i(u_0)$ is above q_0 and so there exists $u_1 < u_0$ such that $q_i(u_1) = q_0$. In the second case, $q_i(1)$ is above q_0 because the distance from $p_i(1)$ to p_0 is less than ρ. Thus, there exists $u_1 < 1$ such that $q_i(u_1) = q_0$. In either case $Z(u_1)$ has a closed orbit through q_0. □

PROOF OF LEMMA 2.5. First we shall prove that, if γ is a nontrivial ω-recurrent orbit, there exists a stable separatrix that accumulates on γ. That is, there exists a stable separatrix γ_2 such that $\alpha(\gamma_2) \supset \gamma$. Consider a point $p \in \gamma$ and a circle C transversal to X through p. Let $P: D \subset C \to C$ be the Poincaré map defined on D. We have $D \ne C$ since, otherwise, X has no singularities. Suppose, if possible, that γ is not accumulated by stable separatrices; that is, suppose that there is an interval in C that contains p and is disjoint from the stable separatrices. Let $I \subset C$ be the maximal interval with this property. As γ is ω-recurrent and $p \in \gamma$, we have $P^k(p) \in I$ for some integer $k > 0$. Now the interval $J = P^k(I)$ is contained in I. This is because, on the one hand, $I \cap J \ne \emptyset$ since $P^k(p) \in I$. On the other hand, if $J \not\subset I$, then J would contain

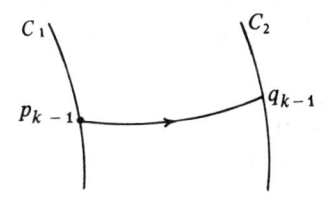

Figure 39

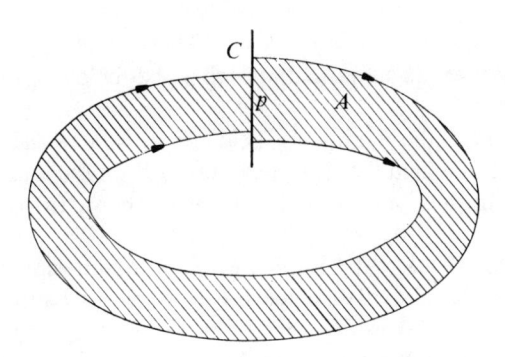

an interval that has one of the end-points of I in its interior. As I is the maximal interval disjoint from the stable separatrices it would follow that J contains points of stable separatrices. As these separatrices are invariant by the flow and, therefore, by P^k, it would follow that I also contains points of stable separatrices of saddles as $J = P^k(I)$. This would contradict the definition of I and so, in fact, $P^k(I) \subset I$. From this we can construct a region A containing p, homeomorphic to an annulus and invariant by X_t, $t > 0$. See Figure 40. This is a contradiction because γ could not be nontrivially ω-recurrent in A. Thus γ is, in fact, accumulated by some stable separatrix γ_2.

We also claim that γ is accumulated by some unstable separatrix or is itself an unstable separatrix. In fact, let us suppose that this claim is false. If the second alternative does not occur we consider, as in the previous case, a point $p \in \gamma$ and a maximal open arc I in C containing p and disjoint from unstable separatrices. As γ is ω-recurrent, there exists an integer $k > 0$ such that $P^k(I) \cap I \neq \emptyset$. It follows from this that, for some $q \in I$, $P^{-k}(q)$ is well defined and $P^{-k}(q) \in I$. There are two possibilities to consider. If P^{-k} is not defined on the whole arc I then there exists a point $z \in I$ whose negative orbit dies in a saddle. In particular, $z \in I \cap \bar{\gamma}$ for some unstable separatrix $\bar{\gamma}$ which contradicts the definition of the arc I. The other possibility is that P^{-k} is well defined on the whole arc I. In this case, as before, we shall have $P^{-k}(I) \subset I$. We leave the reader to complete the proof of the claim from this statement.

We should remark that the property just proved also holds for nonorientable manifolds. The only difference is that the region A used above can be a Möbius band where, again, there cannot be nontrivial recurrence.

Now let γ_1 be an unstable separatrix that either is γ or accumulates on γ. Also let γ_2 be a stable separatrix that accumulates on γ. As in the proof of Lemma 2.4 we consider the circles $C_1 = X_{-\delta}(C)$ and $C_2 = X_\delta(C)$ transversal to X. Let $P: D \subset C_2 \to C_1$ be the Poincaré map and F a flow box containing $p \in \gamma$. As the number of saddles is finite we can take F disjoint from any saddle-connections that X might have.

Let σ_1 and σ_2 be the saddles associated to the separatrices γ_1 and γ_2 which accumulate on γ. Let us consider the family of fields $Z(u) = X + \varepsilon u Y$,

§2 Density of Morse–Smale Vector Fields on Orientable Surfaces 149

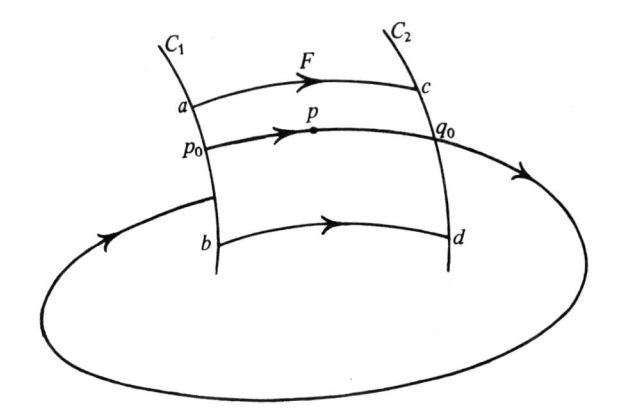

Figure 41

where $\varepsilon > 0$, $0 \leq u \leq 1$ and Y is transversal to X in the interior of F, points upwards and vanishes outside F. If we take ε small then $Z(u)$ is near to X for all $u \in [0, 1]$. We want to show that, for some $0 < u \leq 1$, $Z(u)$ has one saddle-connection more than X.

We fix a small closed interval I in $[a, b]$ containing p_0 in its interior. As before, let $\rho > 0$ be the minimum for $x \in I$ of the vertical distance in F between x and the point where the positive orbit of $Z(1)$ through x first meets C_2. Let x_0 and z_0 be the first times that γ_1 intersects $[a, b]$ and γ_2 intersects $[c, d]$, respectively. Note that the arcs of separatrices $\sigma_1 x_0$ and $\sigma_2 z_0$ are not affected by the perturbations $\varepsilon u Y$ above, see Figure 42.

Now take a point $x \in \gamma_1 \cap I$ near p_0 and a point $z \in \gamma_2 \cap C_2$ such that the vertical distance between x and z is less than ρ. The point x corresponds to the ith intersection of γ_1 with C_1 for some integer $i > 0$. Similarly, the point z corresponds to the jth intersection of γ_2 with C_2 for some $j > 0$. We shall suppose that x is below z in F. If this is not possible then we must take the field Y pointing downwards.

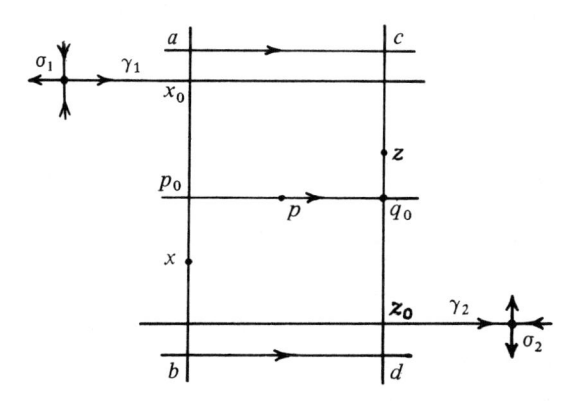

Figure 42

Consider the maps that associate to each u the ith intersection $x(u)$ of the separatrix $\gamma_1(u)$ of $Z(u)$ with C_1 and the jth intersection $z(u)$ of the separatrix $\gamma_2(u)$ of $Z(u)$ with C_2. It is clear that $x(u)$ and $z(u)$ are well defined for small u. As M^2 is orientable, $x(u)$ is monotone increasing on $[a, b]$ and $z(u)$ is monotone decreasing on $[c, d]$.

We have two situations to consider. Suppose first that $x(u)$ and $z(u)$ are well defined for all $u \in [0, 1]$. Then there exists $u_0 \in (0, 1)$ such that the vertical distance between $x(u_0)$ and $z(u_0)$ is zero. This is because $x(u)$ and $z(u)$ are continuous and the vertical distance between $x = x(0)$ and $z = z(0)$ is less than ρ. Thus we have a saddle connection between $\sigma_1(u_0)$ and $\sigma_2(u_0)$.

Now suppose that one of the maps, $x(u)$ for example, is not defined for all $u \in [0, 1]$. This means that, for some $u_0 \in (0, 1)$, $\gamma_1(u_0)$ reaches one of the boundary points of the domain of P; that is, it reaches a point whose positive orbit goes directly to a saddle σ_3. In this case it will be a saddle connection between σ_1 and σ_3. The reasoning is similar for the case where it is $z(u)$ that is not defined for all $u \in [0, 1]$. This completes the proof of Lemma 2.5 and hence of Theorem 2.6. □

To close the section we remark that it follows from Theorem 2.6 that any structurally stable vector field $X \in \mathfrak{X}^r(M^2)$ is Morse–Smale.

§3 Generalizations

Next we make some comments on the density theorem for Morse–Smale fields on orientable surfaces and its partial extension to nonorientable surfaces.

We shall also mention the theorems about the openness and stability of Morse–Smale fields on manifolds of any dimension. In particular, there exist structurally stable fields on any manifold. However, Morse–Smale fields are no longer dense in the space of vector fields on manifolds of dimension three or more. This will be proved in the next section. We should, however, emphasize a useful special case in which Morse–Smale vector fields are dense: namely in the space of gradient fields on any compact manifold.

Our first remark is that in the proof of Lemma 2.5 we cannot guarantee that after the perturbation there is a saddle connection between the first saddles considered. This leads us to formulate the following problem. Let γ_1 be an unstable separatrix and γ_2 a stable separatrix of saddles of $X \in \mathfrak{X}^r(M^2)$. Suppose that $\omega(\gamma_1) \cap \gamma_2 \neq \emptyset$ or $\omega(\gamma_1) \cap \alpha(\gamma_2) \neq \emptyset$. It is a difficult question to know whether it is possible to connect these two separatrices for a C^r perturbation of X. This problem is open whether M^2 is orientable or not and for any $r \geq 1$.

The difficulty in proving the density of Morse–Smale fields in $\mathfrak{X}^r(M^2)$ when M^2 is not orientable lies in the proof of Lemma 2.5. All the other facts

§3 Generalizations

are true in this case. The question is open in this nonorientable case and it is an interesting problem whether the answer is positive or negative, although a negative answer would be a surprise. In this direction some partial results have been obtained as follows.

(1) Morse–Smale fields are dense in $\mathfrak{X}^1(M^2)$ whether M^2 is orientable or not. Pugh obtained this result using the Closing Lemma as we shall show later. The restriction to the C^1 topology comes from the fact that the Closing Lemma has only been proved so far for this case.

(2) It is easy to see that the theorem holds for $\mathfrak{X}^r(P^2)$ and any $r \geq 1$ where P^2 is the projective plane. This is because the vector fields on P^2, as in the case of the sphere S^2, do not have nontrivial recurrence. Density is also true for the Klein bottle K^2 as Markley showed in [56]. Gutierrez [30] simplified the proof for K^2 and showed that in the nonorientable manifold L^2 of genus one more than K^2 that is the torus with a cross cap, the nontrivial recurrences are "orientable". Thus the proof we presented for orientable manifolds also applies to this case. Therefore we have the density of Morse–Smale fields in $\mathfrak{X}^r(M^2)$ for any $r \geq 1$ when M^2 is orientable or $M^2 = P^2$, K^2 or L^2.

We now describe the proof for $\mathfrak{X}^1(M^2)$ using the Closing Lemma.

Closing Lemma [88]. *Let M^n be a compact n-dimensional manifold without boundary. Take $X \in \mathfrak{X}^1(M^n)$ and let γ be a nontrivial recurrent orbit of X. Given $p \in \gamma$ and $\varepsilon > 0$ there exists $Y \in \mathfrak{X}^1(M^n)$, with $|Y - X|_{C^1} < \varepsilon$, having the orbit through p closed.* □

The proof of the Closing Lemma is very delicate even in the case of surfaces. As regards the class of differentiability, the question is open for any $r \geq 2$ and $n \geq 2$.

In the case of surfaces M^2 the closed orbit constructed from a nontrivial recurrent orbit cannot bound a disc. This is because of the existence of a circle transversal to the field and not bounding a disc as constructed in Section 2 of this chapter.

3.1 Theorem. *The subset consisting of Morse–Smale vector fields is dense in $\mathfrak{X}^1(M^2)$ whether or not M^2 is orientable.*

PROOF. We shall show that any field $X \in \mathfrak{X}^1(M^2)$ can be approximated by a Kupka–Smale field with only trivial recurrence. Then the result will follow immediately from the Corollary to Proposition 2.3.

Take a Kupka–Smale field X^* near X. If X^* has only trivial recurrence the proof is finished. Otherwise, consider a nontrivial recurrent orbit γ_1 of X^* and a point $p \in \gamma_1$. By the Closing Lemma there exists X_1 near X^* with a closed orbit σ_1 through p. Now approximate X_1 by a Kupka–Smale field X_1^* that has a hyperbolic closed orbit σ_1^* near σ_1. If X_1^* has only trivial recurrence then X_1^* is a Morse–Smale field. Otherwise, we repeat the above process starting from X_1^* but without changing it on a neighbourhood of σ_1^*. We claim that the number of steps in this process is finite, bounded by 2^g where g is the genus (the number of handles) of the manifold M. In fact each stage of the process is like considering a Kupka–Smale field on a manifold N^2 of lower genus (or, equivalently, of higher Euler–Poincaré characteristic $K(N^2)$ since $K(N^2) = 2 - 2g(N^2)$ for N^2 orientable and $K(N^2) = 2 - g(N^2)$ for N^2 nonorientable, [59], [119]). As $K(N^2) \leq 2$ for any N^2 we shall obtain a Kupka–Smale field with only trivial recurrence after a finite number of steps; this field will then be Morse–Smale as required.

To see this reduction of genus we make a cut in M along the hyperbolic closed orbit σ_1^* that resulted from the nontrivial recurrence. There are two possibilities to consider: either we obtain a manifold M_{10} with boundary or two manifolds M_{11} and M_{12} with boundary [59]. In the first case the boundary of M_{10} consists of one or two copies of σ_1^* depending on whether σ_1^* is orientable or not. In the second case the boundaries of M_{11} and M_{12} are each a copy of σ_1^*. In the first case we glue in one or two discs D_1 and D_2 so that $M_{10} \cup D_1$ or $M_{10} \cup D_1 \cup D_2$ is a manifold without boundary. Thus we have $K(M_{10} \cup D_1) = K(M) + K(D_1)$ or $K(M_{10} \cup D_1 \cup D_2) = K(M) + K(D_1) + K(D_2)$. As $K(D_i) = 1$ for $i = 1, 2$, $K(M_{10} \cup D_1)$ or $K(M_{10} \cup D_1 \cup D_2)$ is greater than $K(M)$. In the second case $K(M) = K(M_{11}) + K(M_{12})$ and $K(M_{11}) < 1$, $K(M_{12}) < 1$ because otherwise σ_1^* would bound a disc. Thus

$$K(M_{11} \cup D_1) = K(M_{11}) + 1 > K(M)$$

and

$$K(M_{12} \cup D_2) = K(M_{12}) + 1 > K(M).$$

Although it is not necessary for our purpose, we can complete the field X_1^* defined on M_{10}, M_{11} and M_{12} by putting in D_1 and D_2 a sink or a source depending on whether σ_1^* is repelling or attracting.

We continue the process with the manifolds M_{ij} ($j = 0, 1, 2$) obtained by glueing in these discs but without altering the field on neighbourhoods of these discs. As the Euler–Poincaré characteristic is bounded by 2 and grows with each cut made, the number of cuts is finite and bounded by 2^g as claimed. This proves the theorem. □

For Morse–Smale fields on M^2 we should also mention that their equivalence classes were described by Peixoto [83] and Fleitas [19]. Also, the connected components of the Morse–Smale fields on M^2 were classified in [32].

Now consider a manifold M of any dimension n endowed with a Riemannian metric. One basic question is whether there exists a structurally stable field on M. Results from [75], [79], [106] show that there exist many Morse–Smale fields and they are structurally stable; see also [57]. These results are as follows:

(1) the set consisting of Morse–Smale fields is open and nonempty in $\mathfrak{X}^r(M^n)$, $r \geq 1$;
(2) if $X \in \mathfrak{X}^r(M^n)$, $r \geq 1$, is Morse–Smale then X is structurally stable;
(3) the set of Morse–Smale gradient fields is open and dense in $\operatorname{Grad}^r(M^n)$, $r \geq 1$.

Here $\operatorname{Grad}^r(M^n)$ denotes the subset of $\mathfrak{X}^r(M^n)$ consisting of the gradient fields of C^{r+1} maps from M to $\mathbb{R}$ with respect to a Riemannian metric on M. Consider the orbit structure of a Morse–Smale gradient field. As we saw in Section 1 of Chapter 1, a gradient field cannot have closed orbits. Moreover, every orbit has its α- and ω-limit sets consisting of singularities. We leave it to the reader to prove that even the nonwandering set consists only of singularities. Thus a Morse–Smale gradient field is just a Kupka–Smale field with nonwandering set a finite number of hyperbolic singularities.

In contrast to what happens in $\mathfrak{X}^r(M^2)$, $r = 1$ or $r \geq 1$ and M^2 orientable, or in $\operatorname{Grad}^r(M^n)$, the Morse–Smale fields are not dense in $\mathfrak{X}^r(M^n)$ for $n \geq 3$. This fact will be seen in the next section together with examples of structurally stable fields in $\mathfrak{X}^r(M^3)$ that are not Morse–Smale since they have infinitely many periodic orbits.

§4 General Comments on Structural Stability. Other Topics

In this section we shall briefly describe Morse–Smale diffeomorphisms, Anosov diffeomorphisms and the diffeomorphisms that satisfy Axiom A and the transversality condition. The first are analogous to the Morse–Smale fields described in previous sections. The last include the first two and constitute the most general class of structurally stable diffeomorphisms known. We shall describe in detail two famous examples that illustrate the last two classes well: one of them is due to Thom (an Anosov diffeomorphism of T^2) and the other is Smale's horseshoe.

Besides its intrinsic importance the study of diffeomorphisms has been of great relevance in understanding the orbit structures of vector fields. This was already emphasized by Poincaré and Birkhoff in their pioneering work on the qualitative theory of dynamical systems. One example is the description of the orbit space of a vector field in a neighbourhood of a closed orbit. As we saw in Chapter 3 this is done by using the Poincaré map (or local diffeomorphism) associated to a transversal section. At the end of Chapter 3 we generalized this idea by the process of suspending a diffeomorphism. Thus any diffeomorphism f of a manifold of dimension n represents the Poincaré map of a field X_f on a manifold of dimension $n + 1$. The field X_f is called the suspension of f and its orbits are in a natural correspondence with the orbits of f. In particular, X_f is a Kupka–Smale field if and only if f is a Kupka–Smale diffeomorphism. Also, X_f is structurally stable if and only if f is structurally stable.

Let $f \in \text{Diff}^r(M)$. A point $p \in M$ is *nonwandering* for f if, for any neighbourhood U of p and any integer $n_0 > 0$, there exists an integer n such that $|n| > n_0$ and $f^n U \cap U \neq \emptyset$. The set $\Omega(f)$ of nonwandering points is closed and invariant, that is, it consists of complete orbits of f. The limit sets $\omega(q)$ and $\alpha(q)$, for any $q \in M$, are contained in $\Omega(f)$. In particular, every fixed or periodic point of f belongs to $\Omega(f)$.

We say that $f \in \text{Diff}^r(M)$ is *Morse–Smale* if

(a) $\Omega(f)$ consists of a finite number of fixed and periodic points, all hyperbolic;
(b) the stable and unstable manifolds of the fixed and periodic points are all transversal to each other.

Next we list some important facts about Morse–Smale diffeomorphisms.

(1) The set of Morse–Smale diffeomorphisms is open (and nonempty) in $\text{Diff}^r(M)$ for any manifold M and any $r \geq 1$, [75].

(2) If $f \in \text{Diff}^r(M)$ is Morse–Smale then f is structurally stable [75], [79].

(3) The set of Morse–Smale diffeomorphisms is dense in $\text{Diff}^r(S^1)$, $r \geq 1$. This fact, due to Peixoto, can be proved directly from the Kupka–Smale Theorem for diffeomorphisms and an argument similar to that in the proof of Lemma 2.4 in this chapter. A more elegant proof is as follows. Let $f \in \text{Diff}^r(S^1)$. Take a C^∞ diffeomorphism $\tilde{f}$ C^r-close to f. Consider the suspension X_f of $\tilde{f}$, which is a C^∞ field defined on T^2 or K^2 depending on whether f preserves or reverses the orientation of S^1. We can consider S^1 as a global transversal section of $X_{\tilde{f}}$ on T^2 or K^2 with $\tilde{f}$ being the associated Poincaré map. If Y is a field on T^2 or K^2 that is C^r-close to $X_{\tilde{f}}$ then S^1 is also a transversal section for Y and the Poincaré map g associated to Y is C^r-close to $\tilde{f}$ and so to f. By the density of Morse–Smale fields in $\mathfrak{X}^r(T^2)$ or $\mathfrak{X}^r(K^2)$ it is possible to choose Y Morse–Smale and C^r-close to $X_{\tilde{f}}$. This implies that g is a Morse–Smale diffeomorphism C^r-close to f.

(4) The set of Morse–Smale diffeomorphisms is not dense in $\text{Diff}^r(M^n)$, $n \geq 2$. We describe next a nonempty open set $\mathscr{U} \subset \text{Diff}^r(S^2)$ such that $\mathscr{U} \cap \text{M-S} = \emptyset$. A similar example can be constructed on any manifold of dimension $n \geq 2$. Consider on S^2 a C^∞ field X with a saddle connection from one saddle to itself as in Figure 43; σ_1 and σ_2 are sinks, σ_4 is a source and σ_3 is a saddle, all hyperbolic. Let X_t be the flow induced by X and $f = X_1$ the time one diffeomorphism. Then σ_3 is a hyperbolic fixed point for f and one of the components of $W^s(\sigma_3) - \sigma_3$ coincides with one of the components of $W^u(\sigma_3) - \sigma_3$. We perturb f so as to obtain a diffeomorphism g that has σ_3 as a hyperbolic fixed point and has orbits of transversal intersection of $W^s(\sigma_3, g)$ with $W^u(\sigma_3, g)$ besides σ_3. To do this we take $p \in W^s(\sigma_3) \cap W^u(\sigma_3)$, $p \neq \sigma_3$, and a small neighbourhood U of p with $U \cap fU = \emptyset$. Let $i: S^2 \to S^2$ be a C^r diffeomorphism supported on U (so that i is the identity on $K =$

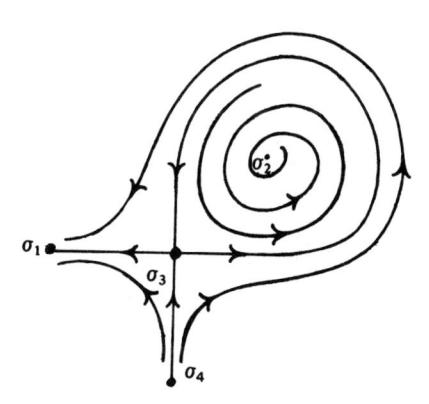

§4 General Comments on Structural Stability. Other Topics

Figure 43

$M - U$) with $i(p) = p$ and $W = i(W^u(\sigma_3))$ transversal to $W^s(\sigma_3)$ at p. Define $g = i \circ f$. We claim that g and any diffeomorphism close enough to g are not Morse–Smale. As $g = f$ outside U, σ_3 is a hyperbolic fixed point of g and the local stable and unstable manifolds of σ_3 for f and g coincide. But $W \subset W^u(\sigma_3, g)$. In fact if $x \in W$ then $i^{-1}(x) \in W^u(\sigma_3)$ and so $(i \circ f)^{-1}(x) = f^{-1}i^{-1}(x) \in W^u(\sigma_3) \cap K$. As i is the identity on K, $(i \circ f)^{-n}(x) = f^{-n}i^{-1}(x) \in W^u(\sigma_3) \cap K$ for $n \geq 1$. Thus, $x \in W^u(\sigma_3, g)$ since $g^{-n}(x) = (i \circ f)^{-n}(x)$ converges to σ_3 as $n \to \infty$ and so $W \subset W^u(\sigma_3, g)$. On the other hand, $W^s(\sigma_3) \cap U \subset W^s(\sigma_3, g)$. In fact if $y \in W^s(\sigma_3) \cap U$ then $f(y) \in W^s(\sigma_3) \cap K$ and so $g^n(y) = (i \circ f)^n(y) = f^n(y) \in W^s(\sigma_3) \cap K$ for $n \geq 1$. Thus $y \in W^s(\sigma_3, g)$ since $g^n(y)$ converges to σ_3 as $n \to \infty$ and this shows that $W^s(\sigma_3) \cap U \subset W^s(\sigma_3, g)$. Thus, $W^s(\sigma_3, g)$ is transversal to $W^u(\sigma_3, g)$ at p. Although it is not necessary we can extend this argument a little to make $W^s(\sigma_3, g)$ and $W^u(\sigma_3, g)$ transversal at all their points of intersection. This corresponds to one of the assertions of the Kupka–Smale Theorem for diffeomorphisms. A point p, as above, of transversal intersection of $W^s(\sigma_3, g)$ and $W^u(\sigma_3, g)$ is called a *transversal homoclinic point*. The reader is challenged

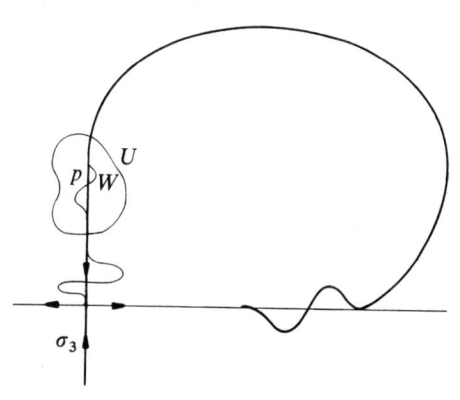

Figure 44

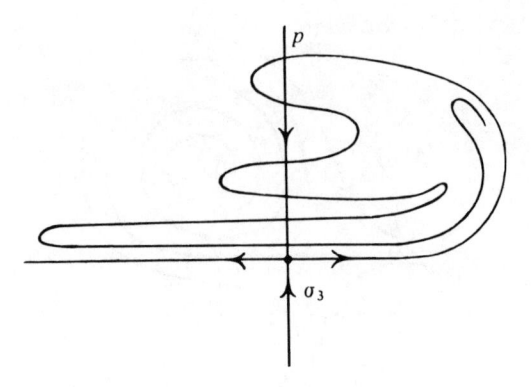

156   4 Genericity and Stability of Morse–Smale Vector Fields

Figure 45

to draw a diagram, with Figure 45 as a rough sketch, of the intersections of the stable and unstable manifolds along a transversal homoclinic orbit.

Birkhoff showed that p is accumulated by hyperbolic periodic orbits of g; Smale generalized this result to higher dimensions [108]; see also [66]. Here we only need the fact that p is nonwandering. To see this consider the arc l of $W^u(\sigma_3, g)$ going from σ_3 to p. For any neighbourhood U of p choose a small arc l_1 of $W^u(\sigma_3, g)$ in U passing through p. Since l_1 is transversal to $W^s(\sigma_3, g)$, $g^n(l_1)$ contains, by the λ-lemma, an arc arbitrarily close to l for all n greater than some $n_0 > 0$. As $l_1 \subset U$ and $l \cap U \neq \emptyset$ we have $g^n U \cap U \neq \emptyset$ for $n > n_0$. Thus $p \in \Omega(g)$ and as p is not periodic g is not Morse–Smale. The same happens for all diffeomorphisms close enough to g since they also have transversal homoclinic points. This follows from the fact that compact parts of the stable and unstable manifolds of a saddle do not change much in the C^r topology when we perturb the diffeomorphism a little. We can thus guarantee that these manifolds still have an orbit of transversal intersection distinct from the perturbed saddle.

From the nondensity of Morse–Smale diffeomorphisms on M^2 we can deduce, using suspension, that Morse-Smale fields are not dense in $\mathfrak{X}^r(M^n)$ for $n \geq 3$.

We shall now give another example, due to Thom, of a diffeomorphism with infinitely many periodic orbits. Then we shall show that this diffeomorphism is structurally stable.

This was one of the examples that motivated the definition given by Anosov of a class of structurally stable dynamical systems with infinitely many periodic orbits [3]. In particular, there exist stable systems that are not Morse–Smale.

Consider a linear isomorphism L of $\mathbb{R}^2$ that is represented with respect to the standard basis of $\mathbb{R}^2$ by a hyperbolic matrix with integer entries and determinant equal to 1. It is easy to see that the eigenvalues of L, λ and $1/\lambda$ with $|\lambda| < 1$, are irrational and that their eigenspaces E^s and E^u have irrational slope. As $\det L = 1$ it follows that L^{-1} has the same properties. If

§4 General Comments on Structural Stability. Other Topics 157

$\mathbb{Z}^2 \subset \mathbb{R}^2$ denotes the set of points with integer coordinates then $L(\mathbb{Z}^2) = \mathbb{Z}^2$. Consider a manifold structure on $T^2 = \mathbb{R}^2/\mathbb{Z}^2$ for which the natural projection $\pi \colon \mathbb{R}^2 \to T^2$ is a local diffeomorphism. This manifold structure can be obtained by identifying $\mathbb{R}^2/\mathbb{Z}^2$ with a torus of revolution as in Example 2, Section 1 of Chapter 1. We recall that $\pi(u, v) = \pi(u', v')$ if and only if $u' - u \in \mathbb{Z}$ and $v' - v \in \mathbb{Z}$. In this case $\pi(L(u, v)) = \pi(L(u', v'))$, which enables us to define a map $f \colon T^2 \to T^2$ by $f(\pi(u, v)) = \pi L(u, v)$. As π is a C^∞ local diffeomorphism it follows that f is of class C^∞. The same argument applies to L^{-1} so f is, in fact, a C^∞ diffeomorphism.

For each $p \in T^2$ and $x \in \mathbb{R}^2$ with $\pi(x) = p$ the curve $W^s(p) = \pi(x + E^s)$ is dense in T^2. Thus $\{W^s(p); p \in T^2\}$ defines a foliation of T^2, called the stable foliation, each leaf of which is dense in T^2. Moreover, this foliation is invariant by f, that is, $fW^s(p) = W^s(f(p))$. Similarly, we define the unstable foliation $\{W^u(p); p \in T^2\}$ by $W^u(p) = \pi(x + E^u)$. If we write E_p^s and E_p^u for the tangent spaces to $W^s(p)$ and $W^u(p)$ at p then $E_p^s = d\pi_x(E^s), E_p^u = d\pi_x(E^u)$ and $E_{f(p)}^s = df_p(E_p^s), E_{f(p)}^u = df_p(E_p^u)$.

On T^2 we consider the metric induced from $\mathbb{R}^2$ by π; that is, if $w_1, w_2 \in T(T^2)_{p=\pi(x)}$ then we define $\langle w_1, w_2 \rangle_p$ to be $\langle d\pi_x^{-1} w_1, d\pi_x^{-1} w_2 \rangle$. In this metric we have

$$\|df_p v\| = |\lambda| \|v\|, \quad \text{if } v \in E_p^s;$$

$$\|df_p w\| = |\lambda|^{-1} \|w\|, \quad \text{if } w \in E_p^u.$$

It follows from this that if $q \in W^s(p)$ then $d(f^n(q), f^n(p)) \to 0$ as $n \to \infty$ and that if $q \in W^u(p)$ then $d(f^{-n}(q), f^{-n}(p)) \to 0$ as $n \to \infty$. Hence every periodic point p of f is hyperbolic and the stable and unstable manifolds of p are the $W^s(p)$ and $W^u(p)$ defined above. Moreover, for any $p, q \in T^2$ we have $W^s(p)$ transversal to $W^u(q)$ and $W^s(p) \cap W^u(q)$ is dense in T^2. In particular, $p = \pi(0)$ is a hyperbolic fixed point of f and its transversal homoclinic points are dense in T^2. As in the previous example this implies that f and any diffeomorphism near f are not Morse–Smale. Birkhoff's result and the density of the transversal homoclinic points imply that the periodic points of f are dense in T^2. We shall now give a direct proof of this.

4.1 Proposition. *The periodic points of $f \colon T^2 \to T^2$ are dense in T^2.*

PROOF. Let $\mathscr{L}$ be the set of points in $\mathbb{R}^2$ with rational coordinates. We shall show that $\pi(\mathscr{L})$ coincides with the set $\text{Per}(f)$ of periodic points of f. As $\mathscr{L}$ is dense in $\mathbb{R}^2$ it follows that $\text{Per}(f)$ is dense in T^2. If $\mathscr{L}_n = \{(m_1/n, m_2/n); m_1, m_2 \in \mathbb{Z}\}$ then $\mathscr{L} = \bigcup_{n \geq 1} \mathscr{L}_n$. As L is a matrix with integer entries we have $L(\mathscr{L}_n) = \mathscr{L}_n$. Therefore $f(\pi \mathscr{L}_n) = \pi \mathscr{L}_n$. As $\pi \mathscr{L}_n = \pi\{(m_1/n, m_2/n); m_1, m_2 \in \mathbb{Z}, 0 \leq m_1 \leq n, 0 \leq m_2 \leq n\}$ we deduce that $\pi \mathscr{L}_n$ is a finite invariant subset of T^2 and so its points are periodic. Hence $\pi(\mathscr{L}) \subset \text{Per}(f)$. On the other hand, let $x \in \mathbb{R}^2$ satisfy $f^n(\pi(x)) = \pi(x)$ for some integer n. We claim that x has rational coordinates. In fact since $\pi(L^n x) = \pi(x)$ the point $y = L^n x - x$ has integer coordinates. As L is a hyperbolic matrix with integer entries it follows that $L^n - I$ is an invertible matrix with integer entries and so $(L^n - I)^{-1}$ is a matrix with rational entries. Thus $x = (L^n - I)^{-1}y$ has rational coordinates and so $\text{Per}(f)$ is contained in $\pi(\mathscr{L})$, which proves the claim. □

At first sight it may seem very difficult to show that the diffeomorphism f, which has infinitely many periodic orbits, is structurally stable. There is, however, one useful property: f has a global hyperbolic structure and is even induced by a linear isomorphism of $\mathbb{R}^2$. In particular, all the periodic orbits of f are saddles with stable manifolds of the same dimension. By contrast, a Morse–Smale diffeomorphism must have sources and sinks and, usually, saddles too. We shall now give a simple and elegant proof, due to Moser [65], that f is structurally stable. This proof is very close to the analytic proof of the Grobman–Hartman Theorem in Chapter 2. We recall that f is induced by an isomorphism L of $\mathbb{R}^2$ where L is defined by a hyperbolic matrix with integer entries and determinant 1.

4.2 Theorem. *The diffeomorphism $f: T^2 \to T^2$ is structurally stable.*

PROOF. Take $g \in \text{Diff}^r(T^2)$ near f. We claim that there exists a diffeomorphism $G: \mathbb{R}^2 \to \mathbb{R}^2$ near L that induces g on T^2. In fact, for each $x \in \mathbb{R}^2$ we can consider $f\pi(x) = \pi L(x)$ where π is the canonical projection from $\mathbb{R}^2$ to T^2. As $g\pi(x)$ is near $f\pi(x)$ there exists a unique point $y \in \mathbb{R}^2$ near $L(x)$ with $\pi(y) = g\pi(x)$. We define $G(x) = y$ and get $\pi G(x) = g\pi(x)$. It is easy to check that G and L are C^r-close. We now write $G = L + \Phi$ where Φ is a C^r small map of $\mathbb{R}^2$. As L is hyperbolic we know from Lemma 4.3 of Chapter 2 that L and $L + \Phi$ are conjugate. That is, there exists a homeomorphism H of $\mathbb{R}^2$ such that $HL = GH$. It is, therefore, enough to check that H induces a homeomorphism h of T^2 with $\pi H = h\pi$ because this will imply that $hf = gh$. In fact $\pi HL = h\pi L = hf\pi$; similarly $\pi GH = g\pi H = gh\pi$ and so $hf\pi = gh\pi$. Since $\pi: \mathbb{R}^2 \to T^2$ is surjective we obtain $hf = gh$. Now we check that the homeomorphism H of $\mathbb{R}^2$ induces a homeomorphism h of T^2. In solving the equation $HL = GH$ we write $H = I + u$ and $G = L + \Phi$ and obtain $uL = Lu + \Phi(I + u)$. We want a solution $u \in C_b^0(\mathbb{R}^2)$ and, for this theorem, we need $I + u$ to project to a map of T^2. This last requirement is equivalent to the following: for each $x \in \mathbb{R}^2$ and each point p with integer coordinates there exists q with integer coordinates such that $(I + u)(x + p) = q + (I + u)(x)$. That is, $u(x + p) = q - p + u(x)$. But u will be constructed with small norm for g near f so we deduce that $u(x + p) = u(x)$ for any $x \in \mathbb{R}^2$ and any p with integer coordinates. Thus we are led to consider the subspace $\mathscr{P}$ of $C_b^0(\mathbb{R}^2)$ consisting of periodic functions $u \in C_b^0(\mathbb{R}^2)$ satisfying $u(x + p) = u(x)$ for any x and p in $\mathbb{R}^2$ where p has integer coordinates. It is immediate that $\mathscr{P}$ is closed in $C_b^0(\mathbb{R}^2)$ and that $\mathscr{L}(\mathscr{P}) \subset \mathscr{P}$ where the operator $\mathscr{L}: C_b^0(\mathbb{R}^2) \to C_b^0(\mathbb{R}^2)$ is defined by $\mathscr{L}(u) = uL - Lu$. Moreover, $\mathscr{L}$ is invertible because L is hyperbolic. On the other hand, as $G = L + \Phi$ does

§4 General Comments on Structural Stability. Other Topics   159

project to a map of T^2 and Φ is C^r small, we have $\Phi \in \mathcal{P}$ for the same reason as above. Therefore, the map $\mu: \mathcal{P} \to \mathcal{P}$, $\mu(u) = \mathcal{L}^{-1}(\Phi(I + u))$ is well defined and is a contraction. The unique fixed point u of μ satisfies the equation $uL - Lu = \Phi(I + u)$ or, equivalently, $(I + u)L = (L + \Phi)(I + u)$. The proof that $H = I + u$ is a homeomorphism is as in Lemma 4.3 of Chapter 2. As $u \in \mathcal{P}$ the homeomorphism $H = I + u$ projects to a homeomorphism h of T^2 and $hf = gh$. This shows that f is structurally stable in Diff$^r(T^2)$, $r \geq 1$. □

This diffeomorphism $f: T^2 \to T^2$ is a good example of an Anosov diffeomorphism and we now give the general definition.

Definition. Let M be a compact manifold. We say that $f \in \text{Diff}^r(M)$, $r \geq 1$, is an *Anosov diffeomorphism* if:

(a) the tangent bundle of M decomposes as a continuous direct sum, $TM = E^s \oplus E^u$;
(b) the subbundles E^s and E^u are invariant by the derivative Df of f; that is, $Df_x E^s_x = E^s_{f(x)}$ and $Df_x E^u_x = E^u_{f(x)}$ for all $x \in M$;
(c) there exists a Riemannian metric on M and a constant $0 < \lambda < 1$ such that $\|Df_x v\| \leq \lambda \|v\|$ and $\|Df_x^{-1} u\| \leq \lambda \|u\|$ for any $x \in M$, $v \in E^s_x$ and $u \in E^u_x$.

Anosov was the first to prove that these diffeomorphisms are structurally stable in Diff$^r(M)$ for M of any dimension (see [3], [65] and the appendix by Mather in [109]). He also defined an analogous class of vector fields and proved their structural stability. The suspension of an Anosov diffeomorphism is an example of such a vector field. Another important case is that of the geodesic flow for a manifold of negative curvature [3].

We remark that the definition of Anosov diffeomorphism imposes strong restrictions on the manifold. For example, among the compact manifolds of dimension two only the torus admits Anosov diffeomorphisms. It is even believed that Anosov diffeomorphisms only exist on very special manifolds such as the torus T^n and nilmanifolds [109], [113]. By contrast there exist Morse–Smale diffeomorphisms on any manifold. It is also known that every Anosov diffeomorphism on T^n is conjugate to one induced by a linear isomorphism of $\mathbb{R}^n$ as in the example described above [55]. It is conjectured that any Anosov diffeomorphism has its periodic orbits dense in the manifold. On these questions various important results were obtained by Franks [20], Manning [55], Newhouse [67] and Farrell–Jones [18]. Recently, a very interesting class of Anosov flows was constructed by Franks and Williams [24] for which the nonwandering set is not all of the ambient manifold. The corresponding question for Anosov diffeomorphisms remains open.

We thus have two classes of structurally stable diffeomorphisms, Morse–Smale and Anosov, that have, as we have already emphasized, very different properties. Smale then introduced a new class of diffeomorphisms encompassing the previous two. These diffeomorphisms, which we shall soon describe, are structurally stable and it is conjectured that they include every structurally stable diffeomorphism.

Consider a compact manifold M. Let $f \in \text{Diff}^r(M)$ and let $\Lambda \subset M$ be a closed invariant set, $f\Lambda = \Lambda$. We say that Λ is *hyperbolic* for f if:

(a) the tangent bundle of M restricted to Λ decomposes as a continuous direct sum, $T_\Lambda M = E_\Lambda^s \oplus E_\Lambda^u$, which is invariant by Df;
(b) there exists a Riemannian metric and a number $0 < \lambda < 1$ such that $\|Df_x v\| \leq \lambda \|v\|$ and $\|Df_x^{-1} u\| \leq \lambda \|u\|$ for any $x \in \Lambda$, $v \in E_x^s$ and $u \in E_x^u$.

Now consider the nonwandering set $\Omega = \Omega(f)$, which is closed and invariant.

Definition. We say that f satisfies *Axiom A* if Ω is hyperbolic for f and $\Omega = \overline{\text{Per}(f)}$, that is, the periodic points of f are dense in Ω.

If f satisfies Axiom A, Smale showed that $\Omega = \Omega(f)$ decomposes as a finite disjoint union of closed subsets which are invariant and transitive, $\Omega = \Omega_1 \cup \Omega_2 \cup \cdots \cup \Omega_k$. For details see [109], [70], [72], [77], [101]. The sets Ω_i are called *basic sets*. *Transitivity* means that there exist dense orbits in each Ω_i. In each Ω_i all the periodic orbits have stable manifolds of the same dimension. In the case of Morse–Smale diffeomorphisms the basic sets are periodic orbits. In the case of the Anosov diffeomorphisms on T^2 there is only one basic set, the whole of T^2. In general these basic sets can have a much more complicated structure as we shall see in examples. As in the Morse–Smale case they can be of attractor, repellor or saddle type. In Examples 3 through 6 below, we exhibit interesting cases of nonperiodic basic sets, which are attractors, repellors and saddles. Examples 5 and 6 can be seen as special instances of a quite large class of Axiom A attractors whose structure was described by Williams [121].

With regard to the above definition of Axiom A it is known that, when $\dim M = 2$, $\Omega(f)$ hyperbolic implies $\overline{\text{Per}(f)} = \Omega(f)$ and that this is not true in higher dimensions [71], [15].

Let us make one last remark about the transitivity requirement for basic sets. Since this is a difficult requirement to check directly, the following criterion may be more useful, as in Examples 4 and 5, below. If Λ is a hyperbolic set for f with periodic points dense and if $\Lambda \supset W^s(p) \cap W^u(q) \neq \emptyset$ for every $p, q \in \text{Per}(f) \cap \Lambda$, then Λ is transitive. In fact, for any $p \in \text{Per}(f) \cap \Lambda$, $W^s(p)$ will accumulate on $\text{Per}(f) \cap \Lambda$ and so $W^s(p)$ is dense in Λ. For each $p \in \text{Per}(f) \cap \Lambda$ choose a base of neighbourhoods $U_1^p, U_2^p, \ldots, U_k^p, \ldots$ of p in Λ. Then $V_k^p = \bigcup_{n \in \mathbb{Z}} f^n U_k^p$ is open and dense in Λ. As Λ is closed it satisfies the Baire property and the periodic orbits of f in Λ are countable since they are hyperbolic. Thus, $D = \bigcap_{p,k} V_k^p$, for $p \in \text{Per}(f) \cap \Lambda$ and $k \in \mathbb{N}$, is dense in Λ. It is easy to see that any $x \in D$ has a

§4 General Comments on Structural Stability. Other Topics

dense orbit in Λ. In fact, given a nonempty open subset A of Λ, there exists some $U_k^p \subset A$ and, as $x \in V_k^p = \bigcup_n f^n U_k^p$, it follows that $f^{-n}(x) \in U_k^p \subset A$ for some $n \in \mathbb{Z}$.

With structural stability in mind we are going to generalize the concepts of stable and unstable manifold to nonperiodic orbits. Let $f \in \text{Diff}^r(M)$ and choose $x \in M$. We define $W^s(x) = \{y \in M; d(f^n x, f^n y) \to 0 \text{ as } n \to \infty\}$ and $W^u(x) = \{y \in M; d(f^n x, f^n y) \to 0 \text{ as } n \to -\infty\}$ where d denotes the metric induced by the Riemannian metric on M. When $\Omega = \Omega(f)$ is hyperbolic and $x \in \Omega$ then $W^s(x)$ and $W^u(x)$ are C^r injective immersions of Euclidean spaces $\mathbb{R}^s$ and $\mathbb{R}^u$ of complementary dimensions [39]. That is the justification in this case for calling the sets $W^s(x)$ and $W^u(x)$ manifolds. In each basic set Ω_i of a diffeomorphism that satisfies Axiom A the stable manifolds of all the orbits in Ω_i have the same dimension. Furthermore, their union coincides with the set of points whose ω-limit set is contained in Ω_i, see [9]. In particular, if Ω_i is an attractor then the union of the stable manifolds of points in Ω_i is a neighbourhood of Ω_i. In the case of an Anosov diffeomorphism of T^2 induced by a linear isomorphism of $\mathbb{R}^2$ the stable manifolds are the projections to T^2 of parallel lines in $\mathbb{R}^2$. The corresponding statements hold for unstable manifolds.

Definition. Let $f \in \text{Diff}^r(M)$ satisfy Axiom A. We say that f satisfies the *transversality condition* if $W^s(x)$ and $W^u(y)$ are transversal for any $x, y \in \Omega(f)$.

Before discussing the structural stability of diffeomorphisms that satisfy Axiom A and the transversality condition we shall give some examples of them.

EXAMPLE 1. Morse–Smale diffeomorphisms.

EXAMPLE 2. Anosov diffeomorphisms of the torus T^2 induced by linear isomorphisms of $\mathbb{R}^2$. It is also true, although we shall not prove it, that any Anosov diffeomorphism satisfies Axiom A and the transversality condition.

EXAMPLE 3. Let $g: S^1 \to S^1$ be a Morse–Smale diffeomorphism with two fixed points, the north pole n is a repellor and the south pole s is an attractor. Let $f: T^2 \to T^2$ be an Anosov diffeomorphism induced by a linear isomorphism of $\mathbb{R}^2$. Consider the product diffeomorphism $g \times f: S^1 \times T^2 \to S^1 \times T^2$. It is easy to see that its nonwandering set has two parts, $\Omega_1 = \{n\} \times T^2$ and $\Omega_2 = \{s\} \times T^2$ and that they are hyperbolic and transitive. Here Ω_1 is a repellor and Ω_2 is an attractor. Let us check the transversality condition. Let $(z, w) \in S^1 \times T^2$ be a point of intersection of $W^s(x)$ and $W^u(y)$ where $x \in \Omega_2$ and $y \in \Omega_1$. Now $W^s(x)$ is the product of W_1^s in $S^1 \times \{w\}$ and W_2^s in $\{z\} \times T^2$. Similarly, let us denote by W_1^u and W_2^u the factors of $W^u(y)$ in $S^1 \times \{w\}$ and $\{z\} \times T^2$. As $W_1^s = (S^1 - n) \times \{w\}$ and $W_1^u = (S^1 - s) \times \{w\}$, W_1^s and W_1^u are transversal in $S^1 \times \{w\}$. Moreover, W_2^s

and W_2^u are transversal in $\{z\} \times T^2$ because they are the projections of lines in $\mathbb{R}^2$ parallel to independent eigenvectors. This proves that $W^s(x)$ and $W^u(y)$ are transversal. If x and y both belong to Ω_1 or both to Ω_2 the transversality of $W^s(x)$ and $W^u(y)$ is immediate. Thus the diffeomorphism $g \times f$: $T^3 \to T^3$ satisfies Axiom A and the transversality condition. It is clear that $g \times f$ is neither Morse–Smale nor Anosov.

EXAMPLE 4. In this example we describe a diffeomorphism f of S^2 satisfying Axiom A and the transversality condition. The nonwandering set $\Omega = \Omega(f)$ consists of three basic sets: Ω_1 is a repelling fixed point, Ω_3 is an attracting fixed point and Ω_2 is a Cantor set in which the periodic saddles are dense. The important part of this example is the "Smale horseshoe". At the north pole of the sphere we put a hyperbolic source Ω_1, and the whole northern hemisphere H_+, including the equator, belongs to its unstable manifold $W^u(\Omega_1)$. Therefore, if H_- denotes the southern hemisphere $f(H_-) \subset$ Int H_-. We now describe f on H_-. The map l is linear, compresses horizontal lines by a compression factor $0 < \lambda < \frac{1}{4}$ and expands vertical lines by an expansion factor $\mu > 4$. The map g twists the rectangle $l(Q)$ at its middle quarter making a horseshoe F and translates it to the position indicated in Figure 46. Thus, $Q \cap fQ$ has two rectangular components $\tilde{R}_1$ and $\tilde{R}_2$ that

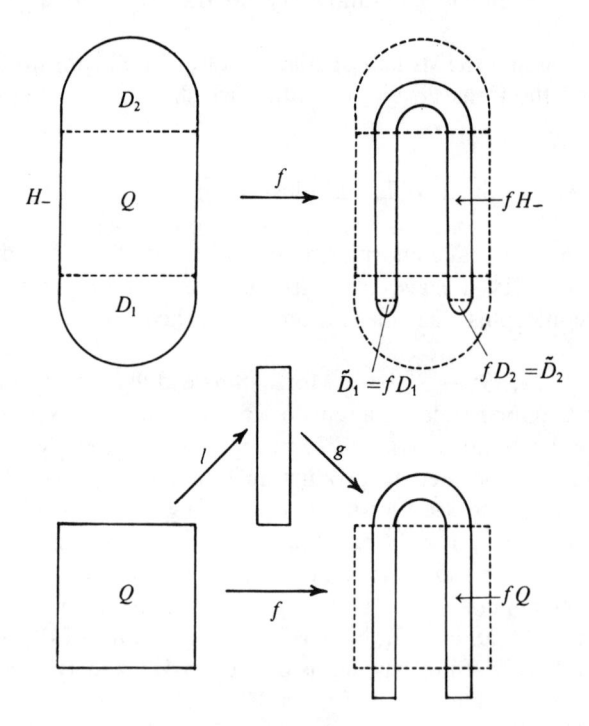

Figure 46

§4 General Comments on Structural Stability. Other Topics 163

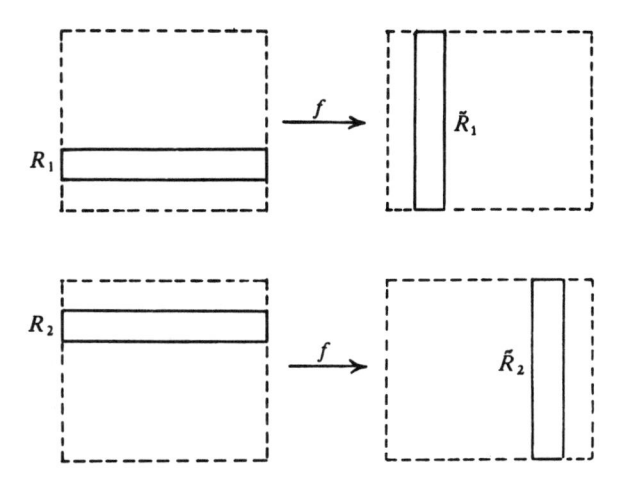

Figure 47

are images of the rectangles R_1 and R_2 in Q. In the rectangles R_1 and R_2 the diffeomorphism f is affine (a linear map composed with a translation), contracting horizontal lines by λ and expanding vertical lines by μ. Finally, at the centre of the disc $\tilde{D}_1 = fD_1$ we put a hyperbolic attracting fixed point Ω_3 and $\tilde{D}_1 \subset W^s(\Omega_3)$. Let us analyse the set $\Omega = \Omega(f)$. If $x \in H_+$ and x is not the north pole then x is wandering, since it belongs to the unstable manifold of the source. If $x \in \tilde{D}_1$ and $x \neq \Omega_3$ then x is wandering, since it belongs to the stable manifold of a sink. If $x \in D_1$ then $f(x) \in \tilde{D}_1$ and x is wandering. If $x \in D_2$ then $f(x) \in D_1$ and x is again wandering. We conclude that, when $x \in H_-$ and $x \neq \Omega_3$, x can only be nonwandering if its orbit is entirely contained in Q. Therefore, if $x \in \Omega$ but $x \neq \Omega_1$ and $x \neq \Omega_3$ we have $x \in \bigcap_{n \in \mathbb{Z}} f^n Q = \Lambda$.

We now describe the set Λ. As $Q \cap fQ$ has two rectangular components, $Q \cap fQ \cap f^2Q$ has four rectangular components and so on. Schematically we have: Q; $Q \cap fQ = \tilde{R}_1 \cup \tilde{R}_2$; $Q \cap fQ \cap f^2Q = \tilde{R}_{11} \cup \tilde{R}_{12} \cup \tilde{R}_{21} \cup \tilde{R}_{22}$; etc. The suffixes are attached to the rectangles in the following way: $f\tilde{R}_1 \cap Q = \tilde{R}_{11} \cup \tilde{R}_{12}$ and $\tilde{R}_{11} \subset \tilde{R}_1$, $\tilde{R}_{12} \cap \tilde{R}_1 = \emptyset$. Similarly, $f\tilde{R}_2 \cap Q = \tilde{R}_{21} \cup \tilde{R}_{22}$ and $\tilde{R}_{21} \cap \tilde{R}_2 = \emptyset$, $\tilde{R}_{22} \subset \tilde{R}_2$. In general, $\tilde{R}_{\sigma_1 \ldots \sigma_p \sigma_{p+1}} \subset f\tilde{R}_{\sigma_1 \ldots \sigma_p}$ and $\sigma_{p+1} = \sigma_p$ if $\tilde{R}_{\sigma_1 \ldots \sigma_p \sigma_{p+1}} \subset \tilde{R}_{\sigma_1 \ldots \sigma_p}$ and $\sigma_{p+1} \neq \sigma_p$ if they are disjoint. Here each σ_i is 1 or 2. Note that, given any integer $k > 0$, $\tilde{R}_1 \supset \tilde{R}_{11 \ldots 1}$ where the suffix 1 is repeated k times. That is, $\tilde{R}_{11 \ldots 1} \subset f^k \tilde{R}_1 \cap Q$ and, in particular, $f^k \tilde{R}_1 \cap \tilde{R}_1 \neq \emptyset$. The same is true for $\tilde{R}_{\sigma_1 \sigma_2 \ldots \sigma_p}$ where each σ_i can be 1 or 2. Consider any horizontal line α such that $\alpha \cap Q \neq \emptyset$ and let $[a, b] = \alpha \cap Q$. Then $[a, b] \cap fQ$ is two closed segments and is obtained from $[a, b]$ by removing three disjoint segments. From each of these two segments we remove three more segments to form $[a, b] \cap fQ \cap f^2Q$, and so on. The reader will recognize that $[a, b] \cap (\bigcap_{n \geq 0} f^n Q)$ is a

Cantor set. If we take a vertical line and follow the same reasoning we find that the intersection of this line with $\bigcap_{n \leq 0} f^n Q$ is also a Cantor set. As f is affine, $\Lambda = \bigcap_{n \in \mathbb{Z}} f^n Q$ is a Cantor set, the product of one Cantor set in a horizontal line and another in a vertical line. The hyperbolicity of Λ is clear: vertical segments are sent by f to vertical segments with an expansion greater than 1 and, dually, horizontal segments remain horizontal and are contracted. Now take $x \in \Lambda$ and let R be a rectangle in Q with two vertical sides of the same height as Q and with $x \in R$. Note that however small the width of R it always contains one of the rectangles $\tilde{R}_{\sigma_1 \sigma_2 \ldots \sigma_p}$. This is because Λ is contained in the intersection of all the rectangles $\tilde{R}_{\sigma_1 \sigma_2 \ldots \sigma_p}$. Let us now show that $x \in \Lambda$ is nonwandering. Let $Q_x \subset Q$ be a square containing x in its interior and let N be a positive integer. We shall show that $f^n Q_x \cap Q_x \neq \emptyset$ for some $n > N$. As f expands vertical segments there exists an integer $m \geq 0$ such that $f^m Q_x \cap Q$ contains a rectangle $R_{f^m(x)}$ of height equal to that of the original square Q. Λ is invariant by f so that $f^m(x) \in \Lambda$. Thus there exists a rectangle $\tilde{R}_{\sigma_1 \sigma_2 \ldots \sigma_p} \subset R_{f^m x}$. As we have already noted, it is possible to choose an integer $k > N + m$ such that $f^k \tilde{R}_{\sigma_1 \sigma_2 \ldots \sigma_p} \cap \tilde{R}_{\sigma_1 \sigma_2 \ldots \sigma_p} \neq \emptyset$. This implies that $f^k R_{f^m x} \cap R_{f^m x} \neq \emptyset$ and so $f^n Q_x \cap Q_x \neq \emptyset$ where $n = k - m > N$. This shows that any $x \in \Lambda$ is nonwandering. Let us next prove that the periodic points of f are dense in Λ. In fact, by the previous argument, for any $x \in \Lambda$ and any square Q_x containing x, we have $f^n Q_x \cap Q_x \neq \emptyset$ for some large n. The map f^n contracts the horizontal sides and expands the vertical sides of Q_x linearly. From this expansion we deduce that there exists a horizontal segment l_h in Q_x for which $f^n l_h \subset l_h$. From the contraction we deduce that there exists a vertical segment l_v in Q_x such that $f^n l_v \supset l_v$. Thus, $l_h \cap l_v$ is a fixed point of f^n, that is, a periodic point of f. This shows that the periodic points are dense in Λ. Before we can call Λ a basic set it remains to prove that there is a dense orbit in Λ. According to the criterion which we gave when we defined basic sets, in order to have transitivity in Λ it is sufficient that the stable and unstable manifolds of any

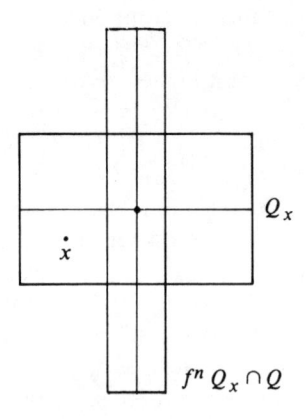

Figure 48

§4 General Comments on Structural Stability. Other Topics

periodic orbits should have nonempty intersection. But this is just what happens in this example because the stable manifolds contain horizontal segments and the unstable manifolds contain vertical segments right across the square Q. We deduce that the basic sets of $f: S^2 \to S^2$ are Ω_1, $\Omega_2 = \Lambda$, Ω_3. The transversality condition is immediate since Ω_1 is a repelling fixed point and Ω_3 is an attracting fixed point. Thus f satisfies Axiom A and the transversality condition.

We remark that a construction like the horseshoe can also be made in dimensions greater than two [108], [66], [70], [72].

EXAMPLE 5. We describe here another significant example of a C^∞ diffeomorphism on the torus T^2 that satisfies Axiom A and the transversality condition. It is due to Smale (see [109], [122]) and known as the DA diffeomorphism, meaning "derived from Anosov". Its nonwandering set is hyperbolic and consists of two basic sets: a repelling fixed point and a one-dimensional attractor which is locally homeomorphic to the product of an interval and a Cantor set. We start with an Anosov diffeomorphism $g: T^2 \to T^2$ induced by a linear isomorphism L of $\mathbb{R}^2$, via the natural projection $\pi: \mathbb{R}^2 \to T^2$ as in Example 2. Let v^s and v^u be a contracting and a repelling eigenvector of L, respectively. Let e^s and e^u be the vector fields on T^2 defined by $e^s(\pi(x)) = d\pi(x) \cdot v^s$ and $e^u(\pi(x)) = d\pi(x) \cdot v^u$. On T^2 we consider a Riemannian metric for which $\{e^s(p), e^u(p)\}$ is an orthonormal basis of $T(T^2)_p$, for each $p \in T^2$. Hence, $dg(p) \cdot e^s(p) = \lambda e^s(g(p))$ and $dg(p) \cdot e^u(p) = \mu e^u(g(p))$, where λ and $\mu = 1/\lambda$ are the eigenvalues of L. Notice that e^s and e^u are C^∞ vector fields and their orbits, which are the leaves of the stable and the unstable foliations, respectively, are dense in T^2. Now we consider a diffeomorphism f of T^2 satisfying the following properties:

(1) f is equal to g in the complement of a small neighbourhood U of the fixed point $p_0 = \pi(0)$ of g;
(2) f preserves the stable foliation of g inducing the same map on the space of leaves; i.e. $f(W^s(p)) = W^s(g(p))$ for each $p \in T^2$;
(3) p_0 is a repelling fixed point for f;
(4) if we define $\alpha, \beta: T^2 \to \mathbb{R}$ by $df(p) \cdot e^s(p) = \alpha(p) e^s(f(p))$ and $df(p) \cdot e^u(p) = \beta(p) e^s(f(p)) + \mu e^u(f(p))$, then $\bar{\beta}^2 < (\mu^2 - 1)(\mu - 1)^2$, where $\bar{\beta} = \sup|\beta(p)|$ for $p \in T^2$, and there exists a neighbourhood V of p_0 in the unstable manifold of p_0 such that $0 < \alpha(p) < \bar{\alpha} < 1$ for some constant $\bar{\alpha}$ and all $p \in T^2 - V$.

Before proving the existence of a diffeomorphism f satisfying the four properties above, we shall describe its dynamics and show that it satisfies Axiom A. Let L_0 denote the leaf of the stable foliation of g through the fixed point p_0. From properties (3) and (4), we have that the restriction of f to L_0 is an expansion near p_0, a contraction on $L_0 - V$ and $V \subset W^u(p_0, f)$.

It follows that f has one and only one fixed point in each connected component of $L_0 - \{p_0\}$. These fixed points, p_1 and p_2, are saddle points, and since $L_0 - \{p_0\} = W^s(p_1, f) \cup W^s(p_2, f)$, we have that $W^s(p_i, f)$ is dense in the torus for $i = 1, 2$. If we set $\Lambda = T^2 - W^u(p_0, f)$ and since $V \subset W^u(p_0, f)$, we conclude that $\Omega(f) \subset \{p_0\} \cup \Lambda$ and that Λ contains the closure of the unstable manifolds of p_i for $i = 1, 2$. We claim that the converse is also true and that, in fact, $W^u(p_0, f)$ is dense in T^2 and thus Λ is the closure of the homoclinic points of p_i for $i = 1, 2$. To prove the claim we start by observing that, since $W^s(p_i)$ is dense in T^2, it intersects $W^u(p_j)$ for $i, j = 1, 2$. Hence, $W^u(p_i)$ accumulates on $W^u(p_j)$ for $i, j = 1, 2$. Let now $p \in \Lambda$ and W be a neighbourhood of p. Let $I \subset W$ be any small interval contained in the stable manifold of some periodic point for g. Because of property (2), the leaf L_1 of the stable foliation which contains I is preserved under a power of f. Since f^{-1} expands L_1 in the complement of V and L_1 is dense, there exists an integer n such that $f^{-n}(I) \cap V \neq \emptyset$. This proves that p is in the closure of $W^u(p_0, f)$. But p does not belong to $W^u(p_0, f)$ and so either $W^u(p_1, f)$ or $W^u(p_2, f)$ intersects W and, therefore, both of them intersect W. This proves that $W^u(p_0, f)$ is dense in T^2 and that both $W^u(p_1, f)$ and $W^u(p_2, f)$ are dense in Λ. From the density of $W^s(p_i, f)$, we conclude that W contains transversal homoclinic points associated to p_i as claimed. In fact, no component of $W^u(p_i, f) \cap W$ can be contained in a stable leaf. Otherwise, by taking negative iterates, we get part of the local unstable manifold of $W^u(p_i, f)$ along the stable foliation, which is clearly impossible. Thus, the homoclinic orbits associated to p_i for $i = 1, 2$ are dense in Λ. In particular, $\Omega(f) = \{p_0\} \cup \Lambda$. Now let us prove that Λ has a hyperbolic structure. For $a, \rho \in \mathbb{R}$ with $-1 \leq a < a + \rho < 1$ and $p \in \Lambda$, consider the cone $C_p(a, \rho) = \{xe^s(p) + ye^u(p); y \neq 0 \text{ and } a \leq x/y \leq a + \rho\}$. From property (4) it follows that the image of $C_p(a, \rho)$ by df_p is the cone $C_{f(p)}(a', \rho')$, where $a' = (\alpha(p)/\mu)a + \beta(p)$ and $\rho' = (\alpha(p)/\mu)\rho$. Since $(\alpha(p)/\mu) < \bar{\alpha}/\mu < 1$, we have that $\bigcap_{n \geq 0} df_{f^{-n}(p)}^n(C_{f^{-n}(p)}(-1, 1))$ is a one-dimensional subspace $E_p^u \subset T(T^2)_p$. Clearly, $df_p(E_p^u) = E_{f(p)}^u$. Thus, for each $p \in \Lambda$, we have a decomposition $E_p^s \oplus E_p^u$ of the tangent space of T^2 at p, where E_p^s is the subspace generated by $e^s(p)$. Such a decomposition is invariant under the derivative of f and E^s is contracted by $\bar{\alpha}$. It remains to show that df_p uniformly expands vectors in E_p^u. To see this, let us estimate the slope of the subspace E_p^u. Let $v = xe^s(p) + ye^u(p)$ be a vector in E_p^u and let $v_n = df_p^{-n}v = x_n e^s(f^{-n}(p)) + y_n e^u(f^{-n}(p))$. From the definition of E_p^u, we get that $|x_n/y_n| \leq 1$ for all $n \geq 0$. On the other hand,

$$\frac{x_{n-1}}{y_{n-1}} = \frac{\alpha(f^{-n}(p))}{\mu} \cdot \frac{x_n}{y_n} + \frac{\beta(f^{-n}(p))}{\mu}$$

because $v_{n-1} = df(f^{-n}(p)) \cdot v_n$. Thus,

$$\left|\frac{x_{n-1}}{y_{n-1}}\right| \leq \frac{\bar{\alpha}}{\mu}\left|\frac{x_n}{y_n}\right| + \frac{\bar{\beta}}{\mu} \quad \text{for all } n \geq 1.$$

§4 General Comments on Structural Stability. Other Topics

By induction, for all $n \geq 1$ we conclude that

$$\left|\frac{x}{y}\right| \leq \left(\frac{\bar{\alpha}}{\mu}\right)^n \left|\frac{x_n}{y_n}\right| + \frac{\bar{\beta}}{\mu}\sum_{i=0}^{n-1}\left(\frac{\bar{\alpha}}{\mu}\right)^i \leq \left(\frac{1}{\mu}\right)^n \left|\frac{x_n}{y_n}\right| + \frac{\bar{\beta}}{\mu}\sum_{i=0}^{n-1}\left(\frac{1}{\mu}\right)^i$$
$$= \left(\frac{1}{\mu}\right)^n \left|\frac{x_n}{y_n}\right| + \frac{\bar{\beta}}{\mu}\frac{1-(1/\mu)^n}{1-(1/\mu)}.$$

Hence, $|x/y| \leq \bar{\beta}/(\mu - 1)$. Now let us prove that df_p uniformly expands vectors in E_p^u. Let $v = xe^s(p) + ye^u(p)$ be a vector in E_p^u and let $\tilde{v} = \tilde{x}e^s(f(p)) + \tilde{y}e^u(f(p))$ be its image under df_p. By property (4) and the expression above, we have that

$$\frac{\|\tilde{v}\|^2}{\|v\|^2} = \frac{\tilde{x}^2 + \tilde{y}^2}{x^2 + y^2} \geq \frac{\mu^2 y^2}{x^2 + y^2} = \frac{\mu^2}{(x^2/y^2) + 1} \geq \frac{\mu^2}{(\bar{\beta}^2/(\mu-1)^2) + 1} > 1.$$

This proves our assertion. We leave the reader to verify that the bundles E^s and E^u are continuous (see Exercise 44). Let us show that the periodic orbits are dense in Λ. This fact follows from the density of transversal homoclinic orbits proved above and Birkhoff's Theorem stating that such homoclinic orbits are accumulated by periodic ones. However, we are going to present a very instructive proof following the so-called Anosov Closing Lemma. Let $p \in \Lambda$ and W be a neighbourhood of p. Let $R \subset W$ be a closed rectangle, which contains p in its interior, whose boundary consists of four intervals: I_1 and I_2 contained in stable leaves, J_1 and J_2 transversal to the stable foliation. Each connected component of the intersection of a stable leaf with R is an interval which we call a stable fibre of R. Since $p \in \Lambda$, $f^{-n}(R)$ intersects R for infinitely many values of $n \in \mathbb{N}$. For n sufficiently big, $f^{-n}(R)$ is a very long (in the stable direction) and very thin rectangle fibred by intervals contained in the stable leaves. We can assume that $f^{-n}(R) \cap R$ is connected for, otherwise, we can shrink R in the stable direction, as indicated in Figure 49. Hence, we may find R and $n \in \mathbb{N}$ such

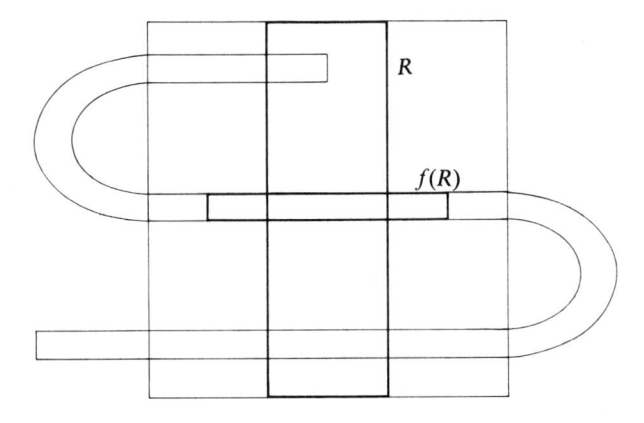

Figure 49

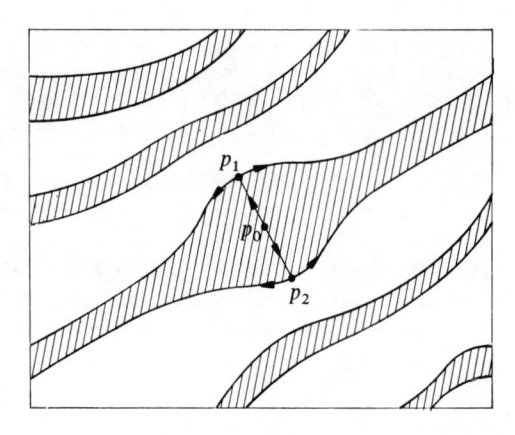

Figure 50

that, for any stable fibre $I \subset R$, $f^{-n}(I)$ contains a stable fibre of R. If I_x denotes the stable fibre through $x \in J_1$, then $f^{-n}(I_x)$ intersects J_1 at a point $\sigma(x)$. Since $\sigma: J_1 \to J_1$ is continuous, it has a fixed point y; i.e. $f^{-n}(I_y)$ contains I_y. The restriction of f^n to the interval $f^{-n}(I_y)$ is a continuous map onto I_y and, therefore, has a fixed point q. Thus, we have found a periodic point of f in W. Using the same criterion as in Example 4 above, the reader can prove the existence of a dense orbit in Λ. Notice that Λ contains the one-dimensional unstable manifolds of the points p_i for $i = 1, 2$ (in fact, it contains the unstable manifold of any point $p \in \Lambda$). Transversally, along the stable foliation, Λ locally contains a Cantor set. Indeed, let $p \in \Lambda$ and let I be a small interval along the stable leaf through p such that $\partial I \subset W^u(p_0, f)$. Finally, we prove the existence of a diffeomorphism $f: T^2 \to T^2$, satisfying the four properties we mentioned at the beginning. Let U be a neighbourhood of p_0 and $\varphi: U \to \mathbb{R}^2$ be the local chart whose inverse is given by $\varphi^{-1}(x_1, x_2) = \pi(x_1 v^s + x_2 v^u)$, where v^s and v^u are the unit eigenvectors of L. Then, we have $\varphi \circ g \circ \varphi^{-1}(x_1, x_2) = (\lambda x_1, \mu x_2)$. Let $\psi: \mathbb{R} \to \mathbb{R}$ be a C^∞ function such that $\psi(\mathbb{R}) \subset [0, 1]$, $\psi(-t) = \psi(t)$ for all t, $\psi(t) = 1$ for $|t| \leq \frac{1}{8}$, $\psi(t) = 0$ for $|t| \geq \frac{1}{4}$ and $\psi(t') \leq \psi(t)$ for $t' \geq t \geq 0$. Now we set $f = g$ in the complement of U and $f = \varphi^{-1} \circ F \circ \varphi$, where $F(x_1, x_2) = (F^1(x_1, x_2), F^2(x_1, x_2)) = (\lambda x_1 + (2 - \lambda)\psi(x_2)\psi(kx_1)x_1, \mu x_2)$. In this expression k is a positive real number chosen so that $|(\partial F^1/\partial x_2)(x)| < (\mu - 1)\sqrt{\mu^2 - 1}$, for all $x = (x_1, x_2)$. Clearly, the origin is a repelling fixed point for F. It remains to show the existence of a neighbourhood $W \subset \varphi(U)$ of the origin such that W is contained in the unstable manifold $W^u(0, F)$ and $\sup_{x \in W} (\partial F^1/\partial x_1)(x) = \bar{\alpha} < 1$. Indeed, since φ is an isometry, after constructing W we just take $V = \varphi^{-1}(W)$. Let $J_t = \{s; (\partial F^1/\partial x_1)(s, t) \geq 1\}$ and $I_{x_2} = \{(x_1, x_2); x_1 \in J_{x_2}\}$. We have that $J_t \subset J_{t'}$ if $t \geq t' \geq 0$ and J_t is either empty or a symmetric interval, say $J_t = [-a_t, a_t]$. Since $(\partial F^1/\partial x_1)(x_1, x_2) \geq 1$ for all $x_1 \in J_{x_2}$ and $F^1(-x_1, x_2) = -F^1(x_1, x_2)$, it follows that $F^{-1}(I_{x_2})$ is a symmetric interval whose length is smaller than or equal to that of I_{x_2}. Therefore, $F^{-1}(I_{x_2})$ is

§4 General Comments on Structural Stability. Other Topics 169

contained in $I_{(1/\mu)x_2}$. Clearly $I_0 \subset W^u(0, F)$ and, thus, $I_{x_2} \subset W^u(0, F)$ for x_2 small. Hence, $I_{x_2} \subset W^u(0, F)$ for all $x_2 \in [-1, 1]$ because $F^{-1}(I_{x_2}) \subset I_{(1/\mu)x_2}$. Since the union of the intervals I_{x_2} for $x_2 \in [-1, 1]$ is a compact set and $W^u(0, F)$ is open, there exists $0 < \bar{\alpha} < 1$ such that $W = \{(x_1, x_2); (\partial F^1/\partial x_1)(x_1, x_2) > \bar{\alpha}\}$ is a neighbourhood of 0 in $W^u(0, F)$. It is now immediate to verify that $f = \varphi^{-1} \circ F \circ \varphi$ satisfies properties (1) to (4) listed above. Notice that we set $V = \varphi^{-1}(W)$ as the neighbourhood mentioned in property (4). Thus, we have constructed a DA diffeomorphism in the torus T^2.

Remark 1. Let p_1 be another periodic point of the Anosov diffeomorphism g considered above. We can modify g simultaneously on neighbourhoods of p_0 and p_1 in order to obtain a diffeomorphism f that satisfies Axiom A with a nonwandering set that consists of three basic sets: a repelling fixed point p_0, a repelling periodic orbit $\mathcal{O}(p_1)$ and a nonperiodic attractor Λ. Performing the same construction along different periodic orbits of g, we can get several examples of nonconjugate hyperbolic attractors (the number of periodic orbits of a given period might be different...).

Remark 2. The same methods can be used to construct Axiom A attractors in higher dimensions. We start with Anosov diffeomorphisms whose stable manifolds have dimension one and perform modifications similar to the ones above.

EXAMPLE 6. We now show that, even for the sphere S^2, it is possible to construct an Axiom A diffeomorphism with a nonperiodic attractor. The result is due to Plykin (see [5] for references and a nice picture). Let Φ: $\mathbb{R}^2 \to \mathbb{R}^2$ be the involution $\Phi(x) = -x$. Since $\Phi(\mathbb{Z}^2) = \mathbb{Z}^2$, Φ induces an involution η of $T^2 = \mathbb{R}^2/\mathbb{Z}^2$. Notice that η has four fixed points: $p_0 = \pi(0, 0)$, $p_1 = \pi(\frac{1}{2}, 0)$, $p_2 = \pi(\frac{1}{2}, \frac{1}{2})$ and $p_3 = \pi(0, \frac{1}{2})$. Let V_i be a small cell neighbourhood of each of the points p_i, $0 \leq i \leq 3$, such that $\eta(V_i) = V_i$ and let $N = T^2 - \bigcup V_i$. The restriction of η to N is an involution without fixed points. Let $\sim$ be the equivalence relation on N that identifies two points if they belong to the same orbit of η. Let M be the quotient space $N/\sim$ and $\rho: N \to M$ be the canonical projection. Considering M with the differentiable structure induced from N by ρ, we can see that M is diffeomorphic to the complement of four disjoint discs in the sphere S^2. M corresponds to the shaded region, with the identification of the corresponding sides, indicated in Figure 51. We have that ρ is a two-sheet covering map. That is, ρ is a local diffeomorphism with each point having two pre-images. Therefore, if $\tilde{f}: N \to N$ is a differentiable mapping such that $\tilde{f}\eta = \eta\tilde{f}$, then it induces a differentiable mapping $f: M \to M$ such that $\rho\tilde{f} = f\rho$. As in Example 5 (see Remark 1), we can construct a DA diffeomorphism $\tilde{f}$ on the torus having p_0 as a repelling fixed point and $\{p_1, p_2, p_3\}$ as a repelling periodic orbit (of period three). Furthermore, it is easy to see that we may construct $\tilde{f}$ so that $\tilde{f}\eta = \eta\tilde{f}$, since we can

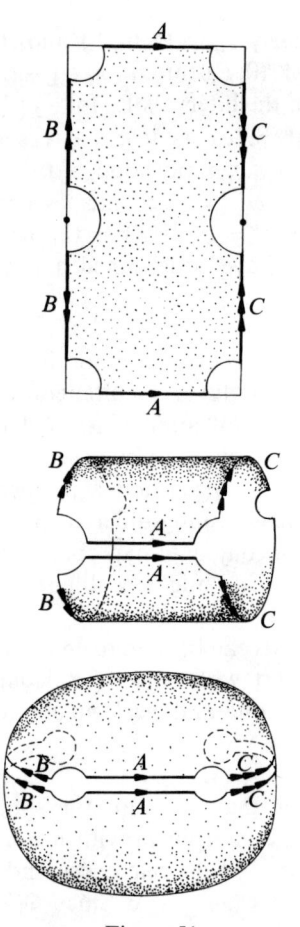

Figure 51

start with an Anosov diffeomorphism g which is induced by a linear isomorphism and, hence, commutes with η. For instance, start with the diffeomorphism of T^2 induced by the linear isomorphism of $\mathbb{R}^2$ given, in canonical coordinates, by $L(x_1, x_2) = (2x_1 + x_2, x_1 + x_2)$. Now, we take closed neighbourhoods V_i of the points p_i so that $\eta(V_i) = V_i$, V_i contained in the unstable manifold of p_i for $0 \leq i \leq 3$. We have that $\tilde{f}$ is a diffeomorphism of $N = T^2 - \bigcup_i V_i$ onto $\tilde{f}(N) \subset N$, and $\tilde{\Lambda} = \bigcap_{n>0} \tilde{f}^n(N)$ is the DA attractor. Thus, $\tilde{f}$ induces a diffeomorphism f of M onto $f(M) \subset M$. We can extend f to the sphere S^2, by putting a source in each of the four discs in the complement of M, one of them being periodic of period three. Then, it is not hard to see that $\Omega(f)$ consists of these repelling fixed or periodic orbits and $\Lambda = \rho(\tilde{\Lambda})$, Λ being a one-dimensional attracting basic set. This last assertion follows from the fact that ρ is a two-sheet covering and $\rho \tilde{f} = f \rho$.

Now we give the results on structural stability of diffeomorphisms satisfying Axiom A and the transversality condition. Let $\mathscr{A}^r(M) \subset \text{Diff}^r(M)$

§4 General Comments on Structural Stability. Other Topics

be the subset of these diffeomorphisms with M compact. First, Robbin [92], [93] proved that any $f \in \mathscr{A}^2(M)$ is structurally stable. Then the same result was proved in [61] for $f \in \mathscr{A}^1(M^2)$. This left the question of the stability of the diffeomorphisms $f \in \mathscr{A}^1(M)$ for M of any dimension. This was settled positively by Robinson [97], and in [95], [96] he proved the corresponding result for vector fields.

An important question, that still has no general solution, is the converse of the above results: if f is structurally stable, does it satisfy Axiom A and the transversality condition? In the particular case where $\Omega(f)$ is finite, $f \in \text{Diff}^r(M)$ is structurally stable if and only if f is Morse–Smale [79]. In the general case where $\Omega(f)$ is not finite partial results were obtained by Pliss [86] and Mañé [53]. Using a very original idea, Mañé [54] has recently solved the question for two-manifolds. Another result in this direction is due to Franks [21], Guckenheimer [26] and Mañé [52]. They introduced the concept below from which they obtained a characterization of stability similar to the one we have just formulated. We say that $f \in \text{Diff}^r(M)$ is *absolutely stable* if there exists a neighbourhood $V(f) \subset \text{Diff}^r(M)$ and a number $K > 0$ such that, for every $g \in V(f)$, there exists a homeomorphism h of M with $hf = gh$ and $\|h - I\|_0 \leq K\|f - g\|_0$. In this expression I is the identity map of M and $\| \ \|_0$ denotes C^0 distance. It was shown that if $f \in \text{Diff}^r(M)$ is absolutely stable then f satisfies Axiom A and the transversality condition. Franks also observed from the proofs of structural stability of the diffeomorphisms $f \in \mathscr{A}^r(M)$ that these diffeomorphisms are absolutely stable. We can therefore say that $f \in \text{Diff}^r(M)$ is absolutely stable if and only if $f \in \mathscr{A}^r(M)$.

In studying (structural) stability of diffeomorphisms and flows, special attention has been devoted to stability restricted to nonwandering sets. The motivation for such a study is based on the idea that the dynamics of a system is ultimately concentrated on the nonwandering sets. There lie all limit sets and in particular the periodic and recurrent orbits. We say that $f \in \text{Diff}^r(M)$ is Ω-*stable* if there exists a neighbourhood $V(f) \subset \text{Diff}^r(M)$ such that, for any $g \in V(f)$, there exists a homeomorphism $h: \Omega(f) \to \Omega(g)$ with $hf(x) = gh(x)$ for all $x \in \Omega(f)$. An important concept here is that of cycles in the nonwandering set. Let f satisfy Axiom A and let $\Omega = \Omega_1 \cup \cdots \cup \Omega_k$ be the decomposition of $\Omega = \Omega(f)$ into basic sets. A *cycle* in Ω is a sequence of points $p_1 \in \Omega_{k_1}, p_2 \in \Omega_{k_2}, \ldots, p_{k_s} \in \Omega_{k_s} = \Omega_{k_1}$ such that $W^s(p_i) \cap W^u(p_{i+1}) \neq \emptyset$ for $1 \leq i \leq s - 1$. In [110] Smale proved that if f satisfies Axiom A and has no cycles in $\Omega(f)$ then f is Ω-stable. The corresponding result for vector fields is due to Pugh–Shub [91]. Later Newhouse [69] showed that it is enough for Ω-stability to require the hyperbolicity and nonexistence of cycles on the limit set. More recently, Malta [50], [51] weakened this hypothesis to the Birkhoff centre, which is defined as the closure of those orbits that are simultaneously α- and ω-recurrent. In the opposite direction, it is known that if f satisfies Axiom A and there exist cycles in $\Omega(f)$ then f is not Ω-stable [76]. There remains the problem: if f is Ω-stable does it satisfy Axiom A? There are results for Ω-stability analogous to those described above characterizing structural stability.

Another natural question is whether the structurally stable diffeomorphisms and vector fields are dense (and, hence, open and dense) in Diff$^r(M)$ and $\mathfrak{X}^r(M)$. The first counter-example to this was given by Smale [111]. Ω-stability is also not generic as is shown by the examples of Abraham–Smale [2], Newhouse [68], Shub and Williams [104] and Simon [105].

Although they are not in general dense in Diff$^r(M)$ or $\mathfrak{X}^r(M)$ the stable systems are still plentiful as we shall indicate. In the first place, they exist on any differentiable manifold; the Morse–Smale diffeomorphisms are typical examples. More than this, Smale [112] showed that there exist stable diffeomorphisms in every isotopy class in Diff$^r(M)$. That is, any diffeomorphism can be joined to a stable one by a continuous arc of diffeomorphisms. Another interesting fact is that any diffeomorphism can be C^0 approximated by a stable diffeomorphism (Shub [102]). Thus, in the space of C^r diffeomorphisms, $r \geq 1$, with the C^0 topology, the stable diffeomorphisms are dense. The isotopy classes in Diff$^r(M)$ that contain Morse–Smale diffeomorphisms were analysed by Shub–Sullivan [103], Franks–Shub [23] and, in the case of surfaces, by Rocha [94]. In this last work a characterization is given in terms of the vanishing of the growth rate of the automorphism induced on the fundamental group. The types of periodic behaviour possible for stable diffeomorphisms in a given isotopy class are studied in [22]. For vector fields, Asimov [8] constructs Morse–Smale systems without singularities on every manifold with Euler characteristic zero and dimension bigger than three. Of course, this is possible for two-dimensional surfaces (the torus and Klein bottle). Somewhat surprisingly, Morgan [60] showed that this cannot be done for certain three-dimensional manifolds.

Perhaps the most important generic property so far discovered in Dynamical Systems is due to Pugh. By combining the Kupka–Smale Theorem with his Closing Lemma, Pugh [89] proved the following theorem: there exists a generic set $\mathscr{G} \subset \text{Diff}^1(M)$ such that, for any $f \in \mathscr{G}$, the periodic points of f are hyperbolic, they are dense in $\Omega(f)$ and their stable and unstable manifolds are transversal. A very important question is to determine whether or not the same result is true in Diff$^r(M)$, $r \geq 2$. This result is known as the General Density Theorem.

Diffeomorphisms that satisfy Axiom A and, in particular, Anosov diffeomorphisms have also been studied using measures that are defined on their nonwandering sets and are preserved by the diffeomorphisms. The branch of Mathematics, related to Dynamical Systems, that uses this technique to describe the orbit structure of a diffeomorphism is called Ergodic Theory. It has its origin in Conservative Mechanics, where the diffeomorphisms that occur usually have the property that they preserve volume.

The Ergodic Theory of diffeomorphisms that satisfy Axiom A begins with the work of Anosov, Sinai and Bowen. The reader will find accounts of this theory in [7], [9], [10], [120].

Another active area of research is Bifurcation Theory, which has been considered by several mathematicians as far back as Poincaré. Roughly speaking, the theory consists of describing how the phase portrait (space of orbits) can change when we consider perturbations of an initial dynamical system. In particular, how can we describe the equivalence classes of systems near an initial one under topological conjugacies or topological equivalences (unfolding). Questions of a similar flavour appear in other branches of Mathematics, like Singularities of Mappings and Partial Differential Equations (see, for instance, [17], [28], [42], [99], [118]), but we will restrict ourselves to a few comments on bifurcations of vector fields and diffeomorphisms. A common point of view is to determine how the phase portrait of a system depending on several parameters evolves when the parameters vary. Bifurcation points are values of the parameters for which the system goes through a topological change in its phase portrait. An increasing number of results is available in this direction, especially for one parameter families (arcs) of vector fields and diffeomorphisms. In order to give an idea of this topic, we will describe two of these results, the first being of a more local nature than the second. Let $I = [0, 1]$ and let M be a compact manifold without boundary. We indicate by $\mathscr{A}$ the space of C^s arcs of vector fields $\xi: I \to \mathfrak{X}^r(M)$ with the C^s topology, $1 \le s \le r$ and $r \ge 4$. From the fact that generically (second category) a vector field is Kupka–Smale, we would expect for a generic arc $\xi \in \mathscr{A}$ that $\xi(\mu) \in \mathfrak{X}^r(M)$ is Kupka–Smale for most values of $\mu \in I$. If $\xi(\mu_0)$ is not Kupka–Smale for some $\mu_0 \in I$, then a periodic orbit of $\xi(\mu_0)$ is not hyperbolic or a pair of stable and unstable manifolds are not transversal. Moreover, if the arc ξ is not "degenerate", either only one periodic orbit of $\xi(\mu_0)$ should be nonhyperbolic or all periodic orbits should be hyperbolic and only one pair of stable and unstable manifolds should be nontransversal along precisely one orbit of $\xi(\mu_0)$. The lack of hyperbolicity of a singularity p of $\xi(\mu_0)$ is due to one eigenvalue (or a pair of complex conjugate ones) of $D\xi(\mu_0)$ at p having real part zero. The lack of hyperbolicity of a closed orbit γ is due to one eigenvalue (or a pair of complex conjugate ones) of DP at $p \in \gamma$ having norm one, where P is the Poincaré map of a cross section through p. In both cases, let us denote this eigenvalue by λ. We now describe the phase portrait of ξ_μ for μ near μ_0. To do this we must assume certain nondegeneracy conditions on the higher order jets of $\xi(\mu)$, which are not discussed here. A crucial fact is the existence for all μ near μ_0 of an invariant manifold for $\xi(\mu)$ associated to λ. It is called the centre manifold and it has dimension one if λ is real and dimension two if not (see [40]). Essentially, the bifurcation occurs along the centre manifold; normally to it we have hyperbolicity (see [80]). In the pictures below hyperbolicity is indicated by double arrows.

Singularity.
(a) $\lambda = 0$. In the central invariant line two saddles collapse and then disappear (or vice versa). This is called a saddle-node bifurcation.

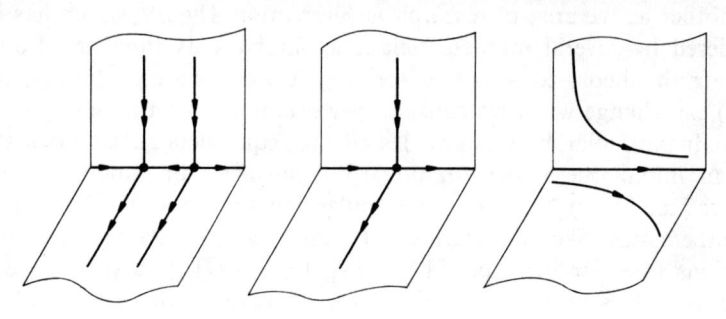

Figure 52

(b) $\lambda = bi$, $b \neq 0$. In the central invariant plane a hyperbolic attracting singularity becomes nonhyperbolic but still attracting, then it changes into a hyperbolic repellor and an attracting closed orbit appears. This is called a Hopf bifurcation.

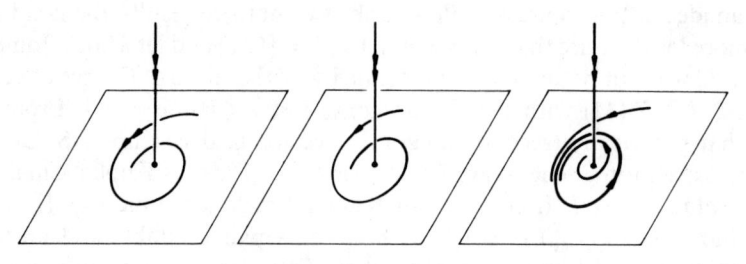

Figure 53

Closed Orbit. We will restrict ourselves to the phase portrait of the associated Poincaré map.
(a) $\lambda = 1$. This is the analogue of the saddle-node singularity. The dots in the picture are just to stress the fact that the orbits of the Poincaré map are discrete.

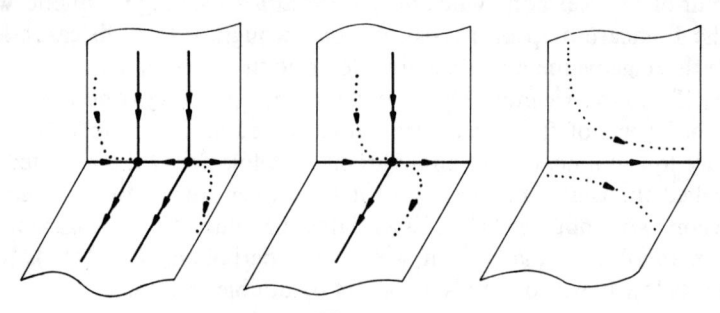

Figure 54

§4 General Comments on Structural Stability. Other Topics 175

(b) $\lambda = e^{i\theta}$, $0 < \theta < \pi$. This is the analogue of the Hopf bifurcation for a singularity: after the bifurcation there appears a circle which is left invariant by the Poincaré map. This circle corresponds to a torus which is left invariant by the flow of the vector field.

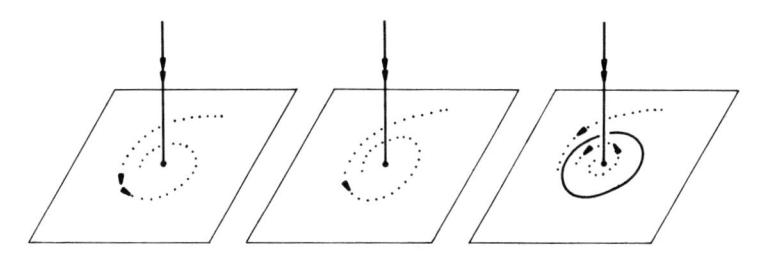

Figure 55

(c) $\lambda = -1$. In the central invariant line, a hyperbolic attracting fixed point becomes nonhyperbolic but still attracting, then it changes into a hyperbolic repellor and an attracting periodic point of period two appears. This is called a flip bifurcation.

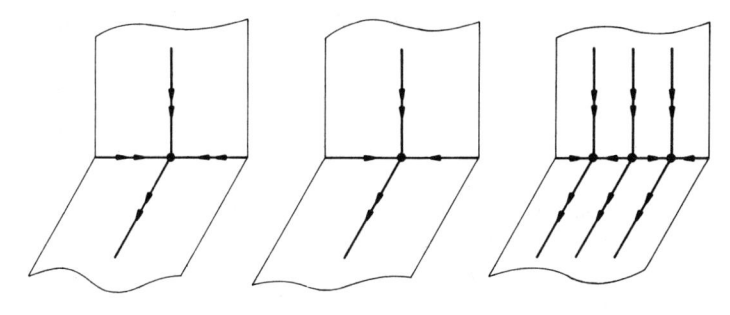

Figure 56

A nonhyperbolic singularity or closed orbit as above is called *quasi-hyperbolic*.

For a nontransversal orbit of intersection of a stable and an unstable manifold, we require the contact to be of second order (quadratic). Figure 57

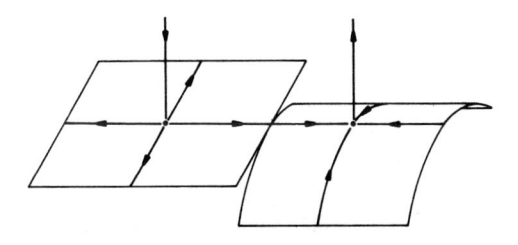

Figure 57

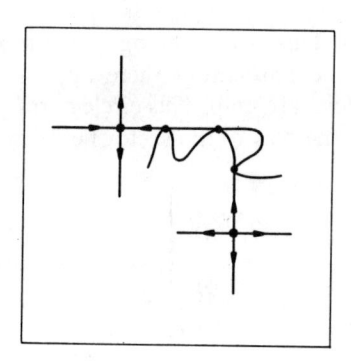

Figure 58

indicates a second-order contact (saddle connection) between a stable and an unstable manifold of two singularities of a three-dimensional vector field. Figure 58 indicates a saddle connection of two closed orbits, drawn in a two-dimensional cross-section.

We can now state the following result, which is mostly due to Sotomayor [115]; parts of it have been considered by several other authors, like Brunowsky [11]. There is a residual subset of arcs $\mathscr{B} \subset \mathscr{A}$ such that if $\xi \in \mathscr{B}$ then $\xi(\mu)$ is Kupka–Smale for $\mu \in I$ except for a countable set $\mu_1, \ldots, \mu_n, \ldots$. For each $n \in \mathbb{N}$ one of the following two possibilities hold. Either $\xi(\mu_n)$ has all periodic orbits hyperbolic except one which is quasi-hyperbolic and their stable and unstable manifolds meet transversally *or* all periodic orbits are hyperbolic and their stable and unstable manifolds meet transversally except along one orbit of second-order contact.

Let us now consider the concept of stability for arcs of vector fields and state one of the results that have recently been obtained in this direction. Two arcs ξ, $\xi' \in \mathscr{A}$ are equivalent if there is a homeomorphism $\rho: I \to I$ such that for each $\mu \in I$ there is an equivalence h_μ between $\xi(\mu)$ and $\xi'(\rho(\mu))$. Moreover, we demand that the homeomorphism h_μ should depend continuously on $\mu \in I$. Let $\mathscr{G} \subset \mathscr{A}$ denote the subset of arcs of gradient vector fields on M. The following recent theorem to appear is due to Palis and Takens: an open and dense subset of arcs in $\mathscr{G}$ are stable.

We refer the reader to [1], [5], [6], [27], [34], [58], [78], [84], [117] for accounts of this topic.

EXERCISES

1. Show that every Morse–Smale vector field on a compact manifold of dimension n has an attracting critical element and a repelling one.

2. Let X be a Kupka–Smale vector field on a compact manifold of dimension n. Show that, if the nonwandering set of X coincides with the union of its critical elements then X is a Morse–Smale field.

Exercises

3. Let $f: M^n \to \mathbb{R}$ be a C^{r+1} Morse function, that is, f has a finite number of non-degenerate critical points. Suppose that two critical points always have distinct images. Show that there exists a filtration for $X = \text{grad } f$.

4. Let X be a gradient field on a compact manifold of dimension n. Show that X can be approximated by a Morse–Smale field.

5. We say that a differentiable function $f: M \to \mathbb{R}$ is a first integral of the vector field $X \in \mathfrak{X}^r(M)$ if $df(p) \circ X(p) = 0$ for all $p \in M$. Show that if X is a Morse–Smale field then every first integral of X is constant.

6. Let M be a compact manifold of dimension n and let X be a C^r vector field on M with two hyperbolic singularities p and q, p being an attractor and q a repellor. Suppose that, for all $x \in M - \{p, q\}$ the ω-limit of x is p and the α-limit is q. Show that M is homeomorphic to S^n.

7. Let $X \in \mathfrak{X}^r(M^2)$ be a Morse–Smale field. We say that X is a polar field if X has only one source, only one sink and has no closed orbits. Let $X \in \mathfrak{X}^r(M^2)$ be a polar field, with sink p and let C_X be a circle that is transversal to X and bounds a disc containing p. Show that, in an orientable manifold of dimension two, polar fields X and Y are topologically equivalent if and only if there exists a homeomorphism $h: C_X \to C_Y$ with the following property: $x, y \in C_X$ belong to the unstable manifold of a saddle of X if and only if $h(x), h(y)$ belong to the unstable manifold of a saddle of Y (G. Fleitas).

8. Show that two polar fields on the torus are topologically equivalent.

9. Show that there exist two polar fields on the sphere with two handles (pretzel) that are not topologically equivalent.

10. Describe all the equivalence classes of polar fields on the pretzel.

    Hint. See Example 10 in this chapter, Section 1.

11. Consider a closed disc $D \subset \mathbb{R}^2$ and let C be its boundary. A chordal system in D is a finite collection of disjoint arcs in D each of which joins two points of C and is transversal to C at these points. Show that, for any chordal system in D there exists a vector field X transversal to C with the following properties:
    (i) the arcs are contained in the unstable manifolds of the saddles of X and each arc contains just one saddle;
    (ii) each connected component of the complement of the union of the arcs contains just one source;
    (iii) the α-limit of a point in C is either a single source or a single saddle.

12. Let X and Y be vector fields on the discs D_1 and D_2 associated to two chordal systems as in Exercise 11. Let $h: \partial D_1 \to \partial D_2$ be a diffeomorphism such that if $p \in \partial D_1$ is contained in the unstable manifold of a saddle of X then $h(p)$ is contained in the stable manifold of a sink of $-Y$. By glueing D_1 and D_2 together by means of h we obtain a field Z on S^2 that coincides with X on D_1 and with $-Y$ on D_2. Show that
    (a) Z is a Morse–Smale field,
    (b) every Morse–Smale field without closed orbits on S^2 is topologically equivalent to a field obtained by this construction.

178   4 Genericity and Stability of Morse–Smale Vector Fields

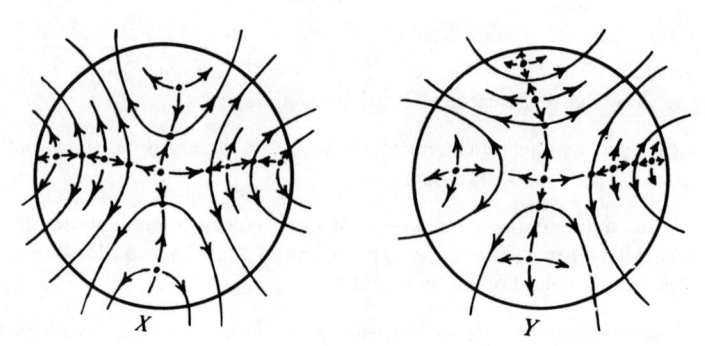

Figure 59

13. Let X and Y be Morse–Smale fields on S^2 with just one sink and such that the orbit structure on the complement of a disc contained in the stable manifold of the sink is as shown in Figure 59. Show that X and Y have isomorphic phase diagrams but are not topologically equivalent.

14. Describe all the equivalence classes of Morse–Smale fields without closed orbits on S^2 with one sink, three saddles and four sources.

15. Describe all the equivalence classes of Morse–Smale fields without closed orbits on the torus T^2 with one sink, two sources and three saddles.

16. Consider a vector field X whose orbit structure is shown in Figure 60. The non-wandering set of X consists of the sources f_1 and f_2, the saddles s_1 and s_2 and the sinks p_1 and p_2. Prove that X can be approximated by a field which has a closed orbit. This shows that X is not Ω-stable.

17. Let X be a Kupka–Smale vector field with finitely many critical elements on a compact manifold of dimension n. Show that, if the limit set of X coincides with the set of critical elements then X is a Morse–Smale vector field.

18. Show that the set of Kupka–Smale fields is not open in $\mathfrak{X}^r(M^2)$ if M^2 is different from the sphere, the projective plane and the Klein bottle.

    Hint. Use Cherry's example (Example 13 of this chapter).

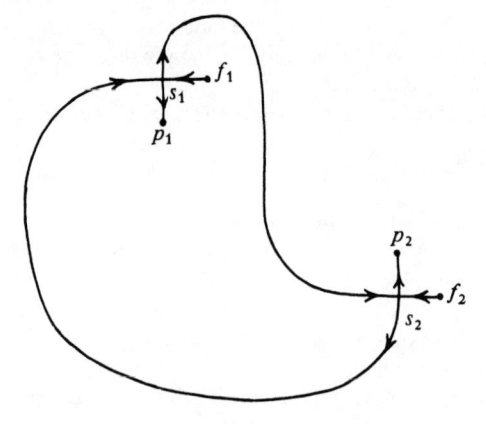

Figure 60

Exercises

19. Let $X \in \mathfrak{X}^r(M^2)$ be a field whose singularities are hyperbolic. If G is a graph of X and $p \in M$ has $\omega(p) = G$ then there exists a neighbourhood V of p such that $\omega(q) = G$ for every $q \in V$.

20. Show that, if M^2 is an orientable manifold and $X \in \mathfrak{X}^r(M^2)$ is structurally stable, then X is Morse–Smale. Show also that if M is nonorientable and $X \in \mathfrak{X}^1(M^2)$ is structurally stable then X is Morse–Smale.

21. Show that, if $f \in \text{Diff}^1(M)$ is structurally stable and has a finite number of periodic points, then f is Morse–Smale.

22. Let $\mathfrak{X}^r(\mathbb{R}^2)$, $r \geq 1$, be the set of vector fields on $\mathbb{R}^2$ with the Whitney topology (see Exercise 17 in Chapter 1). Show that there exist structurally stable vector fields in $\mathfrak{X}^r(\mathbb{R}^2)$.

    Hint. Consider a triangulation of the plane and construct a field with singularities at the centres of the triangles, edges and vertices.

    Remark. In [63], P. Mendes showed that there exist stable vector fields and diffeomorphisms on any open manifold.

23. Show that, if γ is a nontrivial recurrent orbit of a field X on a nonorientable manifold of dimension two, then there exists a transversal circle through a point $p \in \gamma$.

24. Show that, if $h: S^1 \to S^1$ is a homeomorphism that reverses orientation, then h has a fixed point.

25. Show that a vector field without singularities on the Klein bottle has only trivial recurrence.

26. Show that the set of Morse–Smale diffeomorphisms is not dense on any manifold of dimension two.

27. Show that the set of Morse–Smale fields is not dense on any manifold of dimension greater than or equal to three.

28. Show that if the limit set of a diffeomorphism or a vector field consists of finitely many orbits then the Birkhoff centre consists of (finitely many) fixed and periodic orbits. (The Birkhoff centre is the closure of those orbits that are simultaneously ω- and α-recurrent.)

29. Show that if a diffeomorphism or a vector field has finitely many fixed and periodic orbits, all of them hyperbolic, then there can be no cycles between these fixed and periodic orbits along transversal intersections of their stable and unstable manifolds.

30. Show that a structurally stable diffeomorphism or vector field whose limit set consists of finitely many orbits must be Morse–Smale.

31. Show that the sets of Anosov diffeomorphisms and of Morse–Smale diffeomorphisms are disjoint.

32. Show that the only compact manifold of dimension two that admits an Anosov diffeomorphism is the torus.

33. Show that the Anosov diffeomorphism of Proposition 4.1 of this chapter has an orbit dense in T^2.

34. Give an example of an Anosov diffeomorphism of the torus $T^n = S^1 \times \cdots \times S^1$.

35. Using the General Density Theorem show that an Anosov diffeomorphism satisfies Axiom A.

36. Show that on any manifold of dimension two there exists a diffeomorphism satisfying Axiom A and the transversality condition and possessing an infinite number of periodic points.

    Hint. Use Smale's Horseshoe in S^2 and a Morse–Smale diffeomorphism on M^2.

37. Show that if $f_1: M_1 \to M_1$ and $f_2: M_2 \to M_2$ are two diffeomorphisms that satisfy Axiom A and the transversality condition then so does $f: M_1 \times M_2 \to M_1 \times M_2$ where $f(p, q) = (f_1(p), f_2(q))$.

38. Let $f: M \to M$ be a diffeomorphism which is C^1 stable and satisfies Axiom A. Prove that f must satisfy the Transversality Condition.

39. Show that any diffeomorphism satisfying Axiom A has an attracting basic set.

40. Show that, if a diffeomorphism f of a (compact) manifold M satisfies Axiom A, then the union of the stable manifolds of the points in the attracting basic sets of f is open and dense in M.

41. Show that in any manifold there is an open set of C^r ($r \geq 1$) diffeomorphisms isotopic to the identity that do not embed in a flow.

    Recall that a diffeomorphism f is isotopic to the identity if there is a continuous arc of diffeomorphisms connecting f with the identity. A flow of C^r diffeomorphisms is a continuous group homomorphism $\varphi: (\mathbb{R}, +) \to (\text{Diff}^r(M), \circ)$. We say that f embeds in a flow if $f = \varphi(1)$ for some flow φ.

42. Show that two C^r ($r \geq 1$) commuting vector fields on a surface of genus different from zero have a singularity in common (E. Lima).

    Hint. First show that a C^r vector field on a surface can only have finitely many nontrivial minimal sets. For $r \geq 2$, this follows from the Denjoy–Schwartz Theorem since there can be only trivial minimal sets. Notice that, in Chapter 1, we posed the problem above for the 2-sphere. The corresponding question in higher dimensions seems to be wide open.

43. Show that, in an orientable surface M^2, the set of vector fields with trivial centralizer contains an open and dense subset of $\mathfrak{X}^\infty(M^2)$ (P. Sad). The vector field X has trivial centralizer if for any $Y \in \mathfrak{X}^\infty(M^2)$ such that $[X, Y] = 0$ we have $Y = cX$ for some $c \in \mathbb{R}$. Restricting to Morse–Smale vector fields (or, more generally, to Axiom A vector fields), the same result is true for an open and dense subset in higher dimensions. Notice that a pair of commuting vector fields generates an $\mathbb{R}^2$-action. The concept of structural stability for $\mathbb{R}^k$-actions is considered in [12].

44. Let X be a C^∞ vector field on a compact manifold M. Suppose X has a first integral which is a Morse function. Show that X can be C^∞ approximated by a Morse–Smale vector field without closed orbits.

    Recall that $f: M \to \mathbb{R}$ is called a Morse function if all of its critical points are nondegenerate or, equivalently, if all singularities of grad f are hyperbolic. To be a first integral for X means that f is constant along the orbits of X.

45. Let $f \in \text{Diff}^r(M)$ and $\Lambda \subset M$ be a closed invariant set for f. Suppose that there exist a Riemannian metric on M, a number $0 < \lambda < 1$ and for each $x \in \Lambda$ a decomposition

Appendix: Rotation Number and Cherry Flows 181

$TM_x = E_x^s \oplus E_x^u$ such that $Df_x(E_x^s) = E_{f(x)}^s$, $Df_x(E_x^u) = E_{f(x)}^u$, $\|Df_x v\| \le \lambda \|v\|$ for all $v \in E_x^s$ and $\|Df_x^{-1} w\| \le \lambda \|w\|$ for all $w \in E_{f(x)}^u$. Show that Λ is hyperbolic, that is, the subspaces E_x^s and E_x^u of TM_x vary continuously with x.

46. For all $n \ge 2$ there exists a diffeomorphism $f: S^n \to S^n$, satisfying Axiom A and the Transversality Condition, whose nonwandering set contains a repelling fixed point and a nonperiodic attractor.

    Hint. Use Example 6 of Chapter 4, Section 4.

47. Show that the conclusion of Exercise 46 holds for all manifolds M of dimension $n \ge 2$.

48. Show that the set of stable diffeomorphisms of S^2 is not dense in $\text{Diff}^1(S^2)$.

    Hint. Use Exercise 38 and a modification of Example 6 of Section 4.

49. Show that the set of stable diffeomorphisms is not dense in $\text{Diff}^1(M)$ for any manifold M of dimension $n \ge 2$.

Appendix: Rotation Number and Cherry Flows

We shall now give the details of the construction of Cherry's example that we mentioned in Example 13 of this chapter. Firstly we shall show the existence of C^∞ fields on the torus that are transversal to a circle Σ and have exactly two singularities: one sink and one saddle, both hyperbolic. The Poincaré map of such a field is defined on the complement of a closed interval in Σ and extends to a monotonic endomorphism of Σ of degree 1. The concept of rotation number, introduced by Poincaré to study the dynamics of homeomorphisms of the circle, extends to such endomorphisms. We shall show that these fields have nontrivial recurrence if and only if the rotation numbers of the endomorphisms induced on Σ are irrational. As the rotation number varies continuously with the endomorphism, the existence of fields with nontrivial recurrence follows.

Let $\pi: \mathbb{R}^2 \to T^2$ be the covering map introduced in Example 2 of Section 1, Chapter 1. Thus π is a C^∞ local diffeomorphism, $\pi(x, y) = \pi(x', y')$ if and only if $x - x' \in \mathbb{Z}$ and $y - y' \in \mathbb{Z}$ and $\pi([0, 1] \times [0, 1]) = T^2$. If X is a C^∞ vector field on the torus we can define a C^∞ field $Y = \pi^* X$ on $\mathbb{R}^2$ by the expression $Y(z) = (d\pi_z)^{-1} X(\pi(z))$. Clearly the field Y defined like this satisfies the condition

$$Y(x + n, y + m) = Y(x, y), \quad \forall (x, y) \in \mathbb{R}^2, \quad \forall (n, m) \in \mathbb{Z}^2. \qquad (*)$$

Conversely, if Y is a C^∞ vector field on the plane satisfying condition $(*)$ then there exists a unique C^∞ field X on the torus such that $Y = \pi^* X$. We can thus identify the vector fields on the torus with the vector fields on $\mathbb{R}^2$ satisfying condition $(*)$.

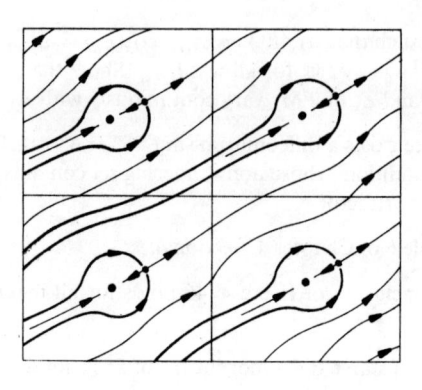

Figure 61

Let $\mathscr{C}$ be the set of vector fields $X \in \mathfrak{X}^\infty(\mathbb{R}^2)$ satisfying the following conditions:

(i) $X(x + n, y + m) = X(x, y); \forall (x, y) \in \mathbb{R}^2, (m, n) \in \mathbb{Z}^2$;
(ii) X is transversal to the straight line $\{0\} \times \mathbb{R}$ and has only two singularities p, s in the rectangle $[0, 1] \times [0, 1]$ where p is a sink and s a saddle, both hyperbolic;
(iii) there exist $a, b \in \mathbb{R}$ with $a < b < a + 1$ such that if $y \in (b, a + 1)$ then the positive orbit of X through the point $(0, y)$ intersects the line $\{1\} \times \mathbb{R}$ in the point $(1, f_X(y))$ while if $y \in (a, b)$ the positive orbit through $(0, y)$ goes directly to the sink without cutting $\{1\} \times \mathbb{R}$;
(iv) $\lim_{y \to b} f'_X(y) = +\infty$ and $\lim_{y \to a+1} f'_X(y) = +\infty$.

In Figure 61 we sketch the orbits of a field $X \in \mathscr{C}$. We remark that it follows from condition (iii) that the ω-limits of the points $(0, a)$ and $(0, b)$ are one and the same saddle. If $(1, c)$ denotes the point where the unstable manifold of this saddle meets the line $\{1\} \times \mathbb{R}$ we can extend f_X continuously

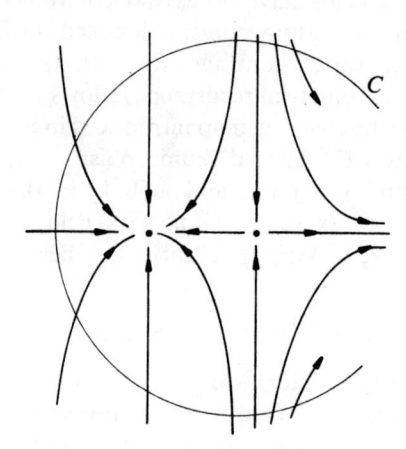

Figure 62

Appendix: Rotation Number and Cherry Flows

to the interval $[a, a + 1]$ by defining $f_X(y) = c$ if $y \in [a, b]$. Condition (i) implies that $f_X(a + 1) = f_X(a) + 1$ so we can extend f_X continuously to $\mathbb{R}$ by defining $f_X(y + n) = f_X(y) + n$ if $y \in [a, a + 1]$ and $n \in \mathbb{Z}$.

Let us denote by $\bar{X}$ the vector field on the torus induced by $X \in \mathscr{C}$, that is $X = \pi^* \bar{X}$. Then $\bar{X}$ has exactly two singularities, both hyperbolic; $\pi(p)$ is a sink and $\pi(s)$ is a saddle. Moreover, $\bar{X}$ is transversal to the circle $\Sigma = \pi(\{0\} \times \mathbb{R})$ and the Poincaré map $P_{\bar{X}}$ defined on $\pi(\{0\} \times (b, a + 1))$ can be extended continuously to Σ. The map f_X is a lifting of $P_{\bar{X}}$. In fact, $\tilde{\pi}: \mathbb{R} \to \Sigma$ defined by $\tilde{\pi}(y) = \pi(0, y)$ is a covering map and $\tilde{\pi} \circ f_X = P_{\bar{X}} \circ \tilde{\pi}$.

Lemma 1. *$\mathscr{C}$ is nonempty.*

PROOF. Consider the vector field $Y(x, y) = (2x(x + \frac{2}{3}), -y)$. The nonwandering set of Y consists of two singularities, a saddle $(0, 0)$ and a sink $(-\frac{2}{3}, 0)$. It is easy to check that Y is transversal to the unit circle at all points of the arc $C = \{(x, y); x^2 + y^2 = 1, x \leq \frac{1}{2}\}$. See Figure 62. Let

$$Z(x, y) = (\varphi(x, y)(2x^2 + \tfrac{4}{3}x) + (1 - \varphi(x, y))(x^2 + 1), -y)$$

where φ is a C^∞ function such that $\varphi(\mathbb{R}^2) \subset [0, 1]$, $\varphi(x, y) = 1$ if $x > \frac{1}{2}$ or if $(x, y) \in U$, $\varphi(x, y) = 0$ if $x < \frac{1}{4}$ and $(x, y) \in \mathbb{R}^2 - V$. Here U and V are small neighbourhoods of C with $\bar{U} \subset V$. If V is sufficiently small, the nonwandering set of Z is empty. Take $T > 0$ such that $Z_t(C) \subset \{(x, y); x > 1\}$ for all $t \geq T$. Using the flow of Z we can define a diffeomorphism $H: (0, 1) \times (0, 1) \to W \subset \mathbb{R}^2$ by $H(x, y) = Z_{xT}(h(y))$ where $h: [0, 1] \to C$ is a diffeomorphism. If $z \in (0, 1) \times (0, 1)$ we define $X(z) = dH^{-1}(H(z)) \cdot Y(H(z))$. As $Z = Y$ in a neighbourhood of C and in $\{(x, y); x > \frac{1}{2}\}$ we have $X(z) = (1, 0)$ if z belongs to a small neighbourhood of the boundary of the rectangle $[0, 1] \times [0, 1]$. We can now extend X to $\mathbb{R}^2$ by defining $X(z) = (1, 0)$ if z belongs to the boundary of $[0, 1] \times [0, 1]$ and $X(x + n, y + m) = X(x, y)$ if $(x, y) \in [0, 1] \times [0, 1]$ and $(n, m) \in \mathbb{Z}^2$. One checks immediately that X satisfies conditions (i)–(iii). Condition (iv) follows from the fact that the trace of $dX(s)$ is positive as the reader can verify. (Although this is not necessary we can assume that X is linear in a small neighbourhood of s.) □

Lemma 2. *There exists a vector field $X \in \mathscr{C}$ such that $df_X(y) > 1$ for all $y \in (b, a + 1)$.*

PROOF. Take $Y \in \mathscr{C}$ with $Y(x, y) = (1, 0)$ if $\frac{2}{3} \leq x \leq 1$. As f_Y takes the interval $(b, a + 1)$ diffeomorphically onto an interval of length 1 and, by condition (iv), $f'_Y(y) > 1$ near b and $a + 1$, it follows that there exists a diffeomorphism $f: (b, a + 1) \to (f_Y(b), f_Y(a + 1))$ such that $f'(y) > 1$, $\forall y \in (b, a + 1)$ and $f(y) = f_Y(y)$ if y is near b or $a + 1$. It remains to show the existence of a field $X \in \mathscr{C}$ with $f_X = f$.

Let $\varphi_t: (c, c + 1) \to (c, c + 1)$ be given by $\varphi_t(y) = (1 - t)y + t\varphi(y)$ where $\varphi = f \circ f_Y^{-1}$ and $c = f_Y(a)$. For each $t \in [0, 1]$, φ_t is a diffeomorphism and $\varphi_t(y) = y$ if y is near c or $c + 1$. Let $\alpha: [\frac{2}{3}, 1] \to [0, 1]$ be a C^∞ function such that $\alpha = 0$ in a neighbourhood of $\frac{2}{3}$ and $\alpha = 1$ in a neighbourhood of 1. Consider the map $H: (\frac{2}{3}, 1) \times (c, c+1) \to (\frac{2}{3}, 1) \times (c, c+1)$ given by $H(x, y) = (x, \varphi_{\alpha(x)}(y))$. Then H is a diffeomorphism. Let X be the vector field defined by

$$X(x, y) = dH(H^{-1}(x, y)) \cdot Y(H^{-1}(x, y)) \text{ if } (x, y) \in (\tfrac{2}{3}, 1) \times (c, c+1),$$

$$X(x+n, y+m) = X(x, y) \text{ if } (x, y) \in (\tfrac{2}{3}, 1) \times (c, c+1) \text{ and } (n, m) \in \mathbb{Z}^2,$$

$$X(x, y) = Y(x, y) \text{ if } ((x, y) + \mathbb{Z}^2) \cap (\tfrac{2}{3}, 1) \times (c, c+1) = \emptyset.$$

It is easy to check that $X \in \mathscr{C}$ and that $f_X = \varphi \circ f_Y = f$. $\square$

Lemma 3. *Let $f, g: \mathbb{R} \to \mathbb{R}$ be monotonic continuous functions such that $f(x+1) = f(x) + 1$ and $g(x+1) = g(x) + 1$ for all $x \in \mathbb{R}$. Then*

(i) $\rho(f) = \lim_{n \to \infty} (f^n(0)/n)$ *exists and* $|(f^n(0)/n) - \rho(f)| < 1/n$;
(ii) $\lim_{n \to \infty} (f^n(x) - x)/n$ *exists for all $x \in \mathbb{R}$ and is equal to $\rho(f)$;*
(iii) $\rho(f) = m/n$ *with $m, n \in \mathbb{Z}$, $n > 0$ if and only if there exists $x \in \mathbb{R}$ such that $f^n(x) = x + m$;*
(iv) *given $\varepsilon > 0$ there exists $\delta > 0$ such that if $\|f - g\|_0 = \sup_{x \in \mathbb{R}} |f(x) - g(x)| < \delta$ then $|\rho(f) - \rho(g)| < \varepsilon$;*
(v) $\rho(f + n) = \rho(f) + n$ *for any integer n.*

PROOF. Let $M_k = \max_{x \in \mathbb{R}} (f^k(x) - x)$ and $m_k = \min_{x \in \mathbb{R}} (f^k(x) - x)$. We claim that $M_k - m_k < 1$. In fact, as $f(x+1) = f(x) + 1$ we have $f^k(x+1) = f^k(x) + 1$. Therefore, $\varphi = f^k - \text{id}$ is periodic with period 1. Consequently there exist $x_k, X_k \in \mathbb{R}$ with $0 \le x_k - X_k < 1$ such that $\varphi(x_k) = m_k$ and $\varphi(X_k) = M_k$. Since f^k is also monotonic nondecreasing we have $f^k(X_k) \le f^k(x_k)$. Hence $M_k + X_k \le m_k + x_k$ and so $M_k - m_k \le x_k - X_k < 1$ which proves our claim.

We are now going to prove that

$$f^k(y) - y - 1 \le f^k(x) - x \le f^k(y) - y + 1, \quad \forall x, y \in \mathbb{R}. \qquad (1)$$

In fact, $f^k(y) - y - 1 \le M_k - 1 < m_k \le f^k(x) - x \le M_k \le m_k + 1 \le f^k(y) - y + 1$. We next put $y = 0$ and $x = f^{k(j-1)}(0)$ in (1) and obtain

$$f^k(0) - 1 \le f^{kj}(0) - f^{k(j-1)}(0) \le f^k(0) + 1.$$

Thus

$$n(f^k(0) - 1) = \sum_{j=1}^{n} (f^k(0) - 1) \le \sum_{j=1}^{n} (f^{kj}(0) - f^{k(j-1)}(0)) \le n(f^k(0) + 1).$$

From this we deduce that

$$nf^k(0) - n \le f^{kn}(0) \le nf^k(0) + n.$$

We now divide by kn to obtain

$$\frac{f^k(0)}{k} - \frac{1}{k} \le \frac{f^{kn}(0)}{kn} \le \frac{f^k(0)}{k} + \frac{1}{k}$$

or
$$\left|\frac{f^{kn}(0)}{kn} - \frac{f^k(0)}{k}\right| \leq \frac{1}{k}. \qquad (2)$$

Similarly
$$\left|\frac{f^{kn}(0)}{kn} - \frac{f^n(0)}{n}\right| \leq \frac{1}{n}.$$

The sequence $f^k(0)/k$ is a Cauchy sequence because $|f^k(0)/k - f^n(0)/n| \leq 1/k + 1/n$ and so it converges to some limit $\rho(f)$. By letting n tend to ∞ in (2) we see that $|\rho(f) - f^k(0)/k| \leq 1/k$ which proves (i).

By putting $y = 0$ in (1) we obtain $f^k(0) - 1 \leq f^k(x) - x \leq f^k(0) + 1$. Thus
$$\frac{f^k(0)}{k} - \frac{1}{k} \leq \frac{f^k(x) - x}{k} \leq \frac{f^k(0)}{k} + \frac{1}{k}.$$

This shows that $(f^k(x) - x)/k$ converges to $\rho(f)$ and proves (ii).

To prove (iii) suppose that there exists $x \in \mathbb{R}$ such that $f^n(x) = x + m$ with $m, n \in \mathbb{Z}$, $n > 0$. It follows easily by induction that $f^{kn}(x) = x + km$. Thus $\rho(f) = \lim_{k \to \infty} (f^{kn}(x) - x)/kn = \lim_{k \to \infty} km/kn = m/n$. Now let $\rho(f) = m/n$ and suppose if possible that $f^n(x) \neq x + m$, $\forall x \in \mathbb{R}$. Thus $f^n(x) - x > m$, $\forall x \in \mathbb{R}$ (or $f^n(x) - x < m$, $\forall x \in \mathbb{R}$). As $f^n - \text{id}$ is periodic there exists $a > 0$ such that $f^n(x) - x \geq m + a$ (or $f^n(x) - x \leq m - a$) $\forall x \in \mathbb{R}$. Thus $f^{kn}(x) - x \geq km + ka$ (or $f^{kn}(x) - x \leq km - ka$) and so $\rho(f) \geq (m + a)/n$ (or $\rho(f) \leq (m - a)/n$) which is a contradiction.

To prove (iv) we remark that
$$|\rho(f) - \rho(g)| \leq \left|\rho(g) - \frac{g^k(0)}{k}\right| + \left|\frac{g^k(0)}{k} - \frac{f^k(0)}{k}\right| + \left|\frac{f^k(0)}{k} - \rho(f)\right|$$
$$\leq \frac{1}{k} + \frac{1}{k}|g^k(0) - f^k(0)| + \frac{1}{k}.$$

Fix an integer k such that $2/k < \varepsilon/2$ and choose $\delta > 0$ such that $|g^k(0) - f^k(0)| < k\varepsilon/2$ if $\|g - f\|_0 < \delta$. Then $|\rho(g) - \rho(f)| < \varepsilon$ if $\|g - f\|_0 < \delta$. It remains to prove that $\rho(f + n) = \rho(f) + n$ if $n \in \mathbb{Z}$. By induction we have $(f + n)^k(x) = f^k(x) + kn$. Thus
$$\rho(f + n) = \lim_{k \to \infty} \frac{(f + n)^k(0)}{k} = \lim_{k \to \infty} \frac{f^k(0) + kn}{k} = \rho(f) + n. \qquad \square$$

Lemma 3 allows us to introduce the rotation number for degree 1 monotonic endomorphisms of the circle. In fact, let $\pi: \mathbb{R} \to S^1$ be the covering map $\pi(t) = \exp(2\pi i t)$. An endomorphism $\alpha: S^1 \to S^1$ is monotonic and has degree 1 if and only if it has a lift $\tilde{\alpha}: \mathbb{R} \to \mathbb{R}$ which is a continuous monotonic function satisfying $\tilde{\alpha}(x + 1) = \tilde{\alpha}(x) + 1$. We then define the rotation number of α as $\rho(\alpha) = \pi\rho(\tilde{\alpha})$. This does not depend on the choice of the lift $\tilde{\alpha}$ because of Lemma 3(v).

Lemma 4. *There exists* $X \in \mathscr{C}$ *such that* $df_X(y) > 1$ *for all* $y \in (b, a + 1)$ *and* $\rho(f_X)$ *is irrational.*

PROOF. Choose $X^0 \in \mathscr{C}$ such that $df_{X^0}(y) > 1$ for all $y \in (b, a+1)$ and $X^0(x, y) = (1, 0)$ if $\frac{2}{3} < x < 1$. We shall construct a family of vector fields $X^\lambda \in \mathscr{C}$ such that $f_{X^\lambda} = f_{X^0} + \lambda$. Let $H: (\frac{2}{3}, 1) \times \mathbb{R} \to (\frac{2}{3}, 1) \times \mathbb{R}$ be given by $H(x, y) = (x, y + \alpha(x)\lambda)$ where $\alpha: [\frac{2}{3}, 1] \to [0, 1]$ is a C^∞ function such that $\alpha(x) = 0$ if x is near $\frac{2}{3}$ and $\alpha(x) = 1$ for x near 1. We define

$$X^\lambda(x, y) = dH(H^{-1}(x, y)) \cdot X^0(H^{-1}(x, y)) \quad \text{if} \quad (x, y) \in (\tfrac{2}{3}, 1) \times \mathbb{R},$$

$$X^\lambda(x + n, y) = X^\lambda(x, y) \quad \text{if} \quad x \in (\tfrac{2}{3}, 1),$$

$$X^\lambda(x, y) = X^0(x, y) \quad \text{if} \quad (x + \mathbb{Z}) \cap (\tfrac{2}{3}, 1) = \emptyset.$$

As $H(x, y + m) = H(x, y) + (0, m)$ for all $m \in \mathbb{Z}$ we have $X^\lambda \in \mathscr{C}$. It follows immediately that $f_{X^\lambda} = f_{X^0} + \lambda$. Consider the map $h: \mathbb{R} \to \mathbb{R}$ given by $h(\lambda) = \rho(f_{X^\lambda})$. By Lemma 3, h is continuous and $h(1) = h(0) + 1$. Thus there exists $\bar{\lambda} \in [0, 1]$ such that $h(\bar{\lambda})$ is irrational. It now suffices to take $X = X^{\bar{\lambda}}$. □

Theorem. *There exists a C^∞ vector field Y on the torus with the following orbit structure*:

(1) *Y has exactly two singularities, a sink P and a saddle S, both hyperbolic;*
(2) *$W^s(P)$ is dense in T^2 and the compact set $\Lambda = T^2 - W^s(P)$ is transitive, that is there exists $q \in \Lambda - \{S\}$ with $\omega(q) = \alpha(q) = \Lambda$;*
(3) *if $q \in T^2 - (\Lambda \cup \{P\})$ then $\omega(q) = P$ and $\alpha(q) \subset \Lambda$;*
(4) *there exists a circle Σ transversal to Y such that $\Sigma \cap \Lambda$ is a Cantor set.*

PROOF. Take $Y \in \mathfrak{X}^\infty(T^2)$ such that $\pi^* Y = X$ satisfies the conditions of Lemma 4. It is clear that Y satisfies (1) with $P = \pi(p)$ and $S = \pi(s)$. The circle $\Sigma = \pi(\{0\} \times \mathbb{R})$ is transversal to Y and f_X is the lift of the Poincaré map $P_Y: \Sigma \to \Sigma$. If Q is the first point of intersection of the unstable manifold of S with Σ then $Q = \pi(1, c)$ and $P_Y^{-1}(Q) = \pi(\{0\} \times [a, b])$. If $\omega(Q) = S$ or $\omega(Q) = P$ there exists $n \in \mathbb{N}$ such that $P_Y^n(Q) = Q$. Thus $\tilde{\pi} f_X^n(c) = P_Y^n \tilde{\pi}(c) = P_Y^n(Q) = Q = \tilde{\pi}(c)$ and so there exists $m \in \mathbb{Z}$ such that $f_X^n(c) = c + m$. This contradicts the fact that $\rho(f_X)$ is irrational. Hence Y has no saddle connection and $\Lambda = \omega(Q)$ does not contain P. A similar argument shows that Y has no closed orbit.

We now show that $W^s(P)$ is dense in T^2. For this it is enough to prove that the compact set $K = \Sigma - W^s(P)$ has empty interior. So we shall suppose, if possible, that the interior of K is nonempty and take a maximal interval $J \subset K$. Let $J_n = P_X^n(J) = P_X(J_{n-1})$. As $J \cap W^s(P) = \emptyset$, J_n is a compact interval longer than J_{n-1}. Since J is a maximal interval and Y has no closed orbit, the intervals J_n are pairwise disjoint. This contradiction proves that $W^s(P)$ is dense.

According to the Denjoy–Schwartz Theorem, Y has only trivial minimal sets. Let us check this fact directly. Suppose L is a nontrivial minimal set for Y. Then $L \cap \Sigma$ is a compact set disjoint from both the stable and the unstable

Appendix: Rotation Number and Cherry Flows

manifolds of the saddle S for otherwise L would contain the saddle. Since $W^s(P)$ is dense and $L \cap W^s(P) = \emptyset$, $W^s(S)$ is dense in L. Thus if J is a maximal interval in $\Sigma - L$ then $P_Y^n(J) \cap J \neq \emptyset$ for some n. As the endpoints of J belong to L we have $P_Y^n(J) = J$ which implies the existence of a closed orbit. This contradiction shows that P and S are the only minimal sets of Y.

The intervals $I_n = P_Y^{-n}(Q)$ are pairwise disjoint and $\bigcup_{n=1}^{\infty} \text{int}(I_n) = W^s(P) \cap \Sigma$ is dense in Σ. The end-points of the interval I_n belong to different components of $W^s(S) - \{S\}$. Hence $W_+^s(S)$ and $W_-^s(S)$ are both dense in $\Sigma - W^s(P)$, where $W_+^s(S)$ and $W_-^s(S)$ denote the components of $W^s(S) - \{S\}$. It follows that $W_+^s(S)$ and $W_-^s(S)$ are both dense in $\Lambda = T^2 - W^s(P)$ and so the α-limit of any point in $W^s(S) - \{S\}$ is Λ. Let $q \in T^2 - (\Lambda \cup P)$. As $\alpha(q)$ is a compact invariant set and the minimal sets of Y are S and P it follows that $\alpha(q)$ contains S. Thus, either $\alpha(q) = S$ or $\alpha(q)$ contains some point of the stable manifold of S in which case $\alpha(q) = \Lambda$. It is easy to see that $\Lambda \cap \Sigma$ is a perfect set with empty interior. Thus $\Lambda \cap \Sigma$ is a Cantor set. Finally, since $(W^s(S) \cup W^u(S)) \cap \Sigma$ is countable and $\Lambda \cap \Sigma$ is uncountable, there exists $q \in \Sigma - (W^s(S) \cup W^u(S))$. Hence $\alpha(q) = \omega(q) = \Lambda$. $\square$

Remark. Although the vector field Y in this theorem is not structurally stable we can describe completely all the topological equivalence classes of vector fields in a small neighbourhood $\mathcal{N}$ of Y. In fact, if $Z \in \mathcal{N}$ and $\mathcal{N}$ is small enough then the circle Σ is still transversal to Z and the Poincaré map P_Z has derivative greater than 1 at all points of its domain. It follows that, if $Z \in \mathcal{N}$ is Morse–Smale, then Z has just one closed orbit and this is a repellor. (There certainly are Morse–Smale fields in $\mathcal{N}$ because they are dense in $\mathfrak{X}^\infty(T^2)$.) If Z is not Morse–Smale there are two possibilities: either Z has a saddle-connection which is a repellor and the nonwandering set of Z reduces to this saddle connection and the sink or else the rotation number of the endomorphism of Σ induced by Z is irrational and in this case Z satisfies conditions (1) to (4) of the theorem. It can be shown that if the endomorphisms of Σ induced by two fields Z_1 and Z_2 near to Y have the same rotation number and this is irrational then Z_1 and Z_2 are topologically equivalent. There is a partial converse: if there is a topological equivalence h near the identity between Z_1 and Z_2 and the rotation number of Z_1 is irrational then the rotation number of Z_2 is equal to that of Z_1. It is actually enough to require that h is homotopic to the identity. Thus, in a neighbourhood of Y we have just one equivalence class of structurally stable fields, one equivalence class of fields that have a saddle-connection and infinitely many equivalence classes of fields that have nontrivial recurrence. These last are characterized by a single real parameter, the rotation number of the endomorphism induced on Σ.

It is good to emphasize the impossibility of describing completely the topological equivalence classes in a neighbourhood of a field without singularities on the torus having a nontrivial recurrent orbit. In fact, if Y is such a field there exists a circle Σ transversal to Y and the Poincaré map P_Y is defined on the whole circle Σ. According to an important theorem of Herman [37], we can approximate Y by a field X that is C^∞ equivalent to an irrational flow. Then we can approximate X by a field with all orbits closed. It follows that the number of closed orbits of the Morse–Smale fields in any neighbourhood of Y is unbounded. Thus, in any neighbourhood of Y there are infinitely many equivalence classes of structurally stable fields.

References

[1] R. ABRAHAM and J. MARSDEN, *Foundations of Mechanics*, rev. ed. Benjamin-Cummings, 1978.
[2] R. ABRAHAM and S. SMALE, Nongenericity of Ω-stability. In: *Global Analysis*. Proc. Symp. in Pure Math., vol. XIV. American Math. Soc., 1970.
[3] D. V. ANOSOV, *Geodesic Flows on Closed Riemannian Manifolds of Negative Curvature*, Proc. Steklov Math. Inst., vol. 90, 1967. American Math. Soc., 1969 (Transl.).
[4] V. ARNOLD, *Equations Différentielles Ordinaires*. Éditions Mir., 1974.
[5] V. ARNOLD, *Chapitres Supplémentaires de la Théorie des Équations Différentielles Ordinaires*. Éditions Mir, 1980.
[6] V. ARNOLD, Lectures on bifurcations and versal families, *Russian Math. Surveys*, 27, 1972.
[7] V. ARNOLD and A. AVEZ, *Théorie Ergodique des Systèmes Dynamiques*. Gauthier-Villars, 1967.
[8] D. ASIMOV, Round handles and nonsingular Morse–Smale flows, *Ann. of Math.*, 102, 1975.
[9] R. BOWEN, *Equilibrium States and the Ergodic Theory of Anosov Diffeomorphisms*, Lecture Notes in Mathematics, vol. 470. Springer-Verlag, 1975.
[10] R. BOWEN, *On Axiom A Diffeomorphisms*, Conference Board Math. Sciences, 35, American Math. Soc., 1977.
[11] P. BRUNOVSKY, On one-parameter families of diffeomorphisms II: generic branching in higher dimensions. *Comm. Math. Univers. Carolinae*, 12, 1971.
[12] C. CAMACHO, On $R^k \times Z^l$-actions. In: *Dynamical Systems*, edited by M. Peixoto. Academic Press, 1973.
[13] T. CHERRY, Analytic quasi-periodic curves of discontinuous type on a torus, *Proc. London Math. Soc.*, 44, 1938.
[14] C. CONLEY, *Isolated Invariant Sets and the Morse Index*, Conference Board Math. Sciences, 38. American Math. Soc., 1980.
[15] A. DANKNER, On Smale's Axiom A diffeomorphisms, *Ann. of Math.*, 107, 1978.
[16] A. DENJOY, Sur les courbes définies par les équations différentielles à la surface du tore, *J. Math. Pure et Appl.*, 11, ser. 9, 1932.
[17] J. DUISTERMAAT, Oscillatory integrals, Lagrange immersions and unfoldings of singularities, *Commun. Pure and Appl. Math.*, 27, 1974.

[18] F. FARRELL and L. JONES, Anosov diffeomorphisms constructed from $\pi_1 \text{Diff}(S^n)$, *Topology*, **17**, 1978.
[19] G. FLEITAS, Classification of gradient like flows on dimensions two and three, *Bol. Soc. Brasil. Mat.*, **6**, 1975.
[20] J. FRANKS, Anosov diffeomorphisms, In: *Global Analysis*. Proc. Symp. in Pure Math., vol. XIV. American Math. Soc., 1970.
[21] J. FRANKS, Differentiably Ω-stable diffeomorphisms, *Topology*, **11**, 1972.
[22] J. FRANKS and C. NARASIMHAN, The periodic behavior of Morse–Smale diffeomorphisms, *Invent. Math.*, **48**, 1978.
[23] J. FRANKS and M. SHUB, The existence of Morse–Smale diffeomorphisms, *Topology*, **20**, 1981.
[24] J. FRANKS and R. WILLIAMS, Anomalous Anosov flows. In: *Global Theory of Dynamical Systems*, edited by Z. Nitecki and C. Robinson. Lecture Notes in Mathematics, vol. 819. Springer-Verlag, 1980.
[25] D. GROBMAN, Homeomorphisms of systems of differential equations, *Dokl. Akad. Nauk. SSSR*, **128**, 1959.
[26] J. GUCKENHEIMER, Absolutely Ω-stable diffeomorphisms, *Topology*, **11**, 1972.
[27] J. GUCKENHEIMER, Bifurcations of dynamical systems. In: *Dynamical Systems*. CIME Lectures-Bressanone (Italy), Birkhäuser, 1980.
[28] M. GOLUBITSKY and V. GUILLEMIN, *Stable Mappings and Their Singularities*. Graduate Texts in Mathematics, vol. 14. Springer-Verlag, 1974.
[29] V. GUILLEMIN and A. POLLACK, *Differential Topology*. Prentice-Hall, 1974.
[30] C. GUTIERREZ, Structural stability for flows on the torus with a cross-cap, *Trans. Amer. Math. Soc.*, **241**, 1978.
[31] C. GUTIERREZ, Smoothing continuous flows and the converse of Denjoy-Schwartz theorem, *Anais Acad. Brasil. Ciências*, **51**, 1979.
[32] C. GUTIERREZ and W. MELO, On the connected components of Morse–Smale vector fields on two manifolds, *Proc. IIIrd Latin American School of Mathematics*, edited by J. Palis and M. do Carmo. Lecture Notes in Mathematics, vol. 597. Springer-Verlag, 1977.
[33] J. HALE, *Ordinary Differential Equations*, rev. ed. Robert E. Krieger Publ. Co., 1980.
[34] J. HALE, *Topics in Dynamic Bifurcation*. Conference Board Math. Sciences. American Math. Soc., 1981.
[35] P. HARTMAN, *Ordinary Differential Equations*. Wiley, 1964.
[36] P. HARTMAN, A lemma in the theory of structural stability of differential equations, *Proc. Amer. Math. Soc.*, **11**, 1960.
[37] M. HERMAN, Sur la conjugaison différentiable des difféomorphismes du cercle à des rotations, *Publ. Math. Inst. Hautes Études Scientifiques*, **49**, 1979.
[38] M. HIRSCH, *Differential Topology*. Graduate texts in Mathematics, vol. 33. Springer-Verlag, 1976.
[39] M. HIRSCH and C. PUGH, Stable manifolds and hyperbolic sets. In: *Global Analysis*, Proc. Symp. in Pure Math., vol. XIV. American Math. Soc., 1970.
[40] M. HIRSCH, C. PUGH, and M. SHUB, *Invariant Manifolds*, Lecture Notes in Mathematics, vol. 583, Springer-Verlag, 1977.
[41] M. HIRSCH and S. SMALE, *Differential Equations, Dynamical Systems and Linear Algebra*. Academic Press, 1974.
[42] G. IOOSS and D. JOSEPH, *Elementary Stability and Bifurcation Theory*, Undergraduate Texts in Mathematics. Springer-Verlag, 1980.
[43] M. IRWIN, On the stable manifold theorem, *Bull. London Math. Soc.*, **2**, 1970.
[44] I. KUPKA, Contribution à la théorie des champs génériques, *Contrib. Diff. Equations*, **2**, 1963.
[45] M. KEANE, Interval exchange transformations, *Math. Z.*, **141**, 1975.

References

[46] S. LANG, *Analysis*, vol. I. Addison-Wesley, 1968.
[47] S. LEFSCHETZ, *Differential Equations: Geometric Theory*. Wiley, 1963.
[48] E. LIMA, *Análise no espaço $\mathbb{R}^n$*. Univ. Brasília and Edgard Blücher, 1970.
[49] E. LIMA, Variedades diferenciáveis, *Monogr. de Mat.*, **15**, IMPA, 1973.
[50] I. MALTA, Hyperbolic Birkhoff centers, *Anais Acad. Brasil. Ciências*, **51**, 1979; and *Trans. Amer. Math. Soc.*, **262**, 1980.
[51] I. MALTA, On Ω-stability of flows, *Bol. Soc. Brasil. Mat.*, **11**, 1980.
[52] R. MAÑÉ, Absolute and infinitesimal stability. In: *Dynamical Systems Warwick 1974*, edited by A. Manning. Lecture Notes in Mathematics, vol. 468. Springer-Verlag, 1975.
[53] R. MAÑÉ, Contributions to the stability conjecture, *Topology*, **17**, 1978.
[54] R. MAÑÉ, An ergodic closing lemma. Preprint IMPA, 1980; to appear in *Ann. of Math.*
[55] A. MANNING, There are no new Anosov diffeomorphisms on tori, *Amer. J. Math.*, **96**, 1974.
[56] N. MARKLEY, The Poincaré–Bendixson theorem for the Klein bottle, *Trans. Amer. Math. Soc.*, **135**, 1969.
[57] L. MARKUS, *Lectures in Differentiable Dynamics*, rev. ed. Conference Board Math. Sciences, 3. American Math. Soc., 1980.
[58] J. MARSDEN and M. McCRACKEN, *The Hopf Bifurcation and Its Applications*, Appl. Math. Sciences, vol. 19. Springer-Verlag, 1976.
[59] W. MASSEY, *Algebraic topology: an Introduction*. Harcourt-Brace-World, 1967.
[60] J. MORGAN, Nonsingular Morse–Smale flows on three-manifolds, *Topology*, **18**, 1979.
[61] W. MELO, Structural stability of diffeomorphisms on two manifolds, *Invent. Math.*, **21**, 1973.
[62] W. MELO and J. PALIS, *Introdução aos Sistemas Dinâmicos*. Projeto Euclides, IMPA-CNPq, 1977.
[63] P. MENDES, On stability of dynamical systems on open manifolds, *J. Diff. Equations*, **16**, 1974.
[64] J. MILNOR, *Topology from the Differentiable Viewpoint*, Univ. Press of Virginia, 1965.
[65] J. MOSER, On a theorem of Anosov, *J. Diff. Equations*, **5**, 1969.
[66] J. MOSER, *Stable and Random Motions in Dynamical Systems with Special Emphasis on Celestial Mechanics*. Princeton Univ. Press, 1973.
[67] S. NEWHOUSE, On codimension one Anosov diffeomorphisms, *Amer. J. Math.*, **92**, 1970.
[68] S. NEWHOUSE, Nondensity of Axiom A(a) on S^2. In: *Global Analysis*, Proc. Symp. in Pure Math., vol. XIV. American Math. Soc. 1970.
[69] S. NEWHOUSE, Hyperbolic limit sets, *Trans. Amer. Math. Soc.*, **167**, 1972.
[70] S. NEWHOUSE, Lectures on dynamical systems. In: *Dynamical Systems*. CIME Lectures-Bressanone (Italy). Birkhäuser, 1980.
[71] S. NEWHOUSE and J. PALIS, Hyperbolic nonwandering sets on two-dimensional manifolds. In: *Dynamical Systems*, edited by M. Peixoto. Academic Press, 1973.
[72] Z. NITECKI, *Differentiable Dynamics*. M.I.T. Press, 1971.
[73] G. PALIS, Linearly induced vector fields and R^2-actions on spheres, *J. Diff. Geometry*, **13** (2), 1978.
[74] J. PALIS, On the local structure of hyperbolic fixed points in Banach spaces, *Anais Acad. Brasil. Ciêncais*, **40**, 1968.
[75] J. PALIS, On Morse–Smale dynamical systems, *Topology*, **8**, 1969.
[76] J. PALIS, A note on Ω-stability. In: *Global Analysis*, Proc. Symp. in Pure Math., vol. XIV. American Math. Soc., 1970.

[77] J. PALIS (Editor). Seminário de Sistemas Dinâmicos, *Monogr. de Mat.*, **4**, IMPA, 1971.
[78] J. PALIS, Moduli of stability and bifurcation theory. *Proc. Int. Congress of Mathematicians, Helsinki*, 1978.
[79] J. PALIS and S. SMALE, Structural stability theorems. In: *Global Analysis.* Proc. Symp. in Pure Math. vol. XIV. American Math. Soc., 1970.
[80] J. PALIS and F. TAKENS, Topological equivalence of normally hyperbolic dynamical systems, *Topology*, **16**, 1977.
[81] M. PEIXOTO, Structural stability on two-dimensional manifolds, *Topology*, **1**, 1962.
[82] M. PEIXOTO, On an approximation theorem of Kupka and Smale, *J. Diff. Equations*, **3**, 1967.
[83] M. PEIXOTO, On the classification of flows on two-manifolds. In: *Dynamical Systems*, edited by M. Peixoto. Academic Press, 1973.
[84] M. PEIXOTO, On bifurcations of dynamical systems. *Proc. Int. Congress of Mathematicians, Vancouver*, 1974.
[85] M. PEIXOTO and M. C. PEIXOTO, Structural stability in the plane with enlarged boundary conditions, *Anais Acad. Brasil. Ciências*, **31**, 1959.
[86] V. PLISS, Properties of solutions of a periodic system of two differential equations having an integral set of zero measure, *Diff. Equations*, **8**, 1972.
[87] V. PLISS, Sur la grossièreté des équations différentielles définies sur le tore, *Vestnik LGU, ser. mat.* **13**, (3), 1960.
[88] C. PUGH, The closing lemma, *Amer. J. Math.*, **89**, 1967.
[89] C. PUGH, An improved closing lemma and a general density theorem, *Amer. J. Math.*, **89**, 1967.
[90] C. PUGH, On a theorem of P. Hartman, *Amer. J. Math.*, **91**, 1969.
[91] C. PUGH and M. SHUB, The Ω-stability theorem for flows, *Invent. Math.*, **11**, 1970.
[92] J. ROBBIN, A structural stability theorem, *Ann. of Math.*, **94**, 1971.
[93] J. ROBBIN, Topological conjugacy and structural stability for discrete dynamical systems, *Bull. Amer. Math. Soc.*, **78**, 1972.
[94] L. ROCHA, Characterization of Morse-Smale isotopy classes on two-manifolds, *Anais Acad. Brasil. Ciências*, **50**, 1978.
[95] C. ROBINSON, Structural stability of vector fields, *Ann. of Math.*, **99**, 1974.
[96] C. ROBINSON, Structural stability of C^1 flows, In: *Dynamical Systems Warwick 1974*, edited by A. Manning, Lecture Notes in Mathematics, vol. 468. Springer-Verlag, 1975.
[97] C. ROBINSON, Structural stability of C^1 diffeomorphisms, *J. Diff. Equations*, **22**, 1976.
[98] C. ROBINSON, C^r structural stability implies Kupka–Smale. In: *Dynamical Systems*, edited by M. Peixoto. Academic Press, 1973.
[99] D. RUELLE, *Dynamical Systems with Turbulent Behavior*. Lecture Notes in Physics, vol. 80. Springer-Verlag, 1978.
[100] A. SCHWARTZ, A generalization of a Poincaré–Bendixson theorem to closed two-dimensional manifolds, *Amer. J. Math.*, **85**, 1963.
[101] M. SHUB, Stabilité globale des systèmes dynamiques, *Astérisque*, **56**, 1978.
[102] M. SHUB, Structurally stable diffeomorphisms are dense, *Bull. Amer. Math. Soc.*, **78**, 1972.
[103] M. SHUB and D. SULLIVAN, Homology theory and dynamical systems, *Topology*, **14**, 1975.
[104] M. SHUB and R. WILLIAMS, Future stability is not generic, *Proc. Amer. Math. Soc.*, **22**, 1969.
[105] C. SIMON, Instability in $\text{Diff}^r(T^3)$ and the nongenericity of rational zeta functions, *Trans. Amer. Math. Soc.*, **174**, 1972.

References

[106] S. SMALE, On gradient dynamical systems, *Ann. of Math.*, **74**, 1961.
[107] S. SMALE, Stable manifolds for differential equations and diffeomorphisms, *Ann. Scuela Sup. Piza*, **17**, 1963.
[108] S. SMALE, Diffeomorphisms with many periodic points. In: *Differential and Combinatorial Topology*. Princeton Univ. Press, 1965.
[109] S. SMALE, Differentiable dynamical systems, *Bull. Amer. Math. Soc.*, **73**, 1967.
[110] S. SMALE, The Ω-stability theorem. In: *Global Analysis*. Proc. Symp. in Pure Math., vol. XIV. American Math. Soc. 1970.
[111] S. SMALE, Structurally stable systems are not dense, *Amer. J. Math.*, **88**, 1966.
[112] S. SMALE, Stability and isotopy in discrete dynamical systems. In: *Dynamical Systems*, edited by M. Peixoto. Academic Press, 1973.
[113] S. SMALE, Essays on dynamical systems, economic processes and related topics. *The Mathematics of Time*. Springer-Verlag, 1980.
[114] J. SOTOMAYOR, Generic one parameter families of vector fields on two-dimensional manifolds, *Publ. Math. Inst. Hautes Études Scientifiques*, **43**, 1973.
[115] J. SOTOMAYOR, Generic bifurcations of dynamical systems. In: *Dynamical Systems*, edited by M. Peixoto. Academic Press, 1973.
[116] J. SOTOMAYOR, *Lições de equações Diferenciais Ordinárias*, Projeto Euclides, IMPA-CNPq, 1979.
[117] F. TAKENS, *Introduction to Global Analysis*, Comm. no. 2, Math. Inst. Univ. Utrecht, 1973.
[118] R. THOM, *Structural Stability and Morphogenesis*. Benjamin-Addison Wesley, 1975.
[119] C. ZEEMAN, Uma introdução informal à topologia das superfícies, *Monogr. de Mat.*, **20**, IMPA, 1975.
[120] P. WALTERS, *An Introduction to Ergodic Theory*. Graduate Texts in Mathematics, vol. 79, Springer-Verlag, 1982.
[121] R. WILLIAMS, Expanding attractors, *Publ. Math. Inst. Hautes Études Scientifiques*, **43**, 1973.
[122] R. WILLIAMS, The "DA" maps of Smale and structural stability. In: *Global Analysis*. Proc. Symp. in Pure Math., vol. XIV. American Math. Soc., 1970.

Index

absolute stability 171
Anosov
  Closing Lemma 167
  diffeomorphism 159, 161, 179
  flow 159
approximating by a C^∞ map 21
atlas, C^r and C^∞ 9
attracting singularity 30, 119
attractor
  basic set 160, 161, 165, 169
  closed orbit 35, 119, 130
automorphism of fundamental group 172
Axiom A diffeomorphism 160, 170, 172, 180

Baire space 20, 37
Baire's Theorem 3
basic set 160
Bifurcation Theory 173
Birkhoff centre 38, 171, 179
Bolzano–Weierstrass Theorem 139
bounded continuous maps 60
bump function 8, 21, 96

canonical form 43, 44, 48, 69
Cantor set 164, 165, 168, 186
centre
  Birkhoff 38
  manifold 173
Chain Rule 2, 6

Characteristic polynomial 44
chart 5, 8
Cherry flow 137, 178, 181
chordal system 177
circle 30, 113, 122, 136, 161, 179
class C^r 2, 6
closed orbit 11, 18
Closing Lemma 151, 167, 172
commuting vector fields 34, 180
complete 20
complex spectrum 44
complexification 42, 46
composition map 22, 36
conjugacy between diffeomorphisms 31, 33, 63
conjugate vector fields 26
constant vector field 68, 94
contraction 48, 61
coordinate neighbourhood 5
critical
  elements (singularities and closed orbits) 104, 106, 118
  point 3
  value 3
cycle 35, 171
cylinder 34, 35

DA diffeomorphism 165
decomposition 160, 166
Denjoy flow 139
Denjoy–Schwartz Theorem 139, 180, 186

195

density
  C^0 of stable diffeomorphisms 172
  of C^∞ maps 22
  of Morse–Smale fields 132, 141
  of periodic points 157, 160
  of transversal homoclinic points 157, 166, 167
derivative 2, 6, 19
derived from Anosov 165
diffeomorphism 2, 6, 31, 48, 59, 68, 153
differentiable
  manifold 4, 9, 112, 169
  map 2, 6
differential equation 4, 26
dimension 52
direct sum decomposition 160

eigenvalues 44, 45, 46, 53
elementary fixed point 59
embed in a flow 89, 180
embedding 7, 9, 73
endomorphism of the circle 185
equator 18, 37
Ergodic Theory 172
Euler–Poincaré characteristic 152
existence and uniqueness of solutions of a differential equation 4
expansion 48
exponential of a linear map 42

family of discs 85
fibration 94
fibre 129
filtration 124, 177
first integral 180
fixed point of a diffeomorphism 59
flow 11
  box 93, 142, 146
foliation 157, 165
fundamental
  domain 81, 86, 130
  group 172
  neighbourhood 82

General Density Theorem 172, 180
generic property 91, 107, 115, 172
genus 152
geodesic flow 159
geometric theory 26
global
  flow 11
  transversal section 111

gradient 12, 28
  vector field 13, 116, 133, 150, 153, 176
graph
  of a map 76, 79
  transform 80
  (union of separatrices and saddles) 139, 142, 179
Grobman–Hartman Theorem 60, 68, 74, 87, 100, 158
Gronwall's inequality 64

Hartman's Theorem 60
Hausdorff 8
height function 15, 116, 133
homoclinic point 155, 166
Hopf bifurcation 174, 175
horseshoe 162
hyperbolic
  closed orbit 95, 98, 104, 106, 116, 119, 130
  fixed point 59, 60, 72, 75
  linear isomorphism 48, 50, 63, 156
  linear vector field 46, 49, 50, 52, 54, 68
  nonwandering set 160
  periodic points 110, 114
  set 160
  singularity 30, 58, 64, 100, 104, 106, 116, 119

immersed submanifold 99
immersion 3, 7, 9, 73
Implicit Function Theorem 3, 55, 78
Inclination Lemma 80
index 46, 51, 52, 68, 119, 195
inner product 13
integral curve 3, 10, 11
invariant
  section 96, 130
  set 31
Inverse Function Theorem 3, 6
irrational flow on the torus 12, 28, 35, 110, 117, 132, 133
isomorphic phase diagrams 124, 126
isotopy class 172

Jordan
  canonical form 44
  Curve Theorem 16

Klein bottle 138, 146, 151, 172, 179
Kupka–Smale
  diffeomorphism 110, 112, 114, 154

Index

Theorem 107, 110
vector field 106, 113, 132, 134, 141, 153, 154, 176, 178

λ-lemma 80, 82, 86, 156
limit set
  α- or ω- 11, 12, 15, 18, 26, 31, 57, 73, 88, 113, 116, 117, 139
  L_α or L_ω 117, 179
linear
  contraction 31, 37, 48
  vector field 27, 41, 46, 54, 88
Lipschitz 60, 64, 80
local
  chart 5, 8
  classification 68
  conjugacy 59, 63, 66, 72, 90
  diffeomorphism 6
  flow 4, 40
  orbit structure 39, 54
  stability 40, 59, 63, 64, 67, 88, 98, 131
  stable manifold 73
  topological behaviour 40, 68
  unstable manifold 73, 166
locally finite cover 8
long tubular flow 93

maps, space of C^r 19
matrix 43, 68
maximal integral curve 11
meagre 3
Mean Value Theorem 78
measure preserved by a diffeomorphism 172
meridians 12
minimal set 138
Möbius band 34, 122, 146, 148
Morse function 89, 177, 180
Morse–Smale
  diffeomorphism 154
  vector field 116, 118, 122, 127, 141, 152, 153, 176, 177, 178
multiplicity of an eigenvalue 45, 53

nilmanifold 159
no cycles 171
nondegenerate bilinear form 89
nondensity
  of Morse–Smale diffeomorphisms 154, 156
  of structurally stable diffeomorphisms 172

nonhyperbolic periodic orbit 173
nonorientable surface 148
nontransversal intersection 116, 173
nontrivial
  minimal set 139
  recurrence 132, 133, 136, 137, 141, 144, 151, 181
nonwandering set 118, 154, 160
north pole 18
  south pole diffeomorphism 161
  south pole field 12, 29, 119

Ω-stability 171
open manifold 179
openness
  of diffeomorphisms 23
  of Morse–Smale fields 153
orbit
  of a diffeomorphism 31
  of a flow 10, 11
  structure 26, 31, 115
orientable surface 141
orientation of orbit 39

parallels of latitude 12
parameter 4
parametrized neighbourhood 8
partial order 123
partition of unity 8
Peixoto's Theorem 115
period 11, 31
periodic
  behaviour possible 172
  points 31, 154, 157
perturbation 52, 57, 79, 173
phase
  diagram 123, 126
  portrait 173
Plykin attractor 169
Poincaré
  compactification 37
  map 92, 100, 111, 113, 130, 145, 173, 174, 183, 188
Poincaré–Bendixson Theorem 18, 132, 138, 143
polar fields 123, 177
polynomial vector fields 34
pretzel 123, 133, 134, 177
projective plane 35

qualitative theory 10, 26, 153
quasihyperbolic 175, 176

rational flow on the torus 12, 28, 117
real
   canonical form 43, 45, 48, 69
   part of eigenvalue 46
recurrent 38, 132
regular
   orbit of a flow 11, 18, 135
   point of a flow 40, 68, 93
   point of a map 3
   value of a map 3, 7, 24
reparametrization 97, 130
repelling
   closed orbit 30, 119, 130
   singularity 30, 119
residual 3, 20, 37, 104, 107, 110
rotation number 181, 185

saddle 119, 166
saddle connection 120, 122, 141, 142
saddle-node bifurcation 174
Sard's Theorem 3, 8, 25
saturation 111
section 92
separable 21
separatrix 128, 133, 142, 148
simple singularity 55
singularity 11, 18, 54
sink 30, 122
Smale horseshoe 162, 180
solution 4
source 30, 137
space
   of bounded maps 60, 158
   of C^r maps 19
spectrum 44, 52
sphere 12, 18, 37, 74, 113, 119, 132, 154, 162, 169, 177, 178, 181
splitting 46, 48, 60, 160
stability of arcs of vector fields 176
stabilized separatrix 142
stable
   foliation 157, 165
   manifold 73, 98, 161
   Manifold Theorem 75
   separatrix 128
structurally stable
   diffeomorphism 32, 154, 158, 170
   linear vector field 54
   vector field 27, 114, 127, 188
submanifold 6
submersion 3, 7
surface 113, 123
suspension 111, 112, 114, 153

tangent
   bundle 8
   vector 9
Thom Tranversality Theorem 25
topological
   embedding 73
   immersion 73
topologically equivalent 26, 39, 187
torus
   T^2 12, 28, 110, 116, 137, 146, 157, 161, 165, 172, 179, 181, 186
   of dimension n 35, 159, 179
   with a cross-cap 151
trajectory 10
transitive 160, 164
transversal
   circle 144, 146, 151, 179, 188
   homoclinic point 155, 166
   intersection 23, 36, 56, 82, 85, 106, 108, 110, 116, 154
   section 92, 96, 111, 130, 142
Transversality
   Condition 161, 170, 180
   Theorem 25, 56
trivial
   centralizer 180
   minimal set 138, 186
   recurrence 132, 139, 141
tubular
   family 129, 130
   flow 92, 125
   Flow Theorem 40, 68, 93, 95, 101, 145
   neighbourhood 9

unstable
   foliation 157, 165
   manifold 73, 98, 161
   separatrix 128
   vector fields, examples 28

vector field 3, 10, 23, 41, 106, 116

wandering point 118
Whitney
   Embedding Theorem 9, 22
   topology 37, 179

Zorn's Lemma 139